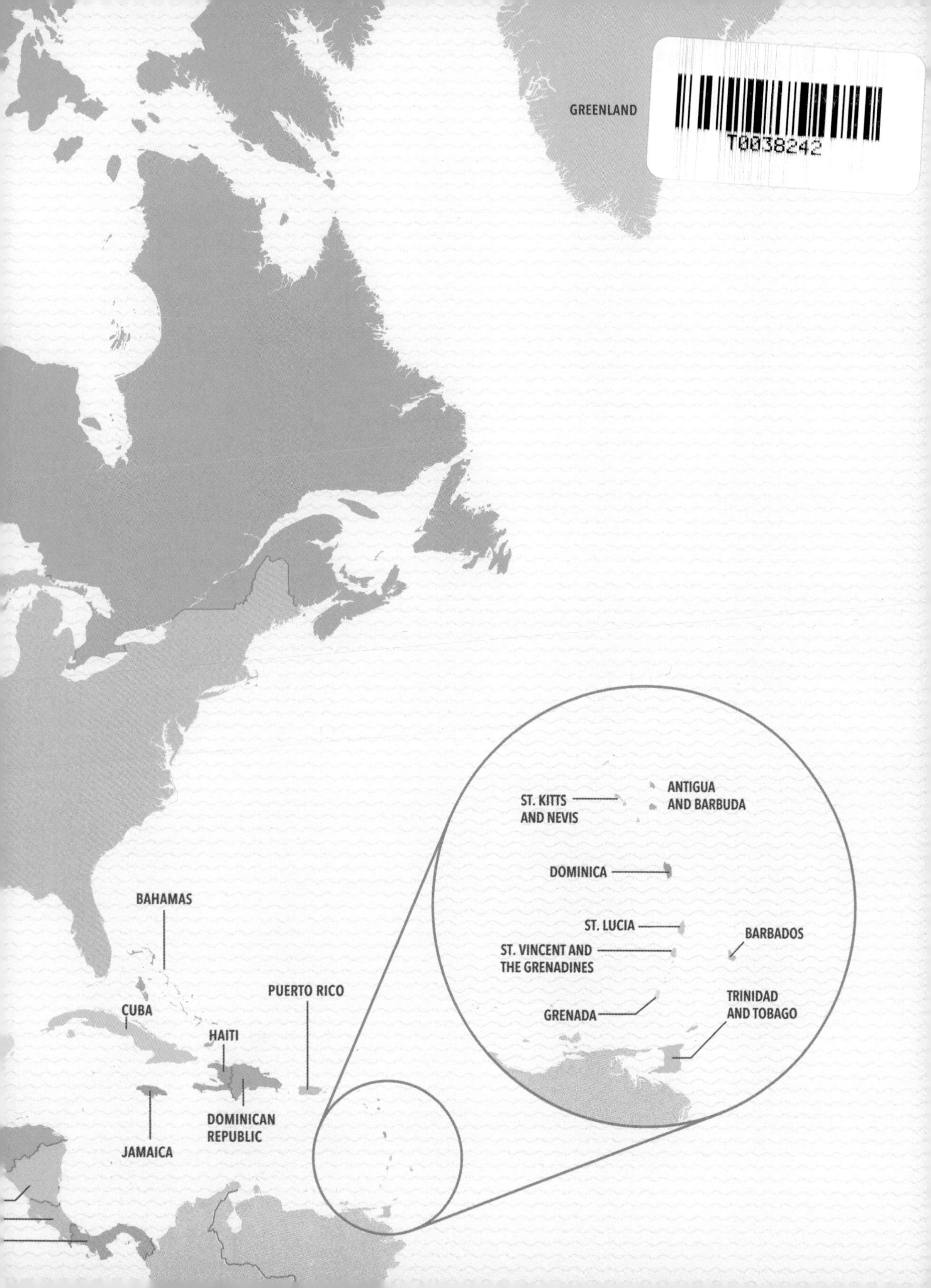
GREENLAND
BAHAMAS
CUBA
HAITI
PUERTO RICO
JAMAICA
DOMINICAN REPUBLIC
ST. KITTS AND NEVIS
ANTIGUA AND BARBUDA
DOMINICA
ST. LUCIA
ST. VINCENT AND THE GRENADINES
BARBADOS
GRENADA
TRINIDAD AND TOBAGO

NORTH AMERICAN MAPS FOR CURIOUS MINDS

100 New Ways to See the Continent

North American Maps for Curious Minds: *100 New Ways to See the Continent*

The Experiment, LLC
220 East 23rd Street, Suite 600
New York, NY 10001-4658
theexperimentpublishing.com

Library of Congress Cataloging-in-Publication Data available upon request

ISBN 978-1-61519-748-4
Ebook ISBN 978-1-61519-749-1

Illustrations, and cover and text design, by Jack Dunnington

Manufactured in Turkey

First printing October 2021
10 9 8 7 6 5 4 3 2 1

NORTH AMERICAN MAPS FOR CURIOUS MINDS

100 New Ways to See the Continent

MATTHEW BUCKLAN & VICTOR CIZEK
Illustrated by Jack Dunnington

Foreword by Ian Wright

THE EXPERIMENT
NEW YORK

CONTENTS

1 GEOGRAPHY

2 POLITICS AND POWER

3

NATURE

4

CULTURE AND SPORTS

5

PEOPLE AND POPULATIONS

6

LIFESTYLE AND HEALTH

7

INDUSTRY AND TRANSPORT

FOREWORD

by Ian Wright

Chances are, you've never wondered what the most common surname is in every state, country, and territory of North America. Or which states have the highest percentage of residents born in-state—or in another state or country. Or how far one day of travel took you from NYC since 1800. Or which state is the most accident prone. Or—here's a really unusual one—how many Waffle Houses there are by latitude.

These are among my favorite maps in *North American Maps for Curious Minds* precisely *because* you likely haven't stopped to wonder about these curiosities. But that's the beauty of this book: It will answer questions that you never knew you had but that you're delighted to know about now. That's a theme very much in line with *Brilliant Maps for Curious Minds*, the first volume of the Maps for Curious Minds series. Here, you'll find the same wide array of topics, from food and health to nature, politics, and power—but with a deeper dive into one continent, allowing for new insights and discoveries across North America. You'll learn what makes Nebraska special, where Americans have second homes, where you'll find the biggest sports stadiums, what day of the year is the hottest across the US, and what about Alaska's geography makes it so weird and wonderful. Most of the maps should stand the test of time, although it was impossible to ignore the singular moment in history during which this book was researched and written: the COVID-19 pandemic. I'm curious about how maps related to commuting times and anxiety and depression, for example, will look in five years.

While I'm on the topic of COVID, I'd like to add that if you happen to be reading this in an independent book shop, make sure you buy a copy from them; like so many industries, the pandemic has been hard on indie booksellers, and they need all the help they can get.

I think the team has done an incredible job making maps that both entertain and educate, and I could write a small essay about each and every one of the brilliant maps you'll find here. But you're better off just reading the book and looking at each one yourself. If you loved *Brilliant Maps for Curious Minds*, you'll love this newest collection, too.

June 2021

Ian Wright runs Brilliant Maps, one of the most popular cartographic sites on the internet, and is the author of *Brilliant Maps for Curious Minds*, the first book in the Maps for Curious Minds series.

INTRODUCTION

There is nothing quite like a good map, and there's no other book of maps quite like this one—none, that is, besides the first book in this series, *Brilliant Maps for Curious Minds*. Here, you'll find the same assortment of thought-provoking maps—by turns surprising, exciting, at times alarming—with one key distinction: Each map in this atlas investigates a curiosity within the confines of North America.

And this one difference makes all the difference. Every map makes meaning by what it leaves out, and in this book, we found that leaving out the six other continents of the world made room for us to dig deeper into previously unexplored realms. In the pages that follow, we revel in the offbeat, overlooked, and beguiling details of the continent: Super-spellers, for example, can discover in "Bees expertise" (page 106) every winning word for every Scripps National Spelling Bee, as well as *where* the speller of each word hailed from and *when* that word was featured in competition. And those with a passion for year-end rollicking good times can let "Don't drop the ball" (page 96) be their travel guide to the many strangely diverse celebrations to ring in the new year. Focusing on North America, it turns out, made us focus on things we had never noticed before.

Creating data-rich maps requires finding good data. And so, while every North American country gets its turn to shine in this book, the bigger players with the most state-funded research institutions—Canada, Mexico, and especially the US (with, among others, its Centers for Disease Control and Prevention and its singularly comprehensive Census Bureau)—hog the spotlight. Still, we've strived to be inclusive: You'll find all kinds of continent-wide contrasts and connections across the chapters of this book. Check out "The price of leadership" (page 42), for one, to see how the salaries of the heads of state of, say, El Salvador and Haiti stack up against that of the US president.

Our benchmark for every map is to teach and to spur interest and further inquiry. And although we're supposed to be the experts here, if we're being honest, while making these maps, we often learned something new and surprising ourselves. And maybe unsurprisingly, the maps we learned the most from tended to be our favorites. Take a peek at "Doing a double take" (page 12), one of our earliest maps; we still find those city comparisons hard to believe. And we were heartened by "'Did Not Vote' ends winning streak" (page 38), where we set out expecting to tell one story, until the data led us to another—one certainly more uplifting for our democracy.

We've done our best to amass a treasure trove of knowledge for the adventurous explorer to mine, and we can't wait for you to share in the discovery of everything we've uncovered about this incredible continent.

MB and VC
June 2021

1

GEOGRAPHY

1 **Identity crisis:** What even is North America?

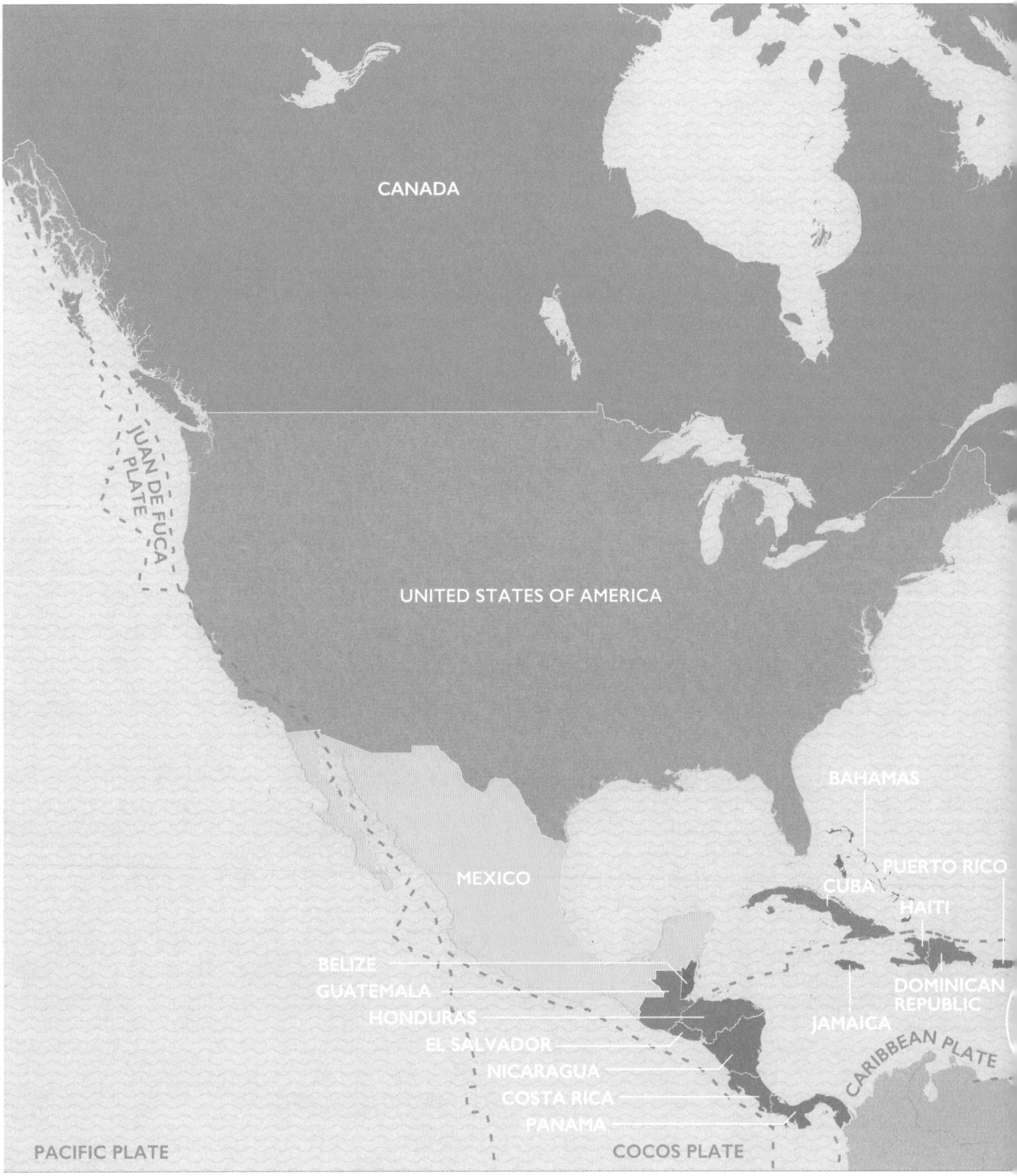

Here's a question more complicated than you may think: What is North America? If we go by physical geography, we might say it's whatever rests on the North American tectonic plate. But where does that leave, say, the majority of Caribbean nations? If you turn to human geography—the history, politics, and culture of places—then Greenland is arguably European. In this book, we'll keep it simple and consider any country marked on this map to be part of North America. With nearly 600 million people divided among 23 countries, it's the fourth-most-populous continent and third-largest by area. There's much to explore!

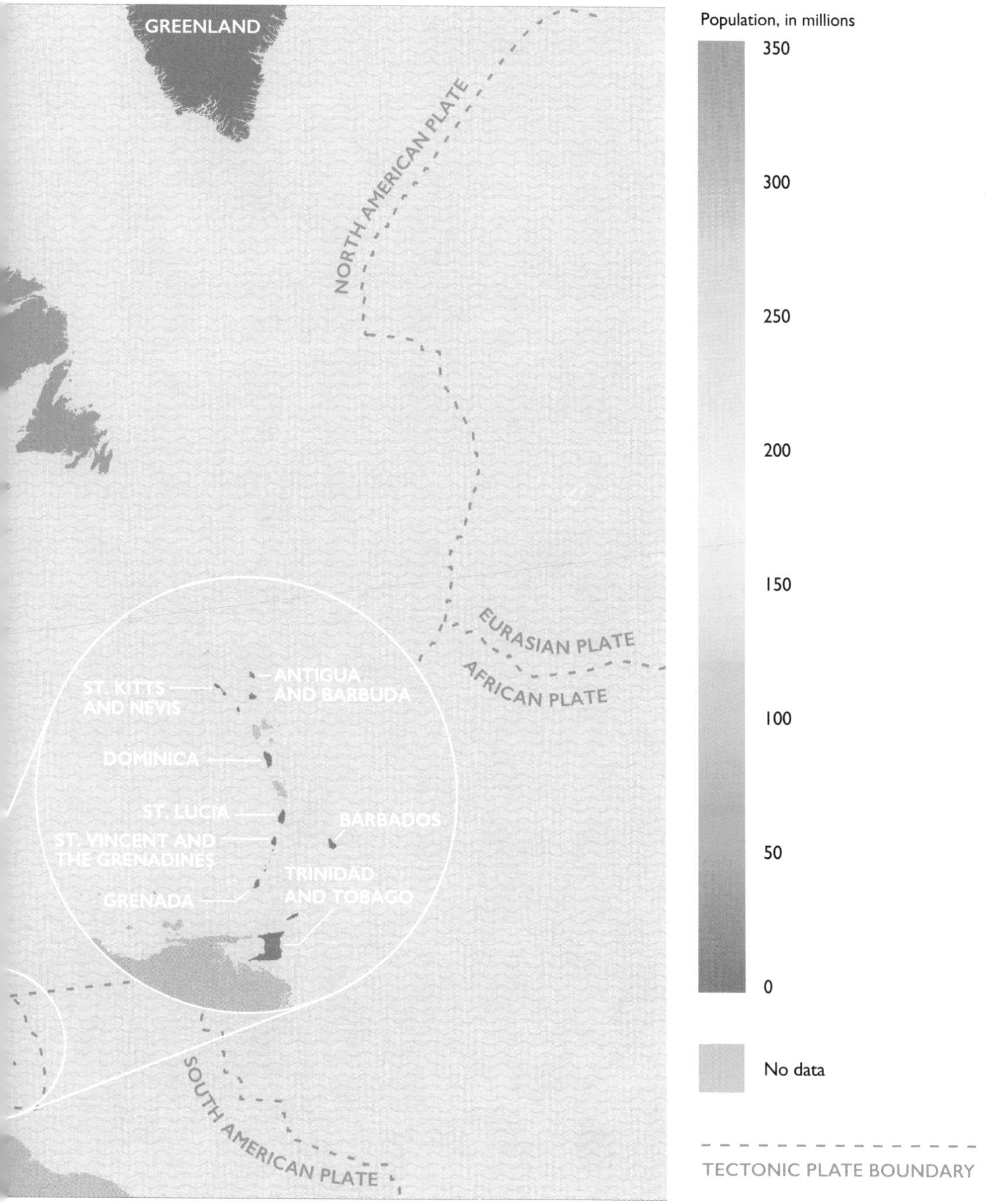

2 **Hawaii is a really long state:** Comparing it to the West Coast puts it in perspective

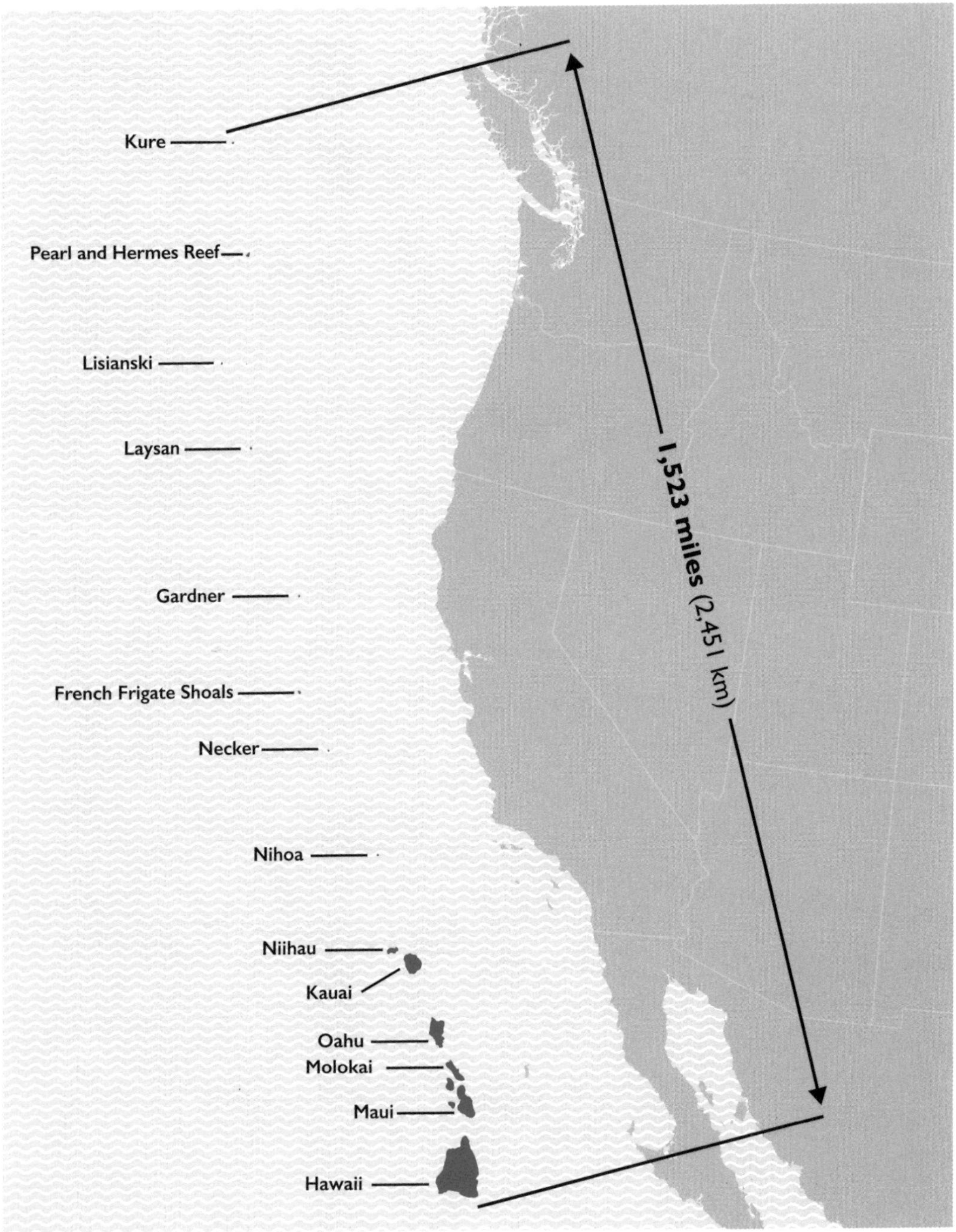

③ **Global scale:** If your state or province were a country, what size by area would it be?

Oklahoma and Cambodia have the exact same area of 69,898 square miles (181,035 km²).

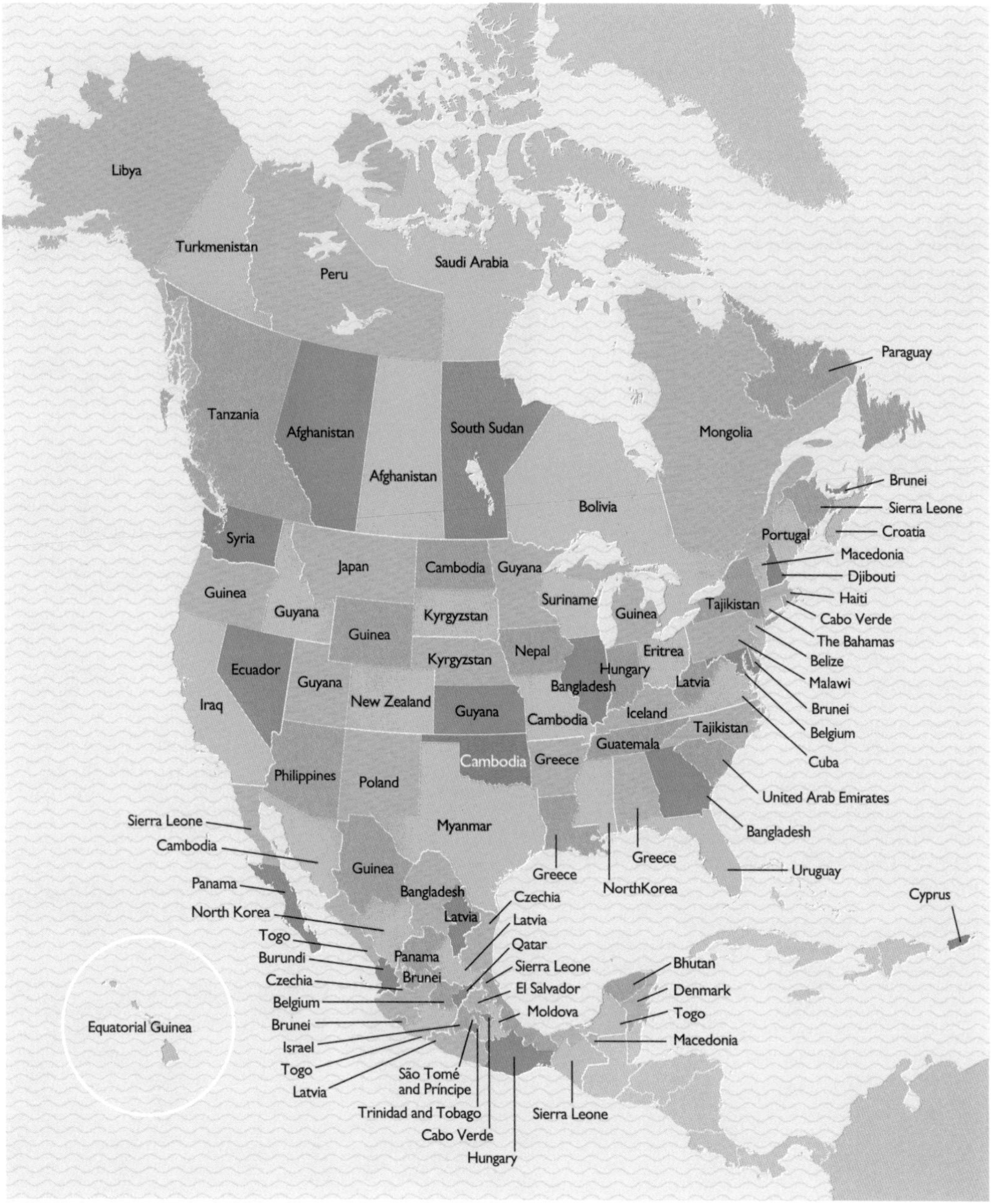

4 **Alas, Alaska:** Big, weird, and wonderful

Alaska is enormous, spanning 570,380 square miles (1,477,300 km²), roughly 2.5 times the size of Texas, the next largest state at 268,596 square miles (695,662 km²). In fact, Alaska's bigger than all but 17 countries. With all that space, it's no wonder the state gets its own map in this book!

Russia's Big Diomede Island **(1)** is only 2 miles (3 km) from Alaska's Little Diomede Island **(2)**.

Alaska contains the easternmost **(3)**, westernmost **(4)**, and northernmost **(5)** points in the US.

Utqiagvik **(6)**, formerly Barrow, is the northernmost town in the US.

Along the Ipnavik River in the National Petroleum Reserve is the most remote spot in the US **(7)**—about 120 miles (190 km) from the nearest habitation.

There are sand dunes north of the Arctic Circle: the Great Kobuk Sand Dunes **(8)**.

The lowest temperature ever recorded in the US was –80°F (–62°C) at Prospect Creek Camp **(9)** on 1/23/71.

Denali **(10)**, formerly Mt. McKinley, is the highest point in North America, at 20,310 ft (6,190 m).

The trans-Alaska pipeline starts in Prudhoe Bay **(11)** and stretches more than 800 miles (1,287 km) to Valdez **(12)**; today it takes a barrel of oil two weeks to cover this span. When finished in 1977, it was the world's largest privately funded construction project ever.

The 1964 Great Alaska Earthquake, magnitude 9.2, at **(13)** was the second-largest earthquake ever recorded (and largest in North America). Alaska has 11 percent of the world's earthquakes, with roughly 1,000 earthquakes every month.

The nation's two largest forests are located in Alaska: the 16.8-million-acre Tongass **(14)** and the 4.8-million-acre Chugach **(15)**. About 60 percent of all National Parks land is in the state.

Juneau **(16)**, Alaska's capital, has no roads connecting to the rest of Alaska or Canada.

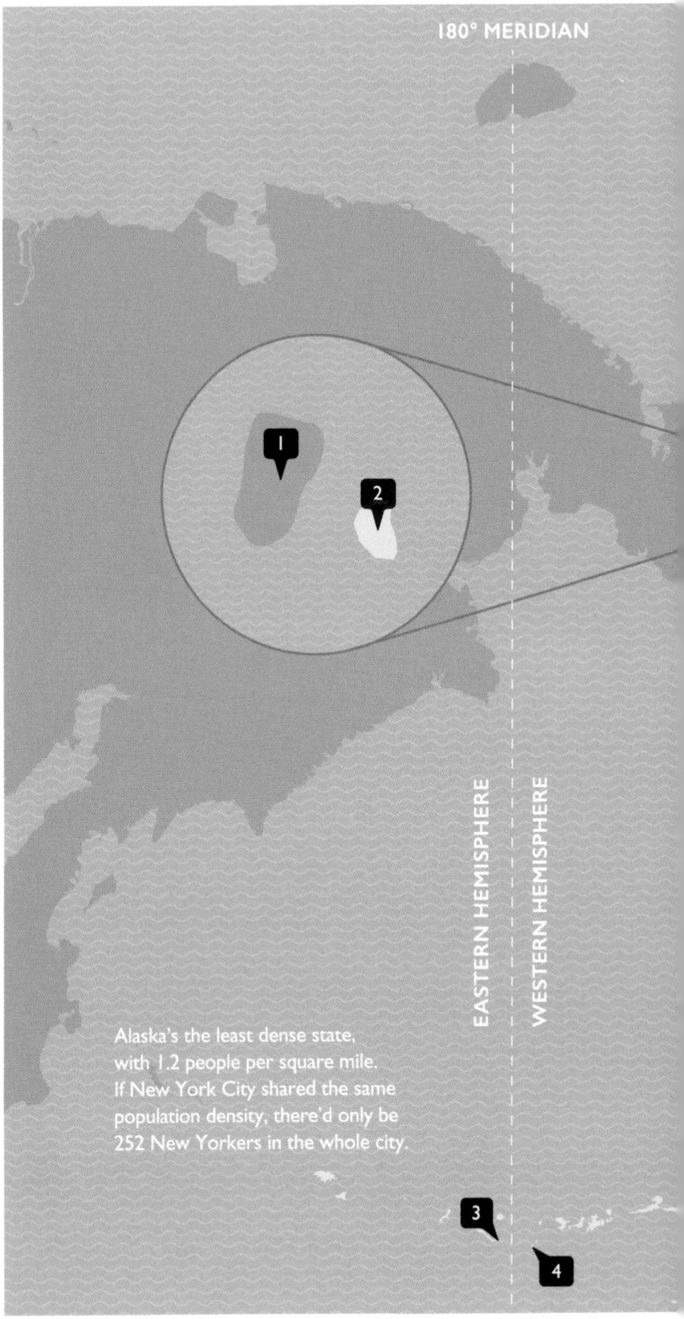

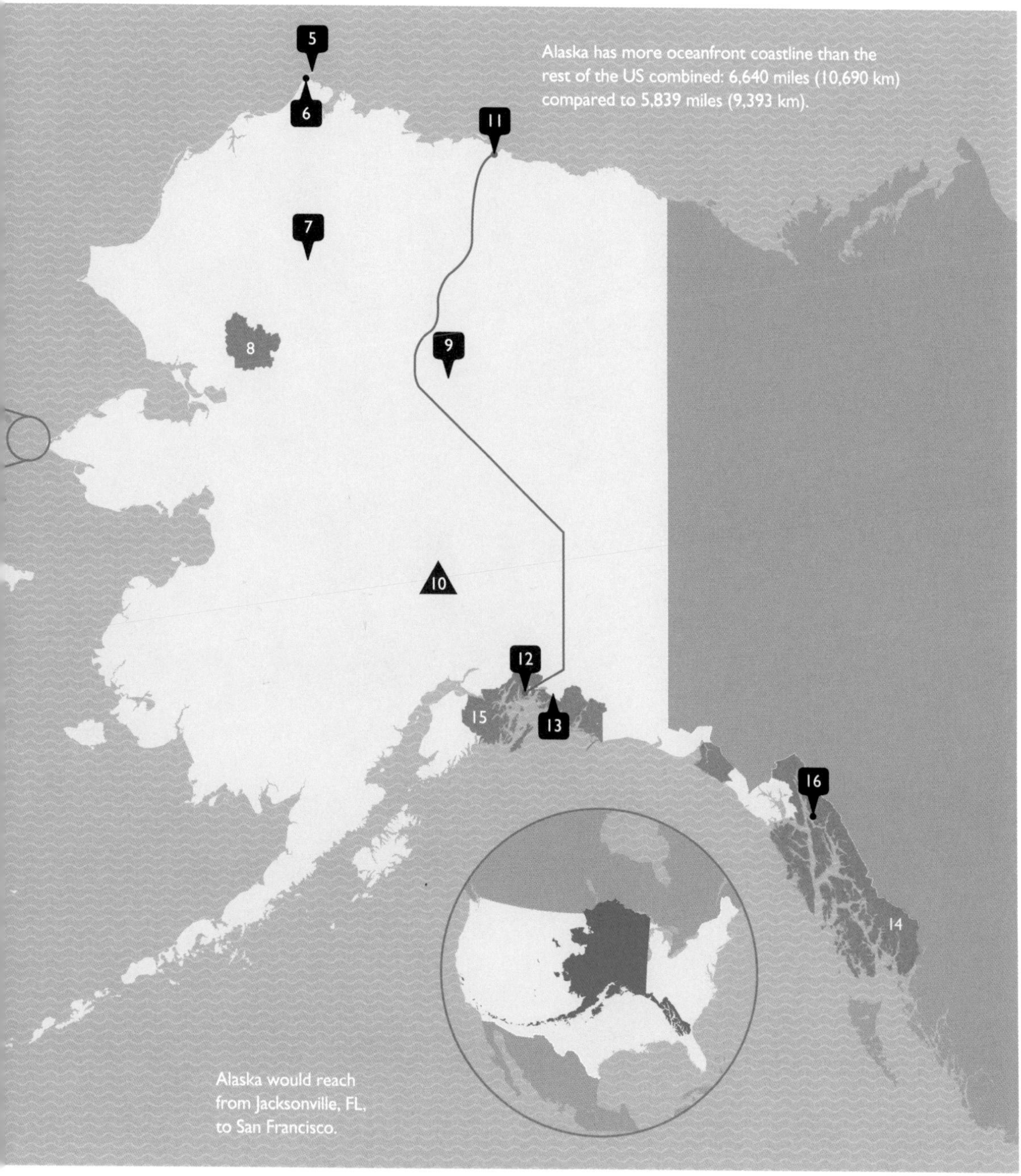
Alaska has more oceanfront coastline than the rest of the US combined: 6,640 miles (10,690 km) compared to 5,839 miles (9,393 km).
5
6
7
8
9
10
11
12
13
14
15
16
Alaska would reach from Jacksonville, FL, to San Francisco.

5 Counties that count:
Giving a few across the US their due

1. **Clallam County, WA**
   Has picked the winning US president since 1976, longer than any other county
2. **Snohomish County, WA**
   First confirmed case of COVID-19
3. **Los Angeles County, CA**
   Most populous (pop. 10,105,518); more people live here than in all but nine states
4. **Kalawao County, HI**
   Smallest (53 mi^2/137 km^2) and least populous (pop. 88); a former leper colony to which no one is allowed to move; last county to get COVID-19
5. **Unorganized Borough, AK**
   An immense wilderness (323,400 mi^2/ 837,600 km^2) covering more than half of Alaska and consisting of eleven "census areas"; bigger than Texas and larger than the smallest sixteen states combined
6. **Lake and Peninsula Borough, AK**
   Least densely populated
7. **North Slope Borough, AK**
   Largest county (94,796 mi^2/ 245,520 km^2); it would be the twelfth largest state
8. **St. John the Baptist Parish, LA**
   Longest name
9. **Hamilton County, NY**
   Least densely populated in eastern US (just over three people per square mile)
10. **New York County (Manhattan), NY**
    Most densely populated
11. **Reedy Creek Improvement District, FL**
    Most famous example of a Special District, a form of local government roughly equivalent to counties; in this case, it regulates Disney World

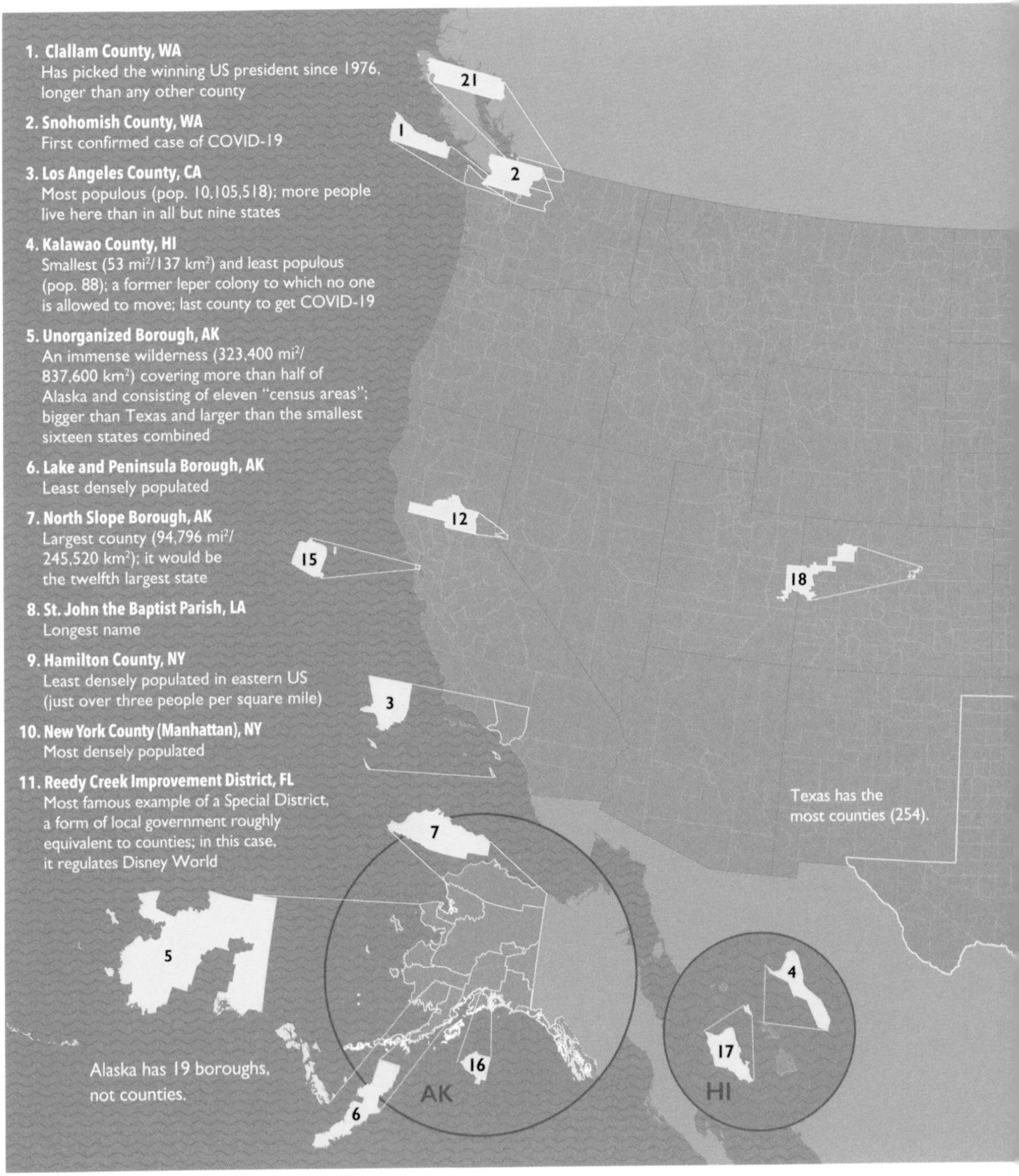

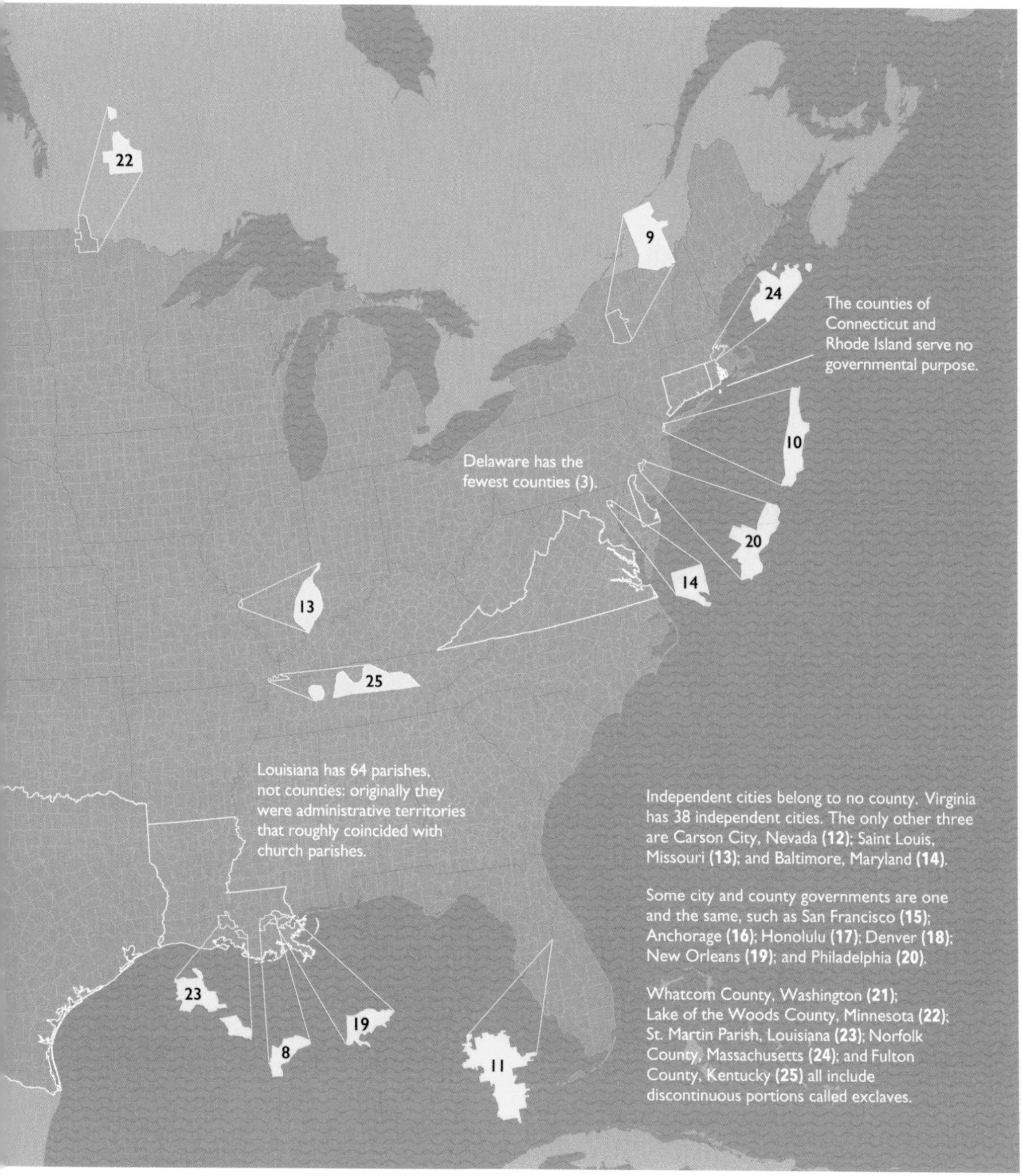
22
9
24
The counties of Connecticut and Rhode Island serve no governmental purpose.
10
Delaware has the fewest counties (3).
20
14
13
25
Louisiana has 64 parishes, not counties: originally they were administrative territories that roughly coincided with church parishes.
Independent cities belong to no county. Virginia has 38 independent cities. The only other three are Carson City, Nevada (12); Saint Louis, Missouri (13); and Baltimore, Maryland (14).
Some city and county governments are one and the same, such as San Francisco (15); Anchorage (16); Honolulu (17); Denver (18); New Orleans (19); and Philadelphia (20).
Whatcom County, Washington (21); Lake of the Woods County, Minnesota (22); St. Martin Parish, Louisiana (23); Norfolk County, Massachusetts (24); and Fulton County, Kentucky (25) all include discontinuous portions called exclaves.
23
19
8
11

6 Drawing outside the lines:
Baffling North American borders

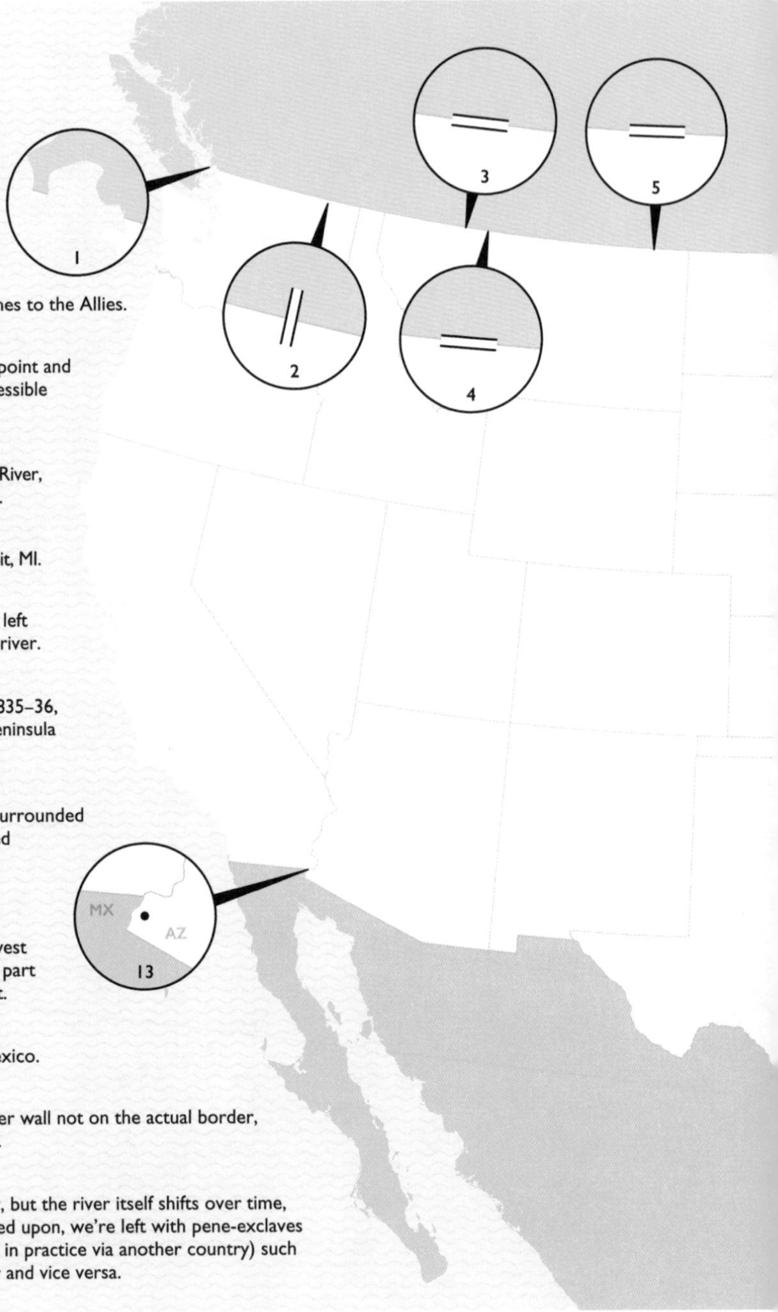

1. Point Roberts, WA
This US location is only reachable via road from Canada.

2. Avey Field State Airport
3. Del Bonita/Whetstone International Airport
4. Coutts/Ross International Airport
5. Coronach/Scobey Border Station Airport
Before the US entered into WWII, it landed aircraft to one of several airports whose runways exactly straddled the border to stay neutral under the Lend-Lease Act but still deliver warplanes to the Allies.

6. Northwest Angle, MN
This is the continental US's northernmost point and its only land north of the 49th parallel, accessible by a single road in Canada.

7. Carter Lake, IA
Formed by a flood that altered the Missouri River, this Iowa city lies entirely within Omaha, NE.

8. Detroit, MI
You can head south into Canada from Detroit, MI.

9. Kaskaskia, IL
A change of course in the Mississippi River left this the only portion of Illinois west of the river.

10. Lost Peninusla, MI
After the Toledo War border dispute of 1835–36, Michigan was left with a nub on an Ohio peninsula cut off from the rest of the state.

11. Kentucky Bend, KY
A stranded piece of Kentucky completely surrounded by Missouri and Tennessee, this was created by earthquakes shifting the course of the Mississippi River two centuries ago.

12. Tennessee Chimney, TN
This notch in the otherwise straight east–west Tennessee-Kentucky border was caused in part by human error and surveyor disagreement.

13. Yuma, AZ
You can head north from Yuma, AZ, to Mexico.

14. Brownsville, TX
There are portions of the US-Mexico border wall not on the actual border, thus cutting off pieces of the US from itself.

15. Rio Grande Pene-Exclaves
The Rio Grande is the international border, but the river itself shifts over time, and until new boundaries are officially agreed upon, we're left with pene-exclaves (areas not fully separate but only reachable in practice via another country) such as this one—bits of the US across the river and vice versa.

The US-Mexico border gets all the attention for sociopolitical reasons, but we'd like to take a closer look at the lesser-known *geographic* highlights and oddities of North America's boundary lines. The US-Canada divide—the world's longest international border, at roughly 5,500 miles (8,900 km)—is especially teeming with liminal quirks.

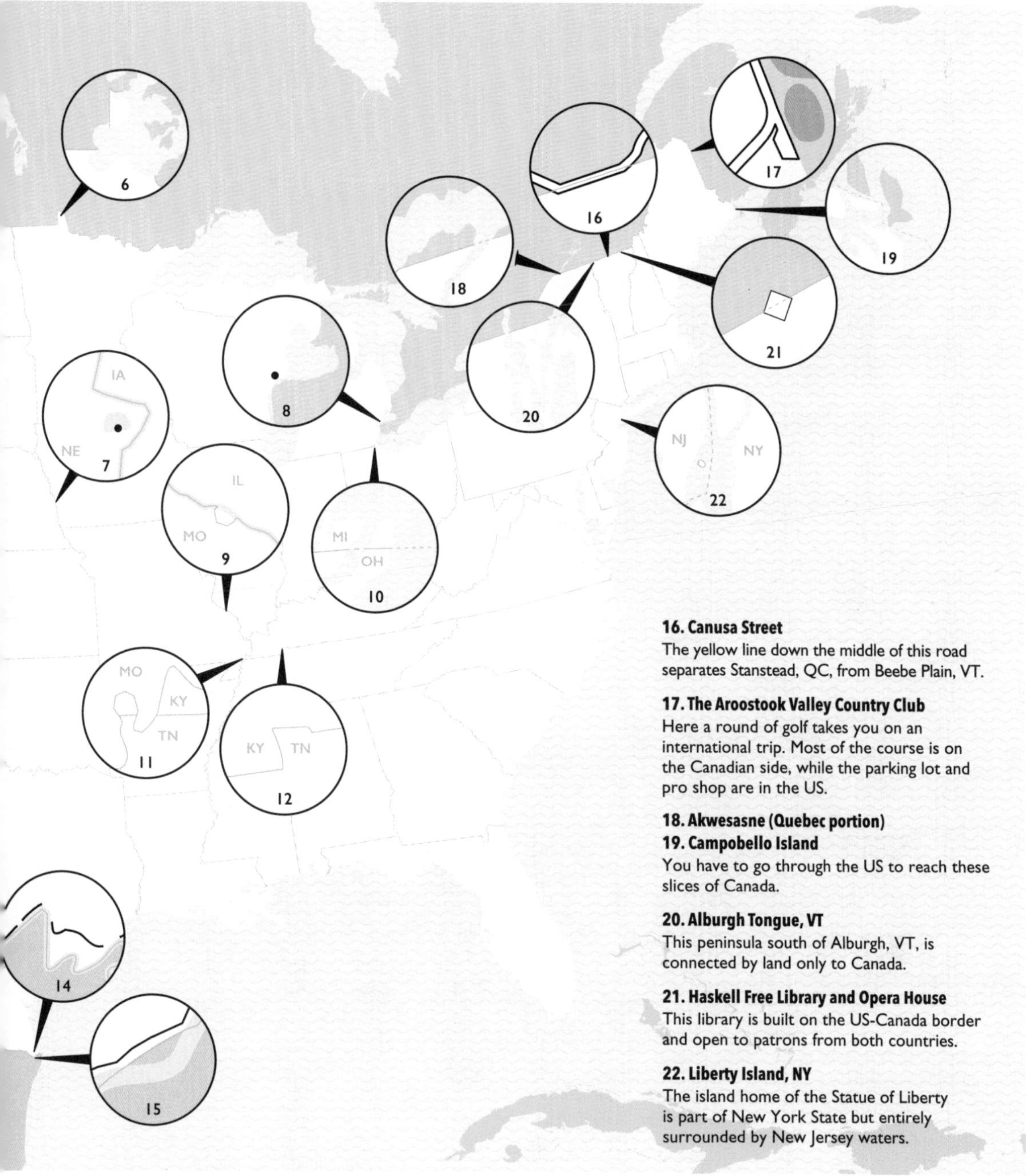

16. Canusa Street
The yellow line down the middle of this road separates Stanstead, QC, from Beebe Plain, VT.

17. The Aroostook Valley Country Club
Here a round of golf takes you on an international trip. Most of the course is on the Canadian side, while the parking lot and pro shop are in the US.

18. Akwesasne (Quebec portion)
19. Campobello Island
You have to go through the US to reach these slices of Canada.

20. Alburgh Tongue, VT
This peninsula south of Alburgh, VT, is connected by land only to Canada.

21. Haskell Free Library and Opera House
This library is built on the US-Canada border and open to patrons from both countries.

22. Liberty Island, NY
The island home of the Statue of Liberty is part of New York State but entirely surrounded by New Jersey waters.

7 Doing a double take:
City pairs to leave you second-guessing

Detroit is farther east than **Atlanta.**

Reno is farther west than **Los Angeles.**

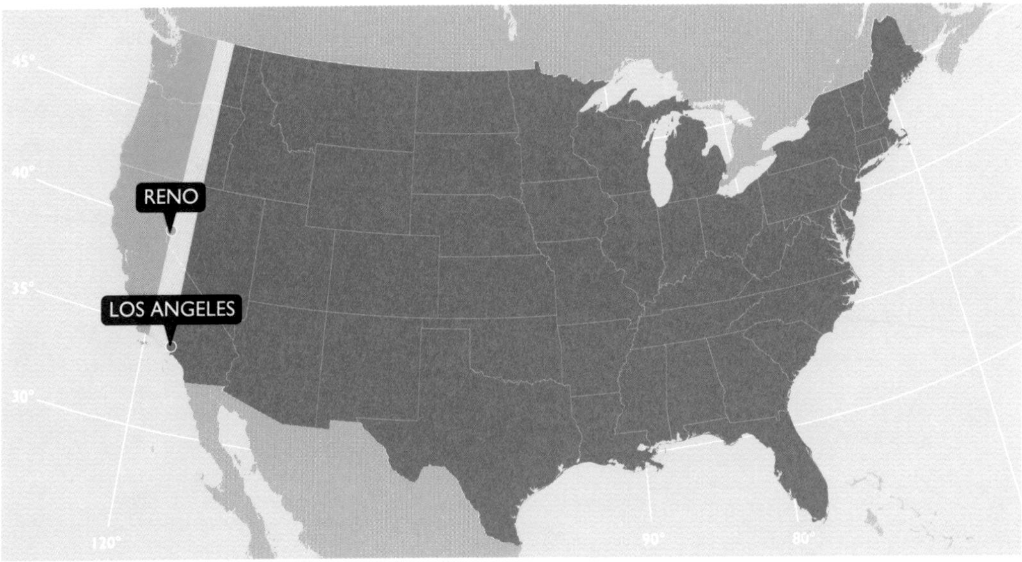

Portland, ME, is south of **Portland, OR.**

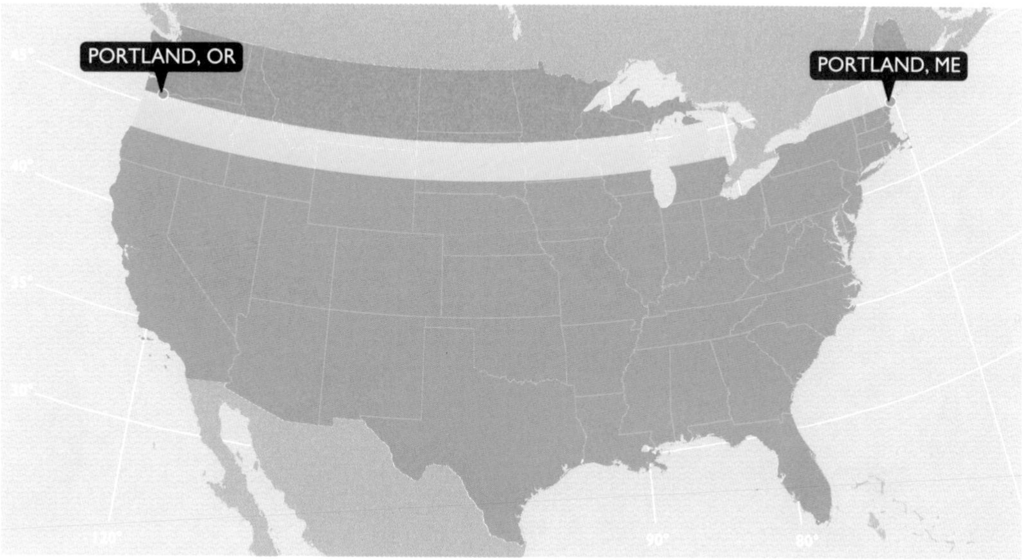

Tijuana is north of **Savannah.**

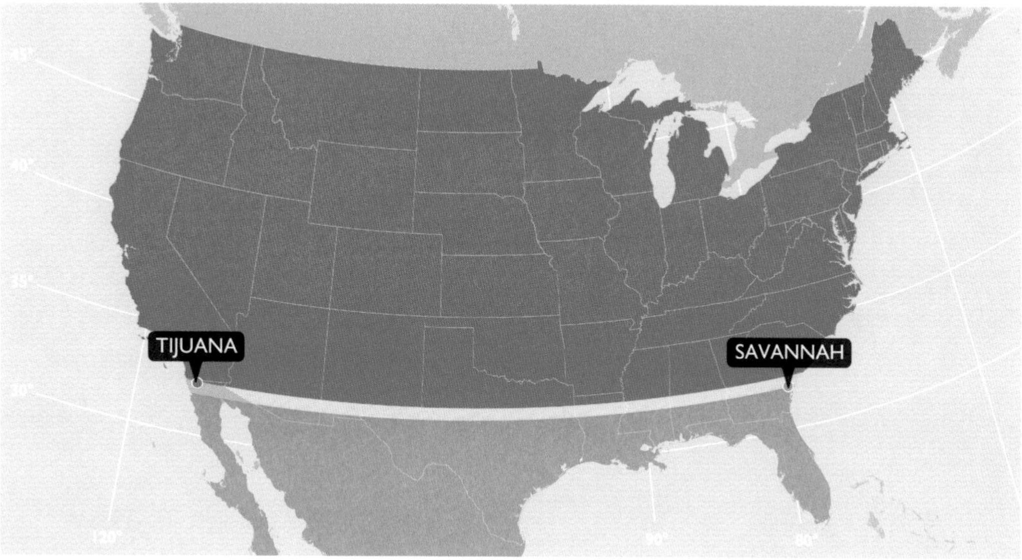

8 Just keep swimming:
The first country you'll reach, coast to coast

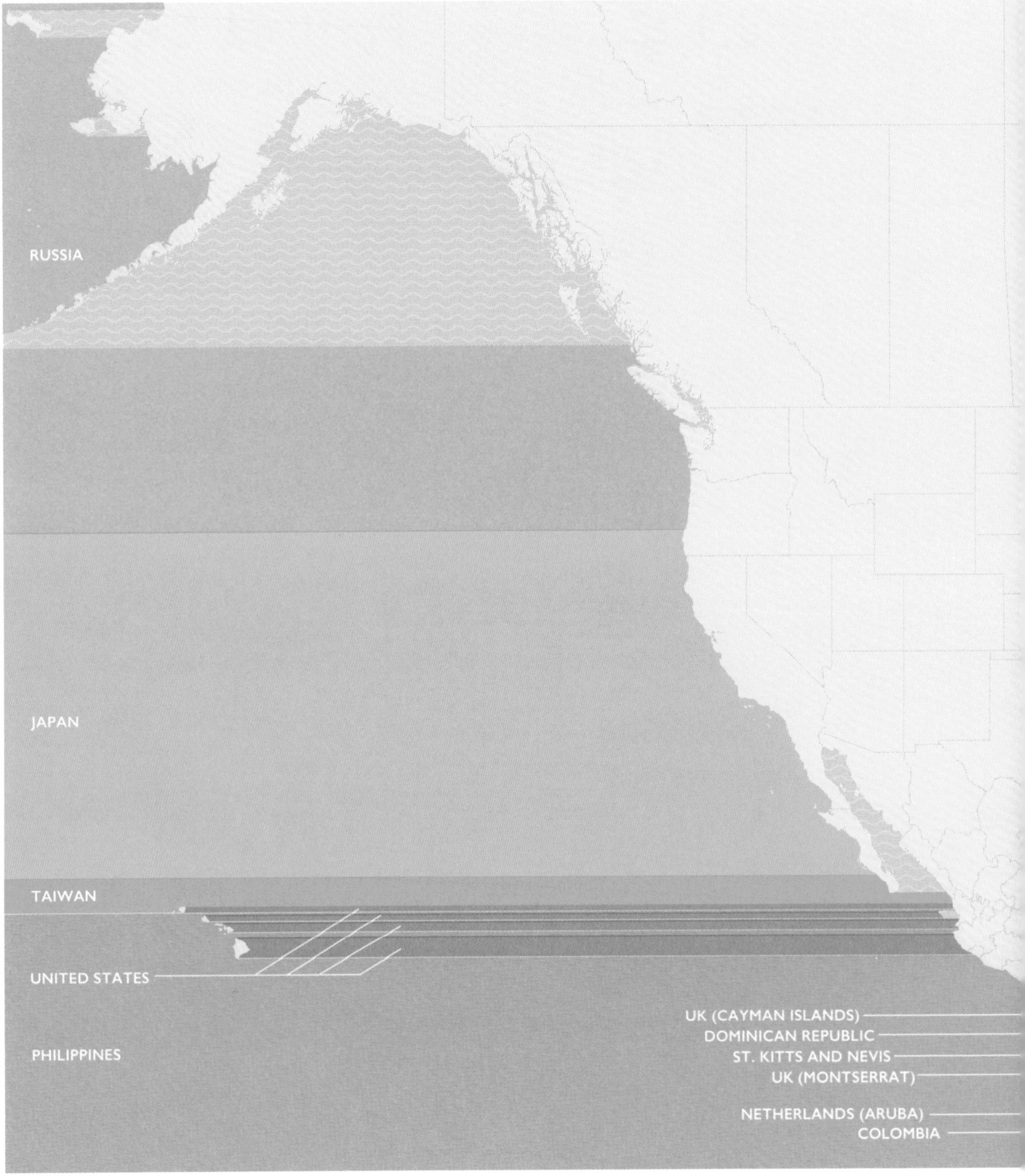

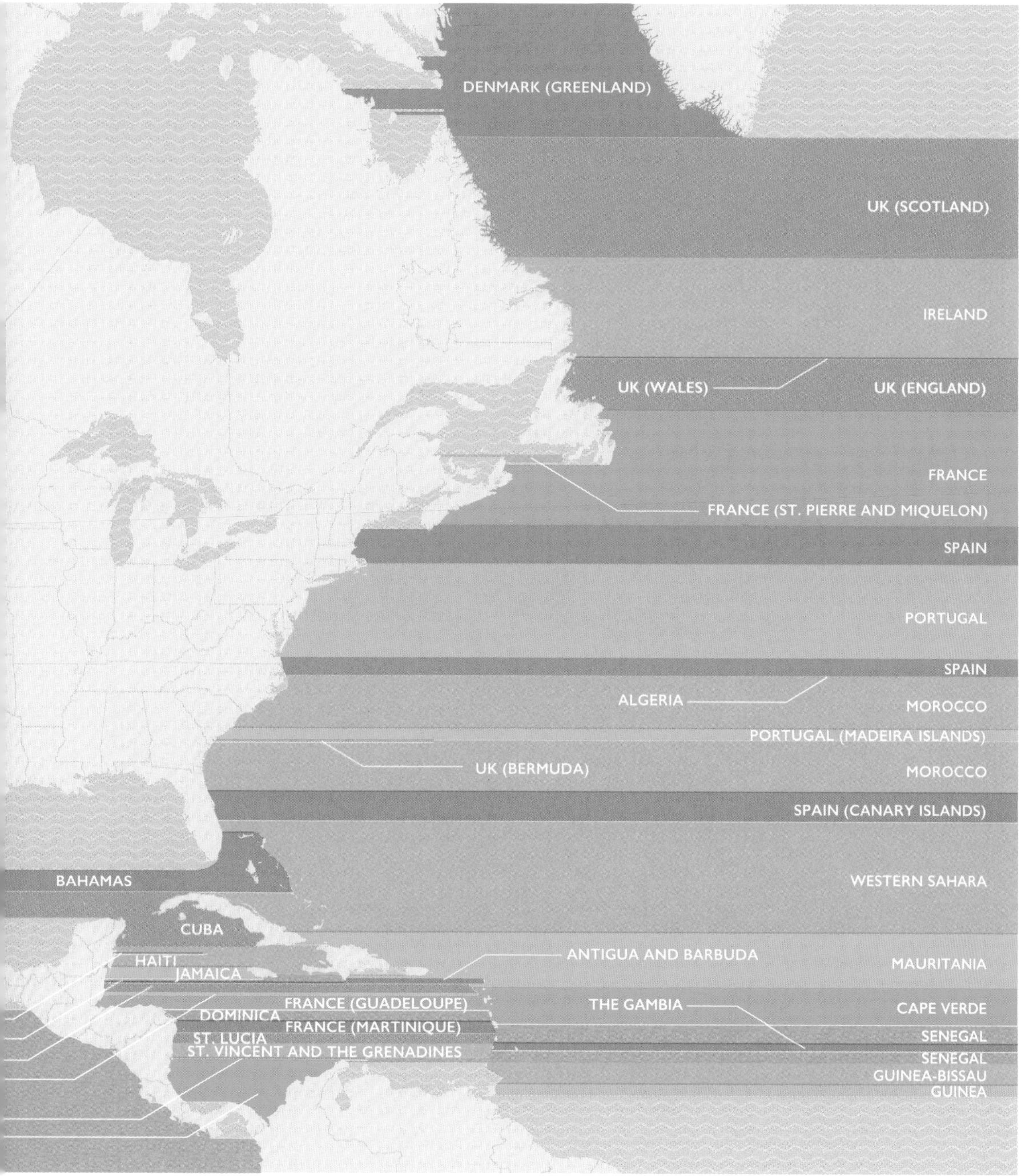
DENMARK (GREENLAND)
UK (SCOTLAND)
IRELAND
UK (WALES)
UK (ENGLAND)
FRANCE
FRANCE (ST. PIERRE AND MIQUELON)
SPAIN
PORTUGAL
SPAIN
ALGERIA
MOROCCO
PORTUGAL (MADEIRA ISLANDS)
UK (BERMUDA)
MOROCCO
SPAIN (CANARY ISLANDS)
BAHAMAS
WESTERN SAHARA
CUBA
ANTIGUA AND BARBUDA
MAURITANIA
HAITI
JAMAICA
FRANCE (GUADELOUPE)
THE GAMBIA
CAPE VERDE
DOMINICA
FRANCE (MARTINIQUE)
SENEGAL
ST. LUCIA
ST. VINCENT AND THE GRENADINES
SENEGAL
GUINEA-BISSAU
GUINEA

9 **The inner banks:** Canada's great island lakes and lake islands

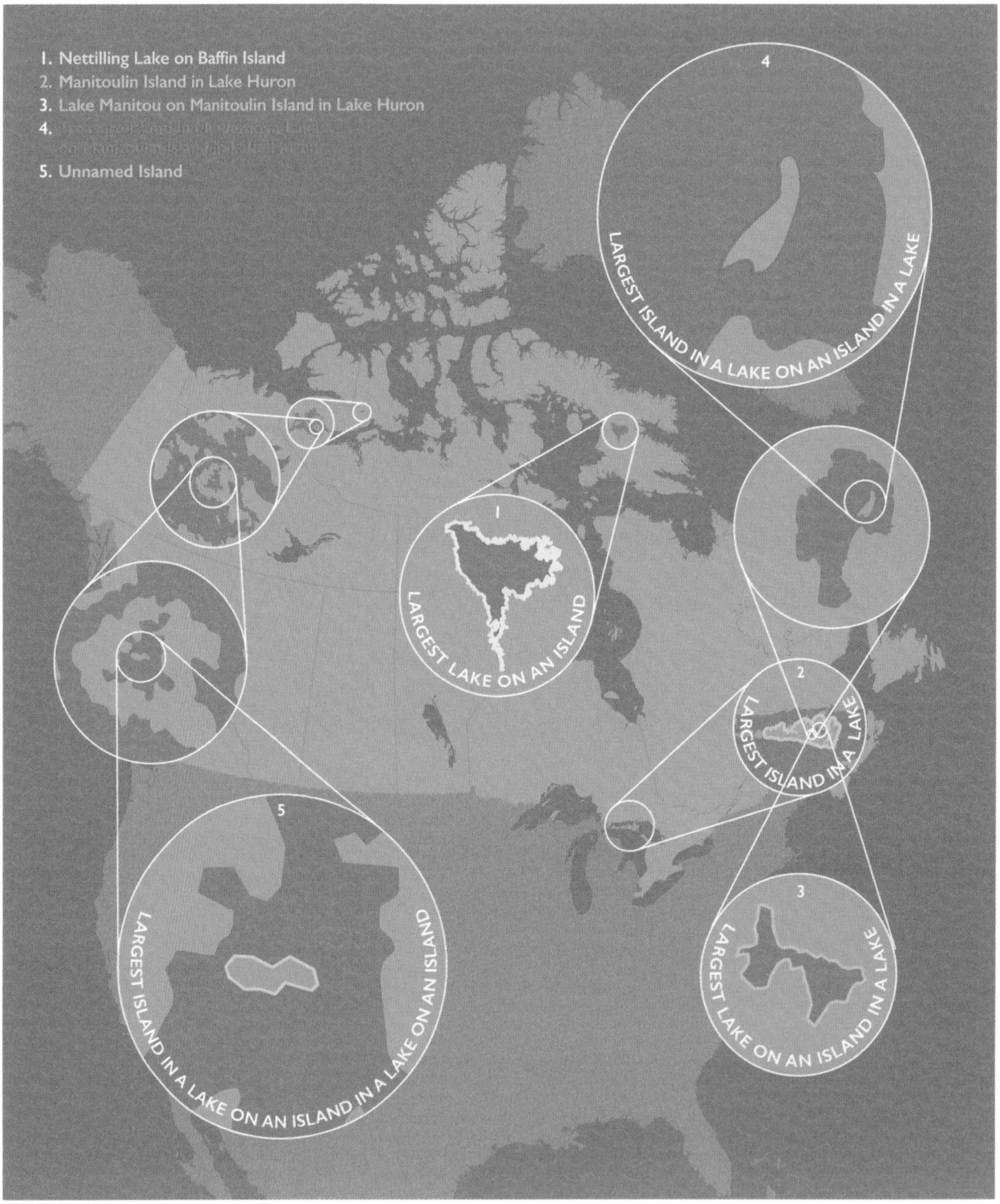

The Canadian Arctic is home to three of the ten largest islands in the world—Victoria Island (third), Baffin Island (fifth), and Ellesmere Island (tenth). All the above features are the world's largest of their kind.

10 **On dry land:** States and provinces not touching an ocean, gulf, or bay

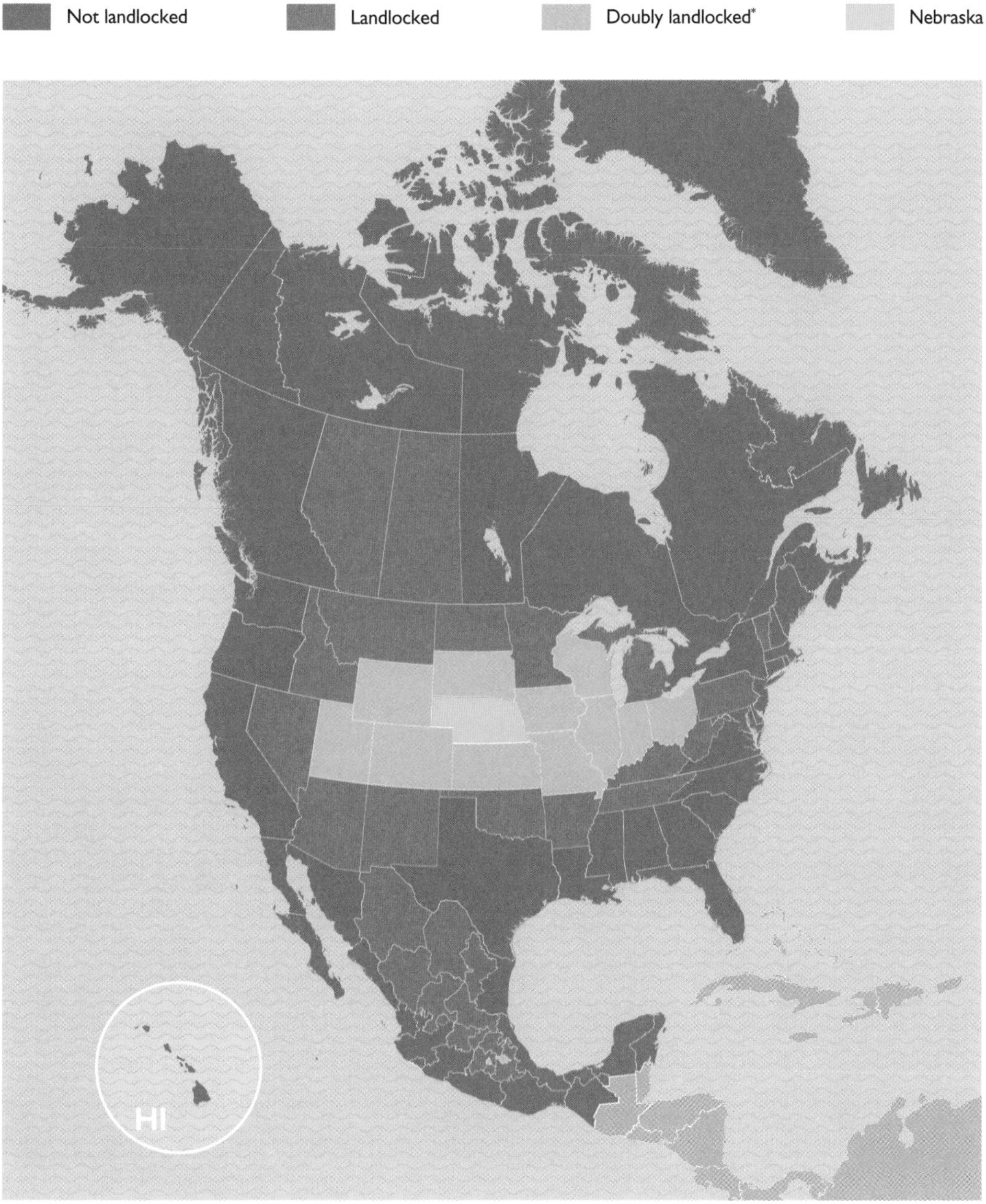

* Doubly landlocked means two steps removed from states or provinces touching an ocean, gulf, or bay.

11 The 8 greatest lakes in North America:
Plus 7 pretty good ones

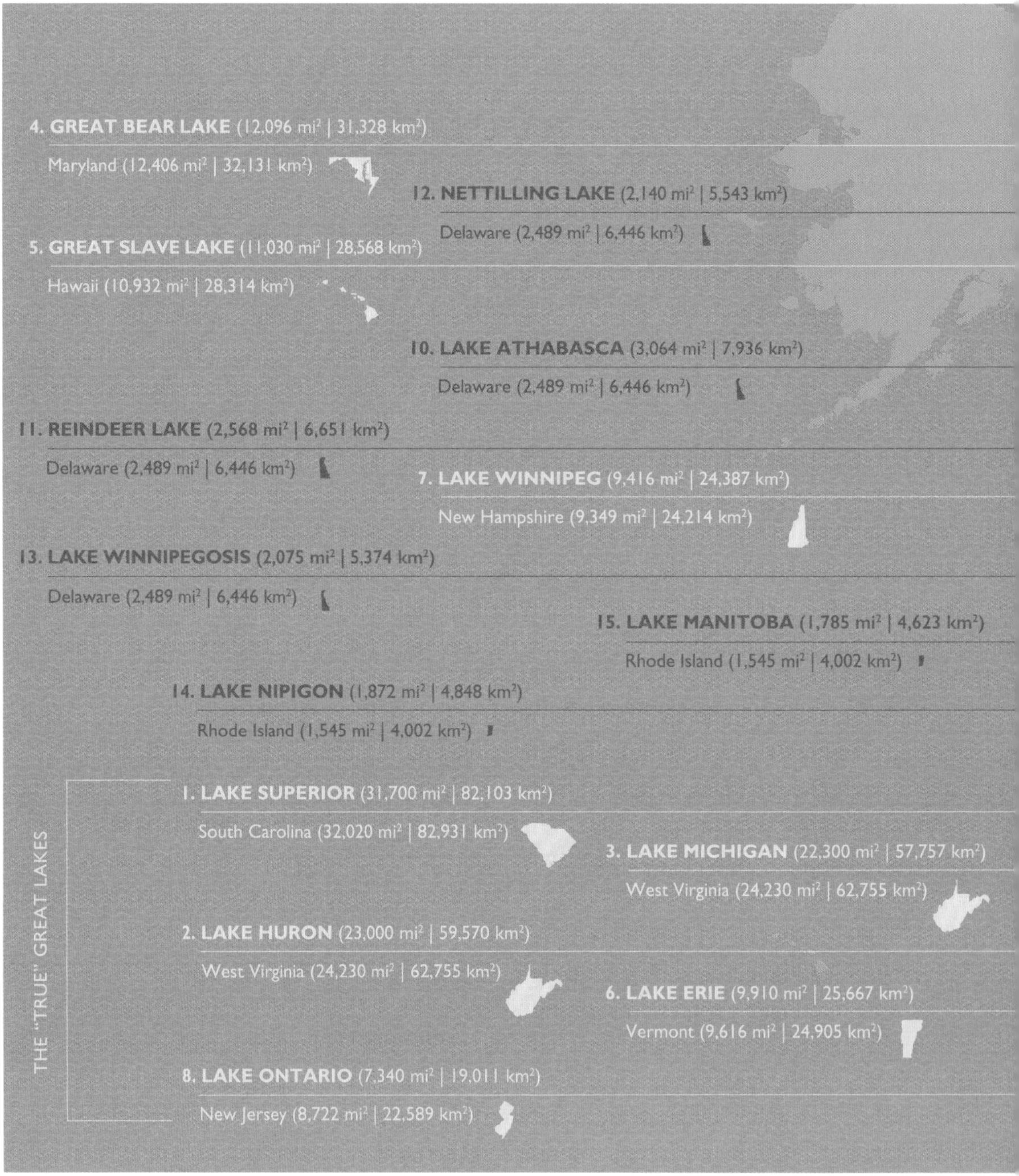
4. GREAT BEAR LAKE (12,096 mi^2 | 31,328 km^2)
Maryland (12,406 mi^2 | 32,131 km^2)
12. NETTILLING LAKE (2,140 mi^2 | 5,543 km^2)
Delaware (2,489 mi^2 | 6,446 km^2)
5. GREAT SLAVE LAKE (11,030 mi^2 | 28,568 km^2)
Hawaii (10,932 mi^2 | 28,314 km^2)
10. LAKE ATHABASCA (3,064 mi^2 | 7,936 km^2)
Delaware (2,489 mi^2 | 6,446 km^2)
11. REINDEER LAKE (2,568 mi^2 | 6,651 km^2)
Delaware (2,489 mi^2 | 6,446 km^2)
7. LAKE WINNIPEG (9,416 mi^2 | 24,387 km^2)
New Hampshire (9,349 mi^2 | 24,214 km^2)
13. LAKE WINNIPEGOSIS (2,075 mi^2 | 5,374 km^2)
Delaware (2,489 mi^2 | 6,446 km^2)
15. LAKE MANITOBA (1,785 mi^2 | 4,623 km^2)
Rhode Island (1,545 mi^2 | 4,002 km^2)
14. LAKE NIPIGON (1,872 mi^2 | 4,848 km^2)
Rhode Island (1,545 mi^2 | 4,002 km^2)
THE "TRUE" GREAT LAKES
1. LAKE SUPERIOR (31,700 mi^2 | 82,103 km^2)
South Carolina (32,020 mi^2 | 82,931 km^2)
3. LAKE MICHIGAN (22,300 mi^2 | 57,757 km^2)
West Virginia (24,230 mi^2 | 62,755 km^2)
2. LAKE HURON (23,000 mi^2 | 59,570 km^2)
West Virginia (24,230 mi^2 | 62,755 km^2)
6. LAKE ERIE (9,910 mi^2 | 25,667 km^2)
Vermont (9,616 mi^2 | 24,905 km^2)
8. LAKE ONTARIO (7,340 mi^2 | 19,011 km^2)
New Jersey (8,722 mi^2 | 22,589 km^2)

What makes the five "true" Great Lakes great? Not their size, apparently. Three North American lakes—all in Canada—are larger than Lake Ontario, two of which also edge out Lake Erie for top-five status. Great Bear Lake—the greatest of the not-actually Great lakes—is bigger than eight US states.

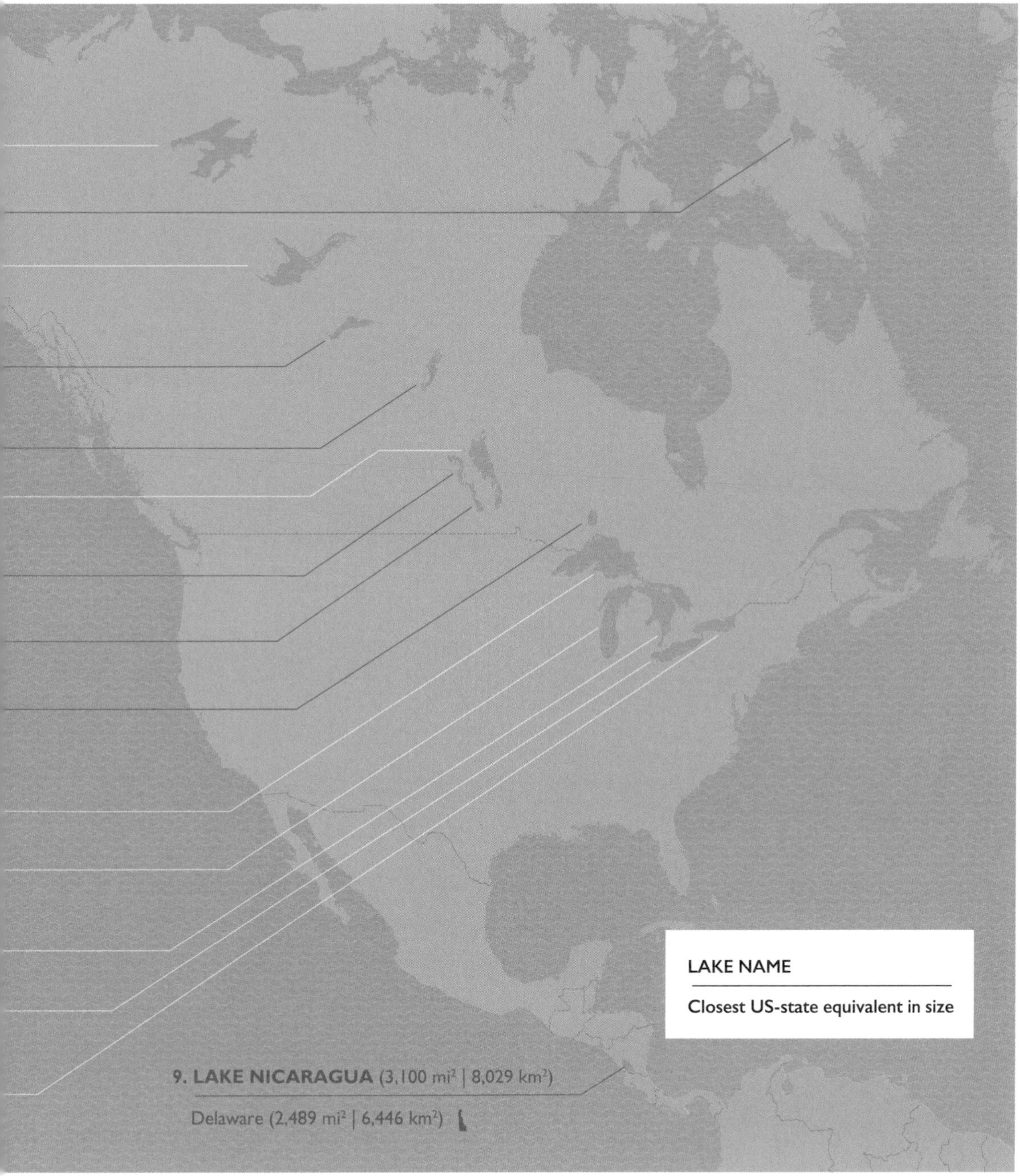

12 The United States of Washington:
Every place named after the first president

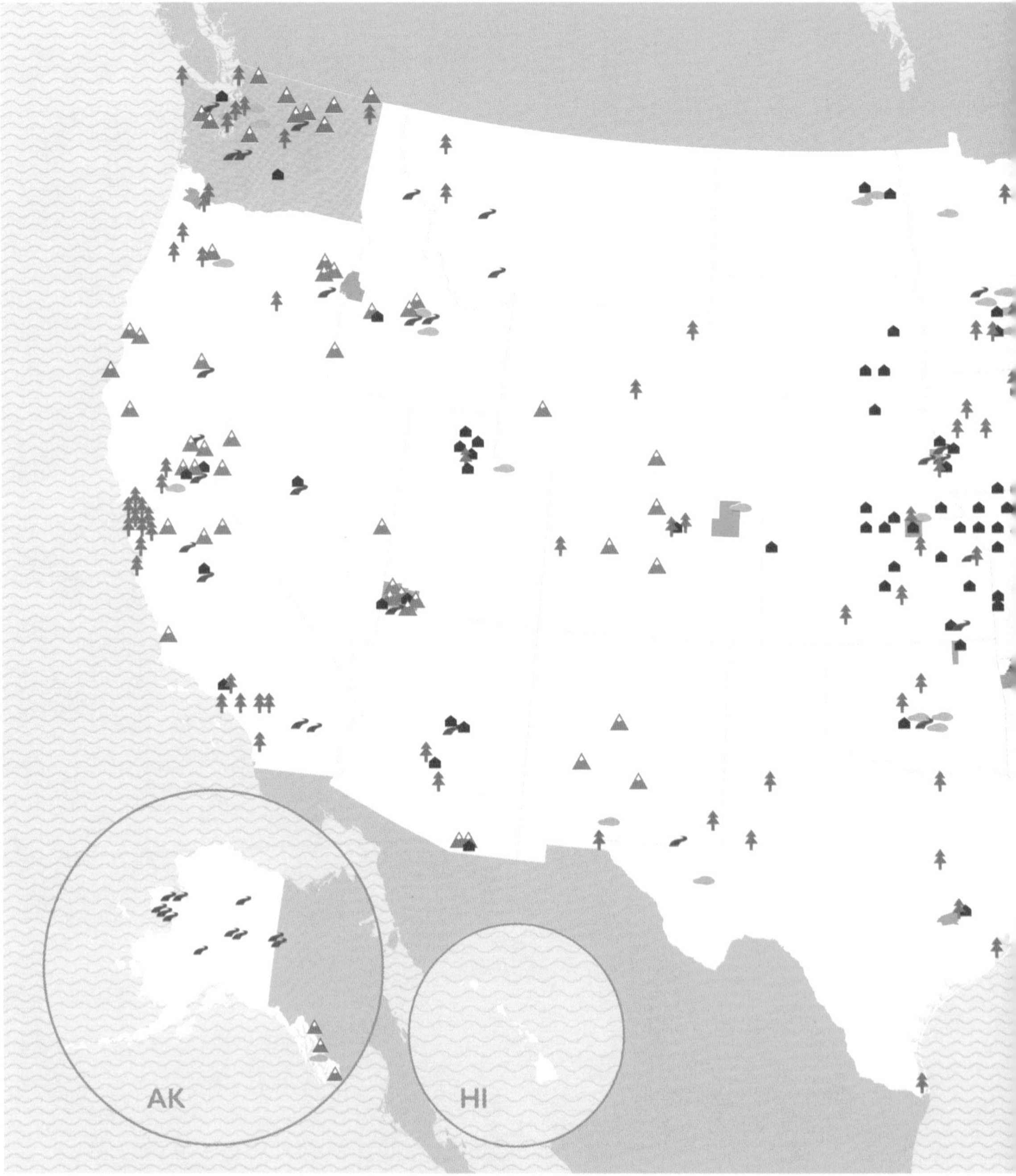

The US Census Bureau recently found Washington to be the most common place name in the country. Here, we've mapped every single state, county, city, town, and other census-recognized place, as well as every natural feature named to honor the first US president. There are a *lot* of them. And this isn't even counting the many thousands of human-made places, including airports, bridges, fire and police stations, courthouses, libraries, churches, hospitals, schools, canals, cemeteries, dams, military facilities, oil fields, post offices, reservoirs, radio towers, and more. We'd need a bigger map to show them all.

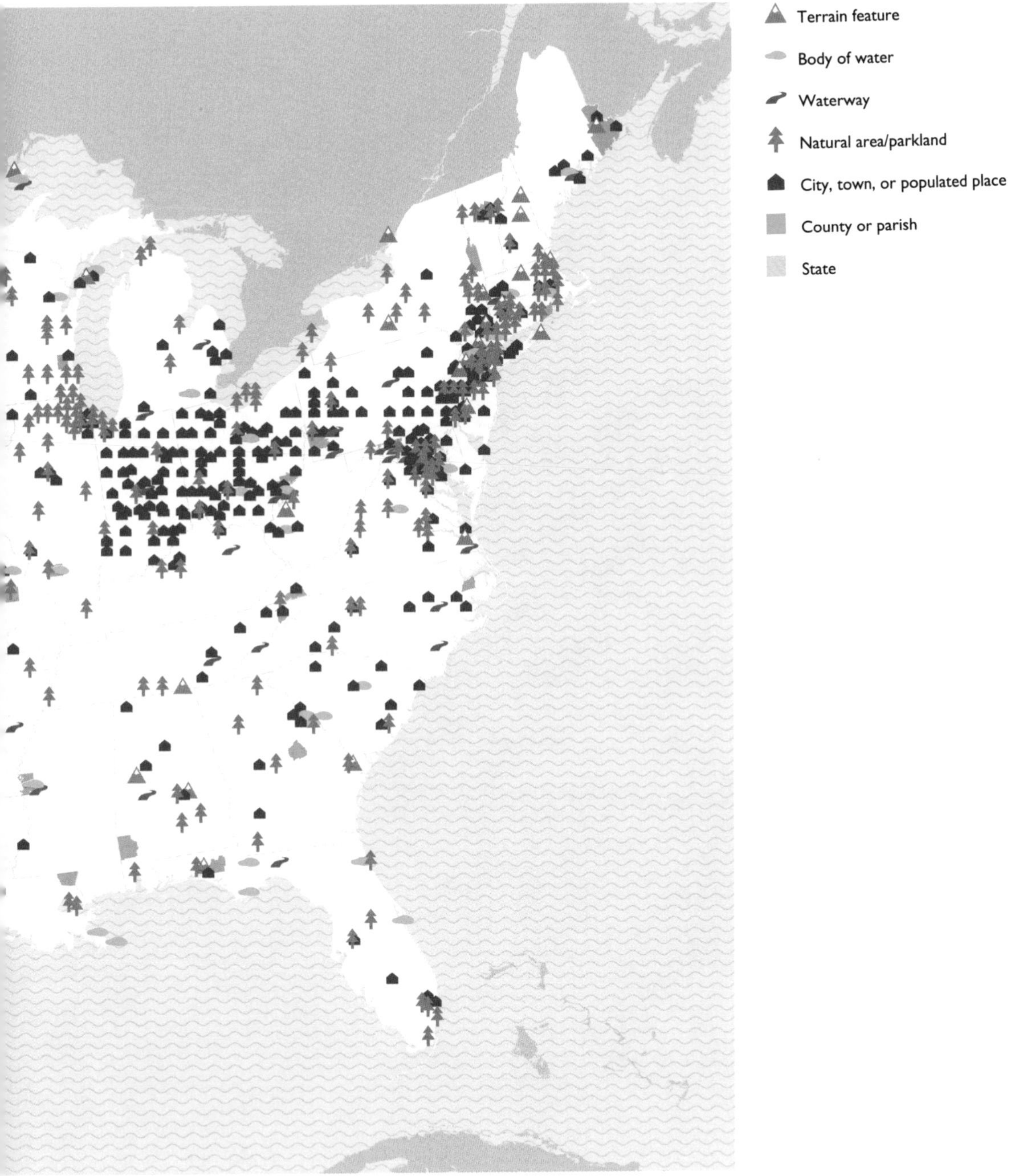

13 **Rhode nation:** If every state were the size of Rhode Island, there would be 2,457 states

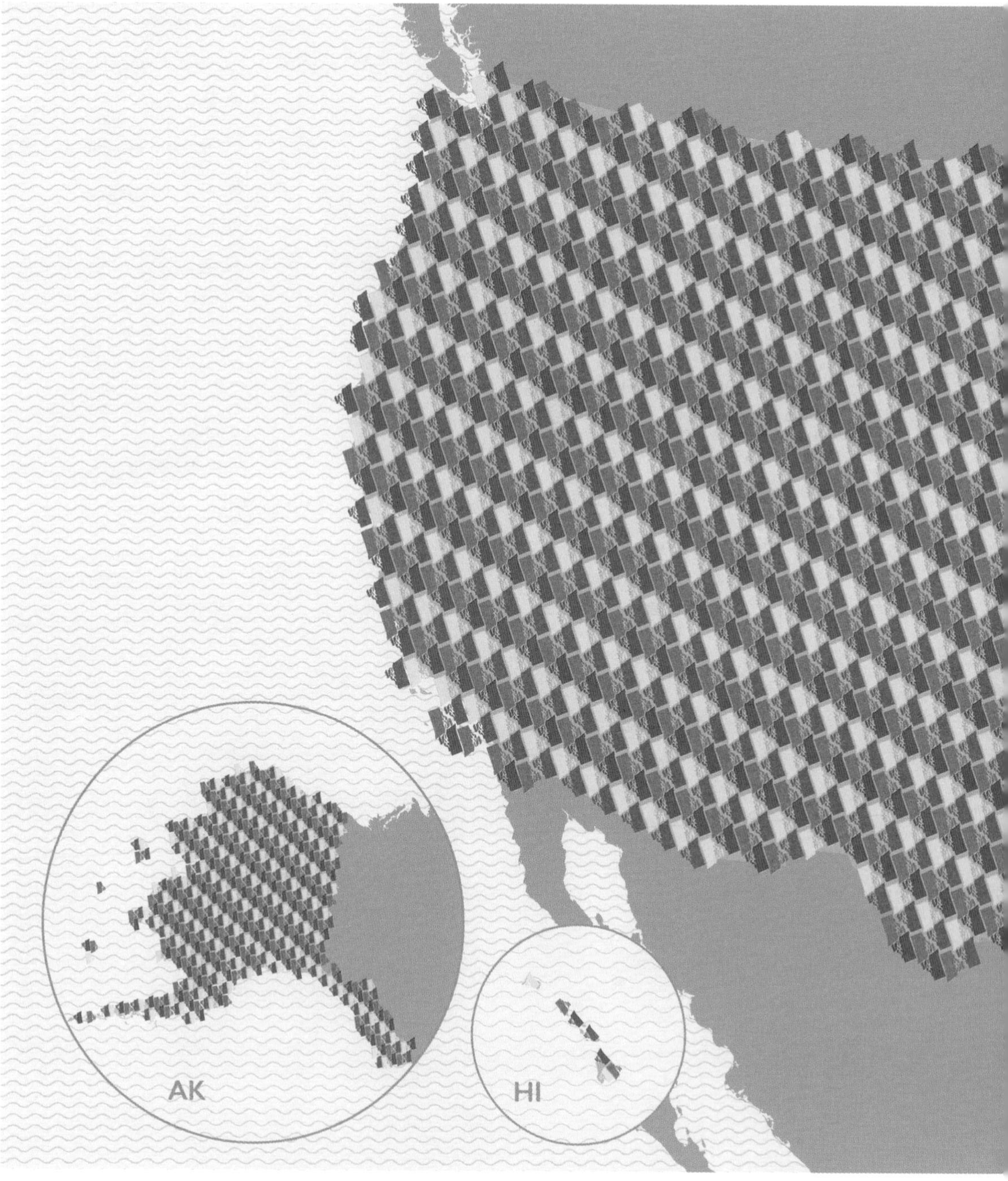

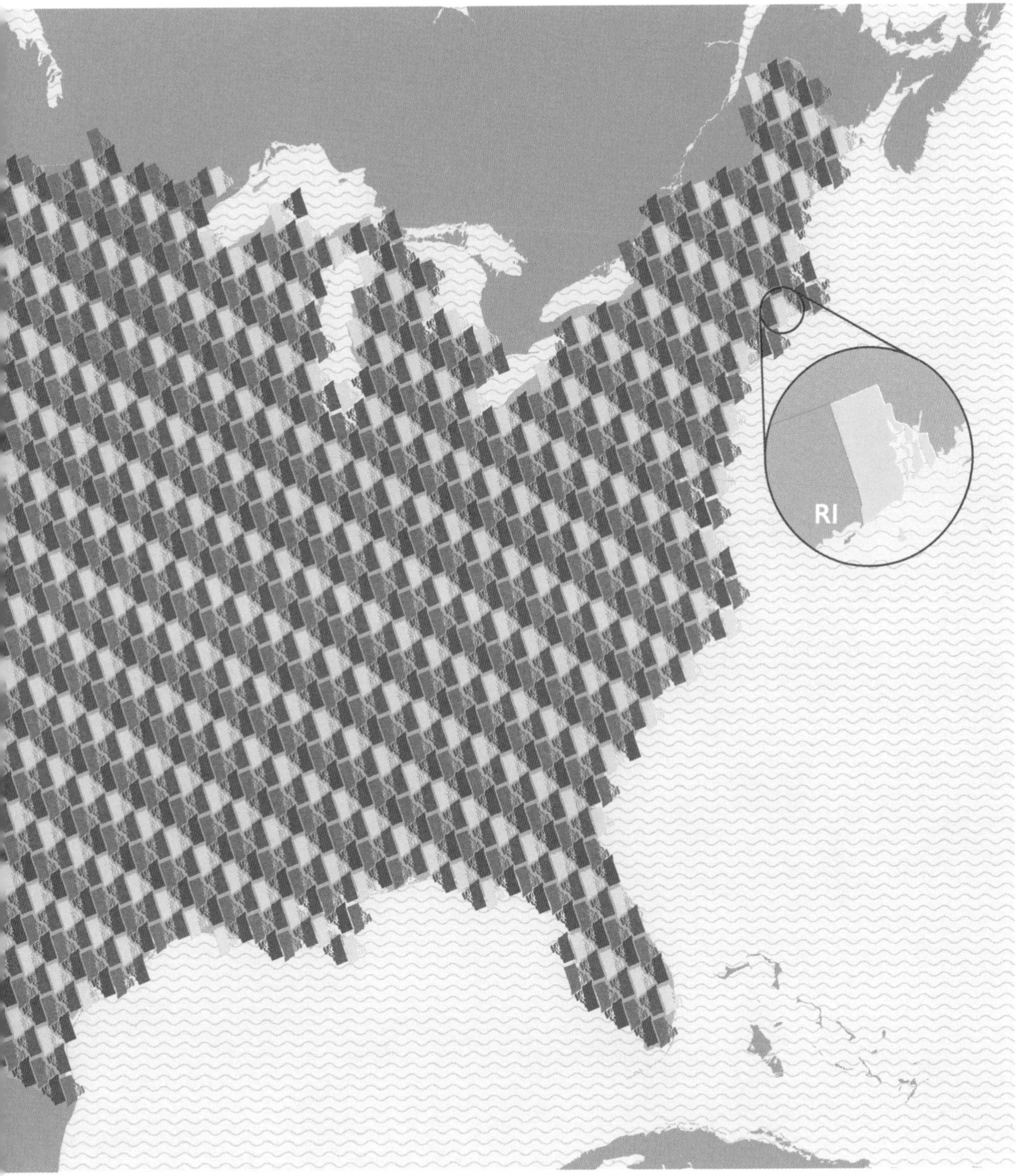
RI

14 **Name dropping:** City names, then and now

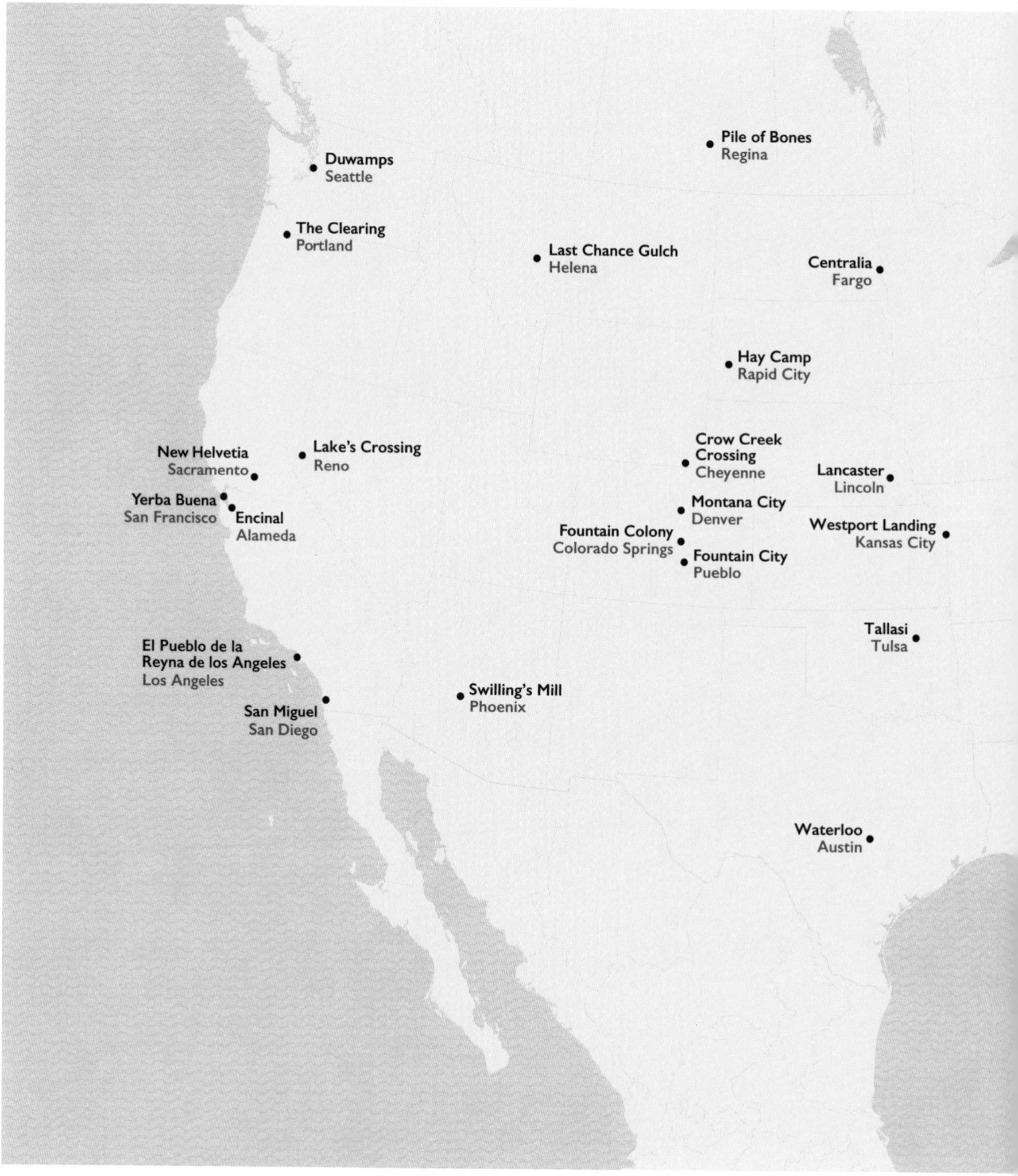
Pile of Bones
Regina
Duwamps
Seattle
The Clearing
Portland
Last Chance Gulch
Helena
Centralia
Fargo
Hay Camp
Rapid City
New Helvetia
Sacramento
Lake's Crossing
Reno
Crow Creek Crossing
Cheyenne
Lancaster
Lincoln
Yerba Buena
San Francisco
Encinal
Alameda
Montana City
Denver
Fountain Colony
Colorado Springs
Westport Landing
Kansas City
Fountain City
Pueblo
Tallasi
Tulsa
El Pueblo de la Reyna de los Angeles
Los Angeles
Swilling's Mill
Phoenix
San Miguel
San Diego
Waterloo
Austin

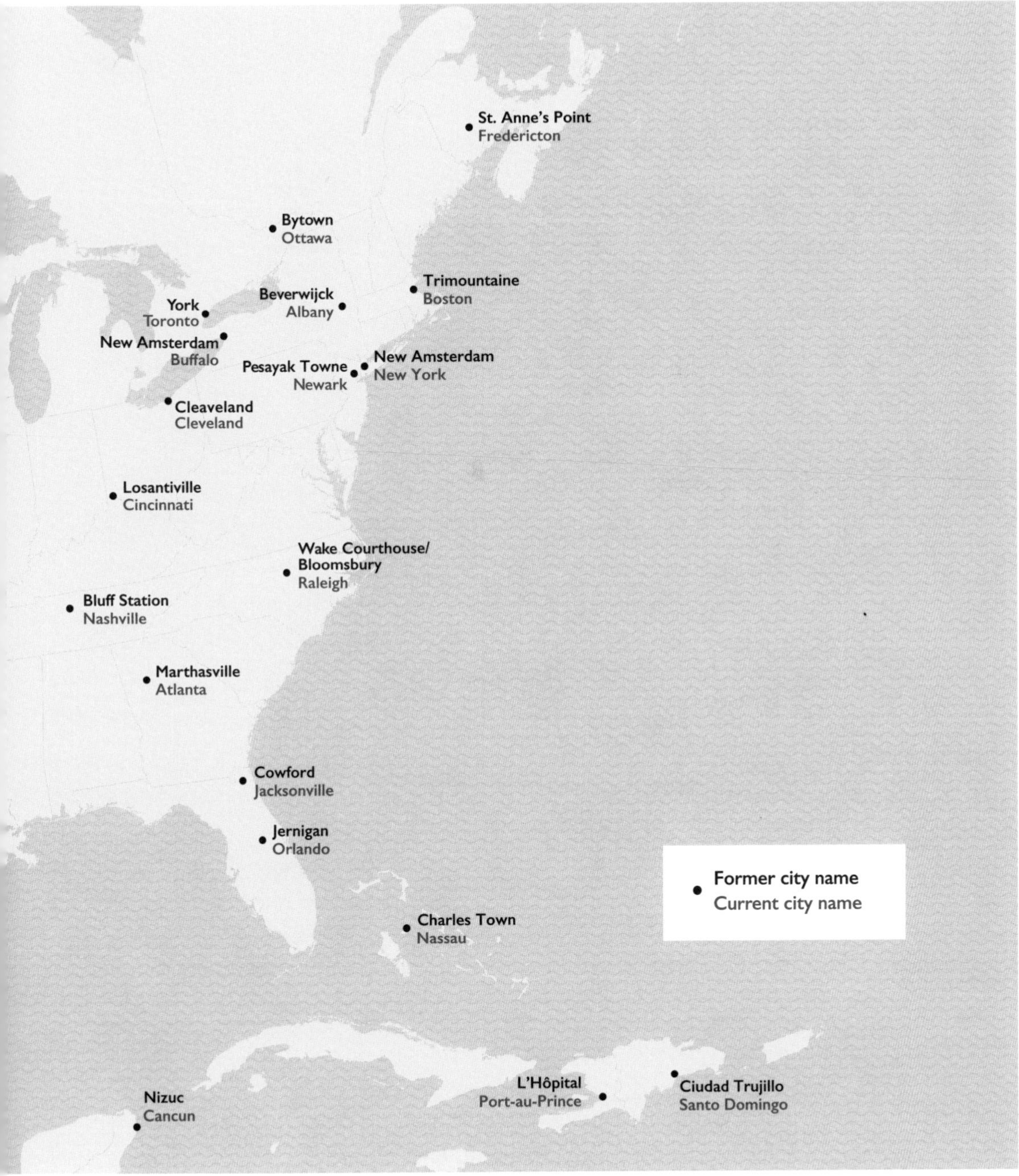
St. Anne's Point
Fredericton
Bytown
Ottawa
Trimountaine
Boston
York
Toronto
Beverwijck
Albany
New Amsterdam
Buffalo
Pesayak Towne
Newark
New Amsterdam
New York
Cleaveland
Cleveland
Losantiville
Cincinnati
Wake Courthouse/
Bloomsbury
Raleigh
Bluff Station
Nashville
Marthasville
Atlanta
Cowford
Jacksonville
Jernigan
Orlando
Former city name
Current city name
Charles Town
Nassau
Nizuc
Cancun
L'Hôpital
Port-au-Prince
Ciudad Trujillo
Santo Domingo

15 **A river runs beside it:** US natural borders

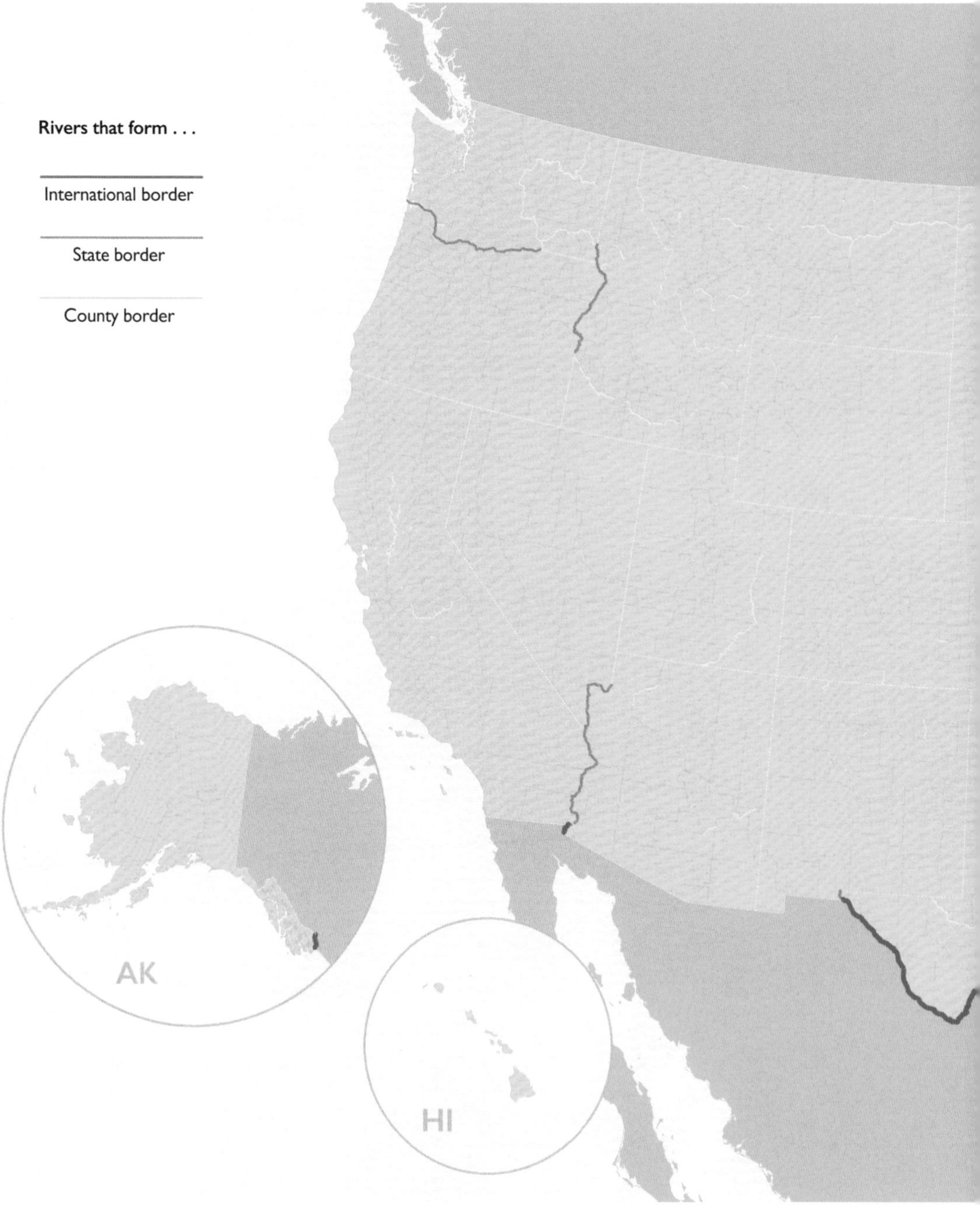

Rivers form natural boundaries, and they account for a great deal of the US's political borders that resemble meandering squiggles rather than straight lines. Only five US states—Hawaii, Colorado, Wyoming, Utah, and Montana—have no river borders; the rest are shaped at least in part by the rivers that run along their state lines. This map is largely based on satellite imagery that counts rivers about 100 feet (30 m) or wider.

2

POLITICS AND POWER

16 **Madam Governor:** Number of women governors by state or province*

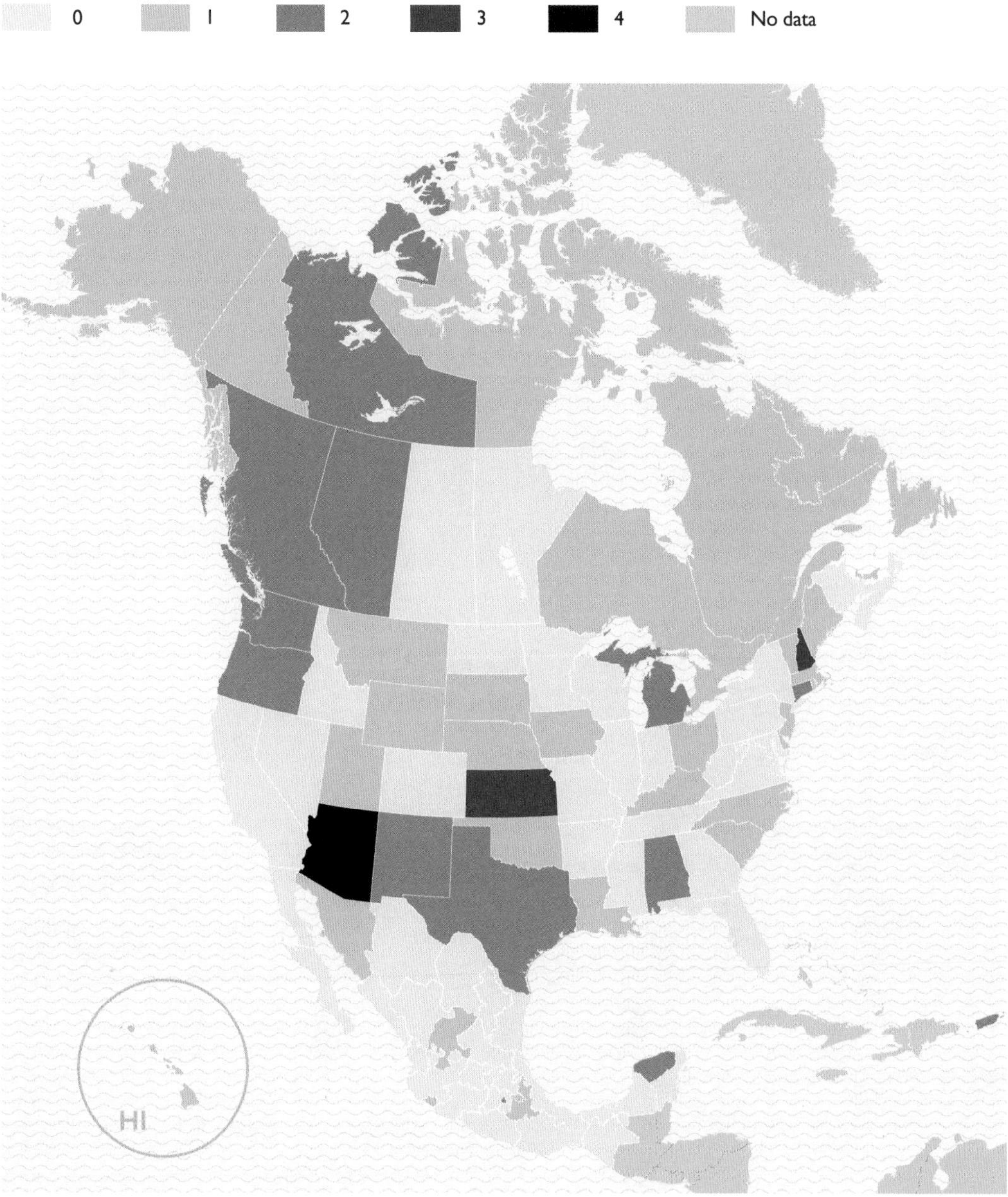

* Canada's equivalent of a governor, the head of government of a province, is called a premier. Also note that the head of government of Mexico City, though included here, functions somewhat differently from a governor of a Mexican state (gobernador).

17 **Belles of the ballot:** When women gained the right to vote

The US's ratification of the 19th Amendment in 1920, a crucial moment in the history of women's suffrage, is far from the whole story in North America, with some states and provinces granting women the right to vote well before 1920—and many countries well after. Canada granted women the right to vote across the country in 1918, while farther south, the first federal election in which women could cast a vote in Mexico took place in 1955.

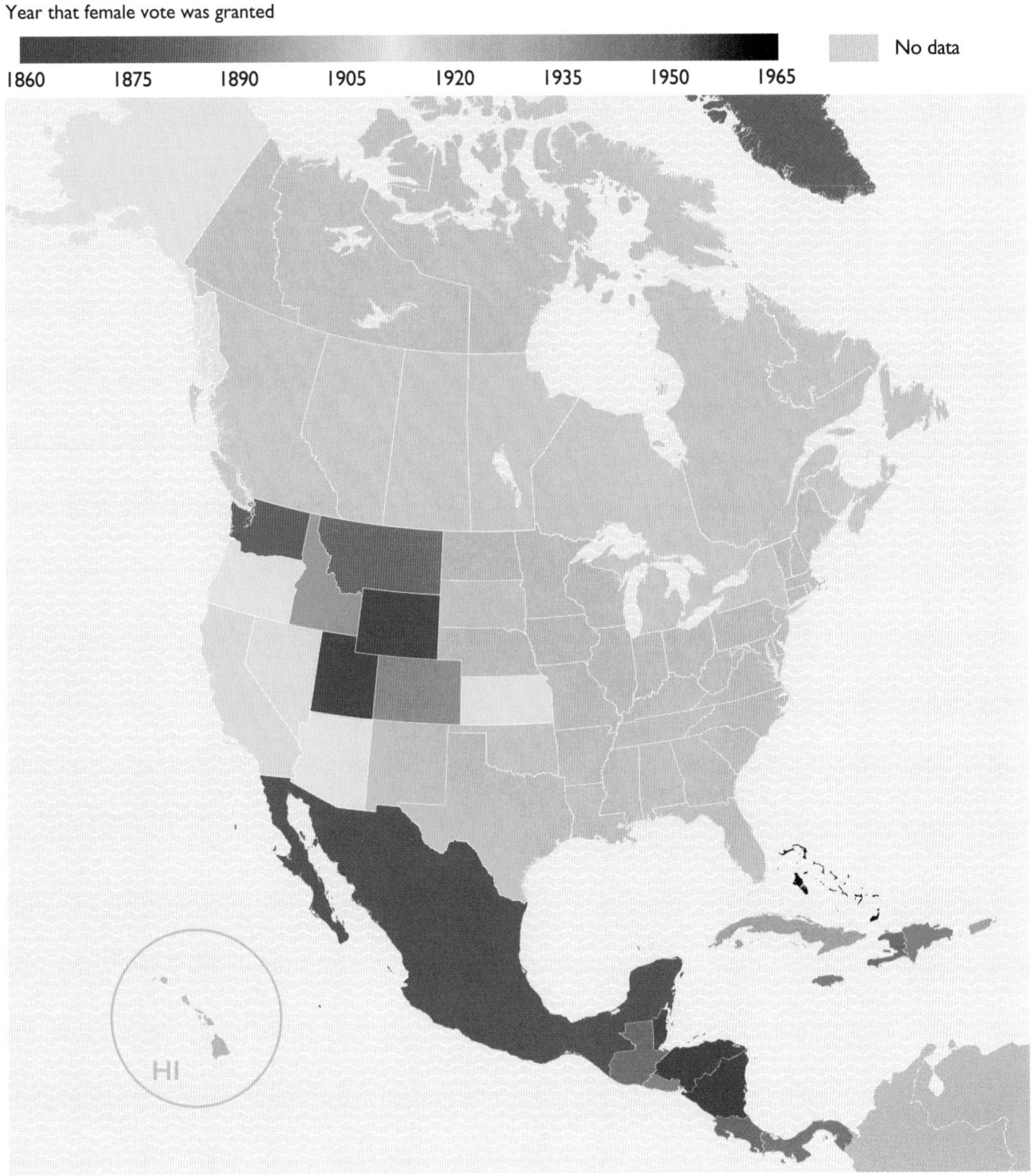

18 **Engendering equality:** The ongoing struggle to pass the Equal Rights Amendment

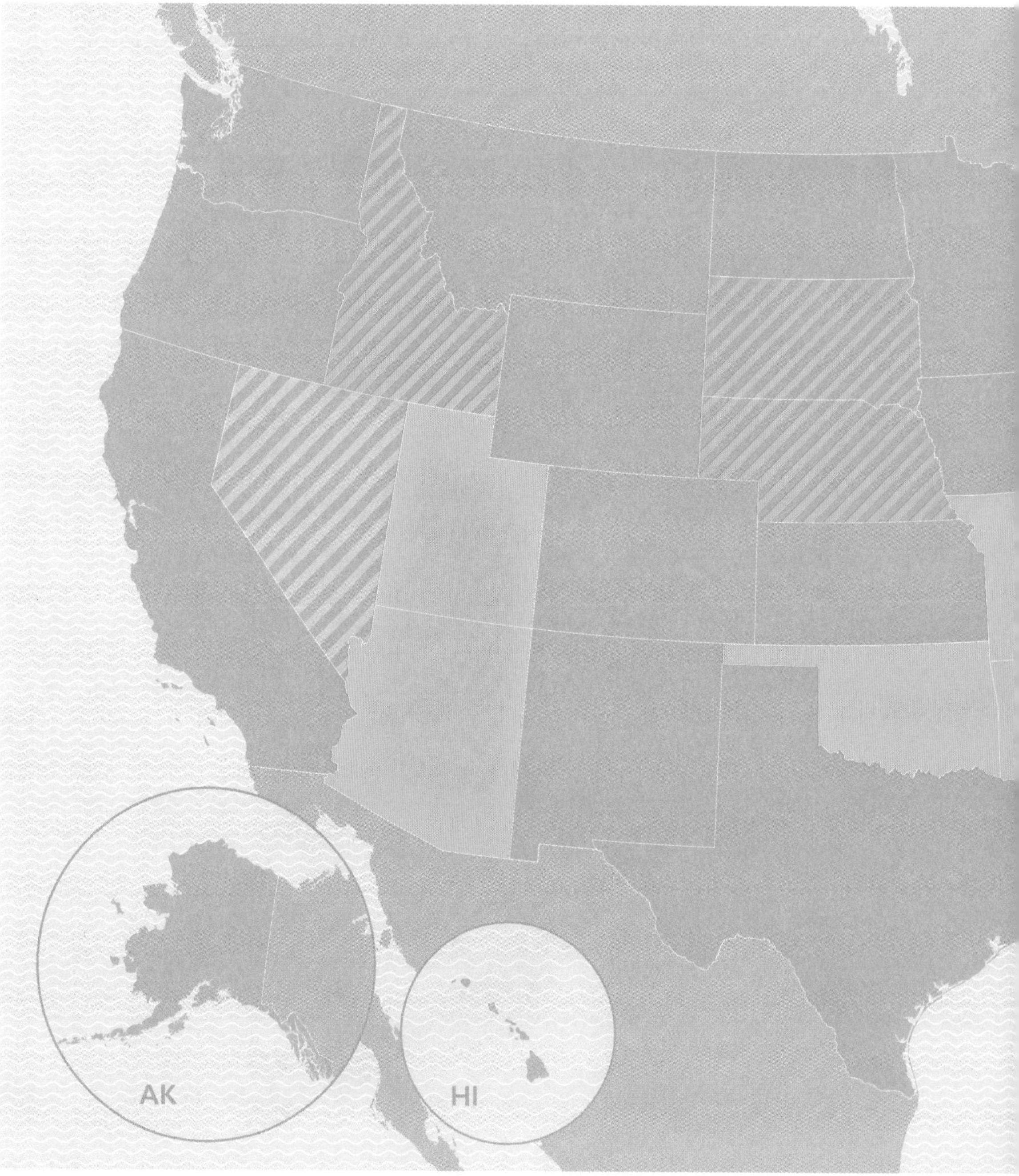

Nothing in the constitution explicitly prohibits gender discrimination; the Equal Rights Amendment (ERA) was passed by Congress in 1972 to fill what many consider a gaping hole in our laws. It then went to the states for ratification . . . and fell just short of the required 38 states by a 1982 deadline. And while a handful of states have backed out (the legality of which is disputed), the struggle to ratify the ERA is alive and well. Three states have ratified it in the past few years, most recently Virginia in 2020. And in 2021, the House voted to cancel the original deadline. A half-century later, the ERA may yet pass.

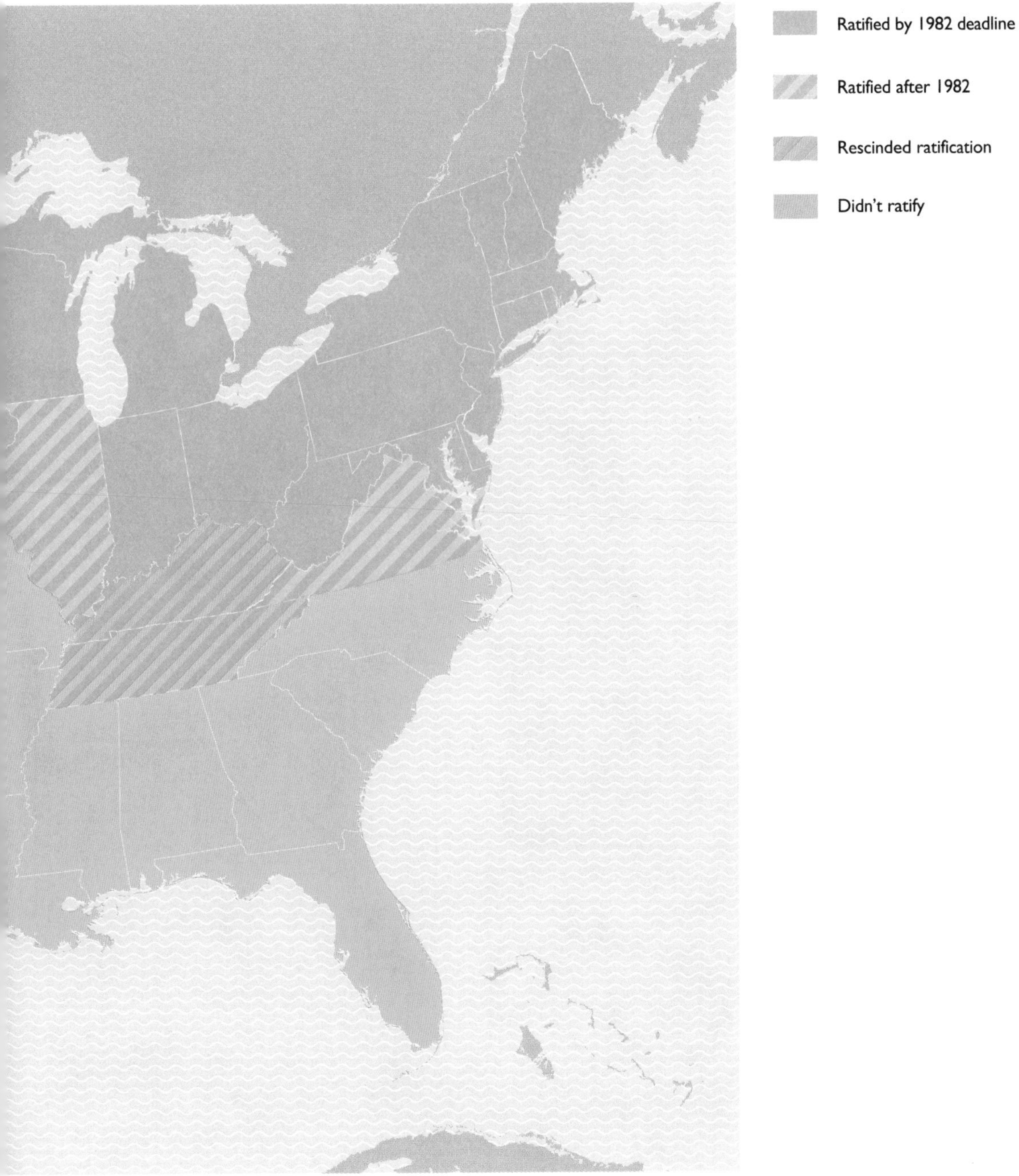

19 Red California and blue Texas?: Presidential election results since 1856

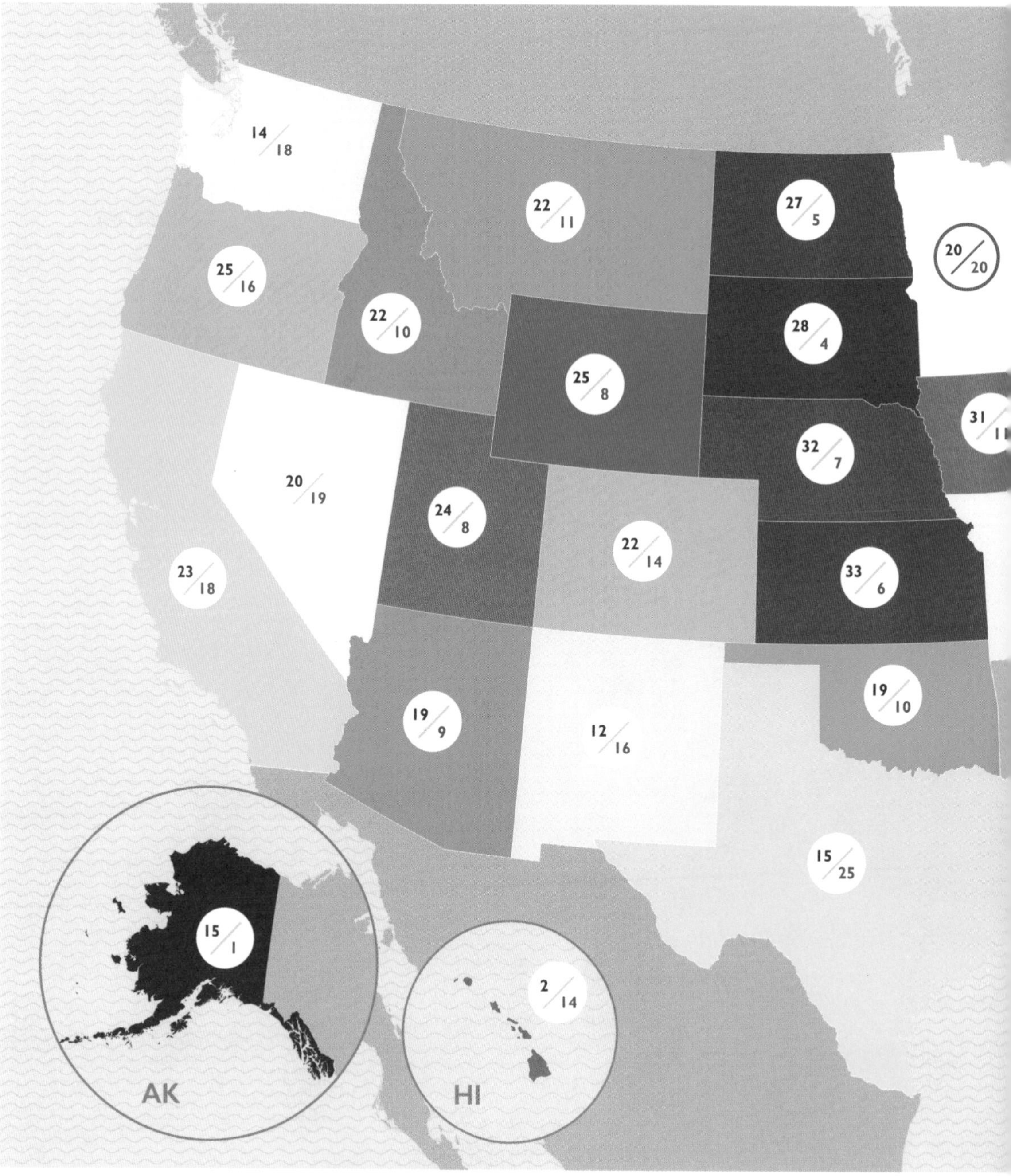

We've grown accustomed to a political map where all but a few states are dyed-in-the-wool Republican or Democrat. But since the establishment of these two modern US political parties in the 1850s, states' party allegiance may not be as one-sided as you might expect. Yes, the parties themselves have undergone something of a platform switch over time, but even this ideological transformation makes the point: Nothing is inevitable.

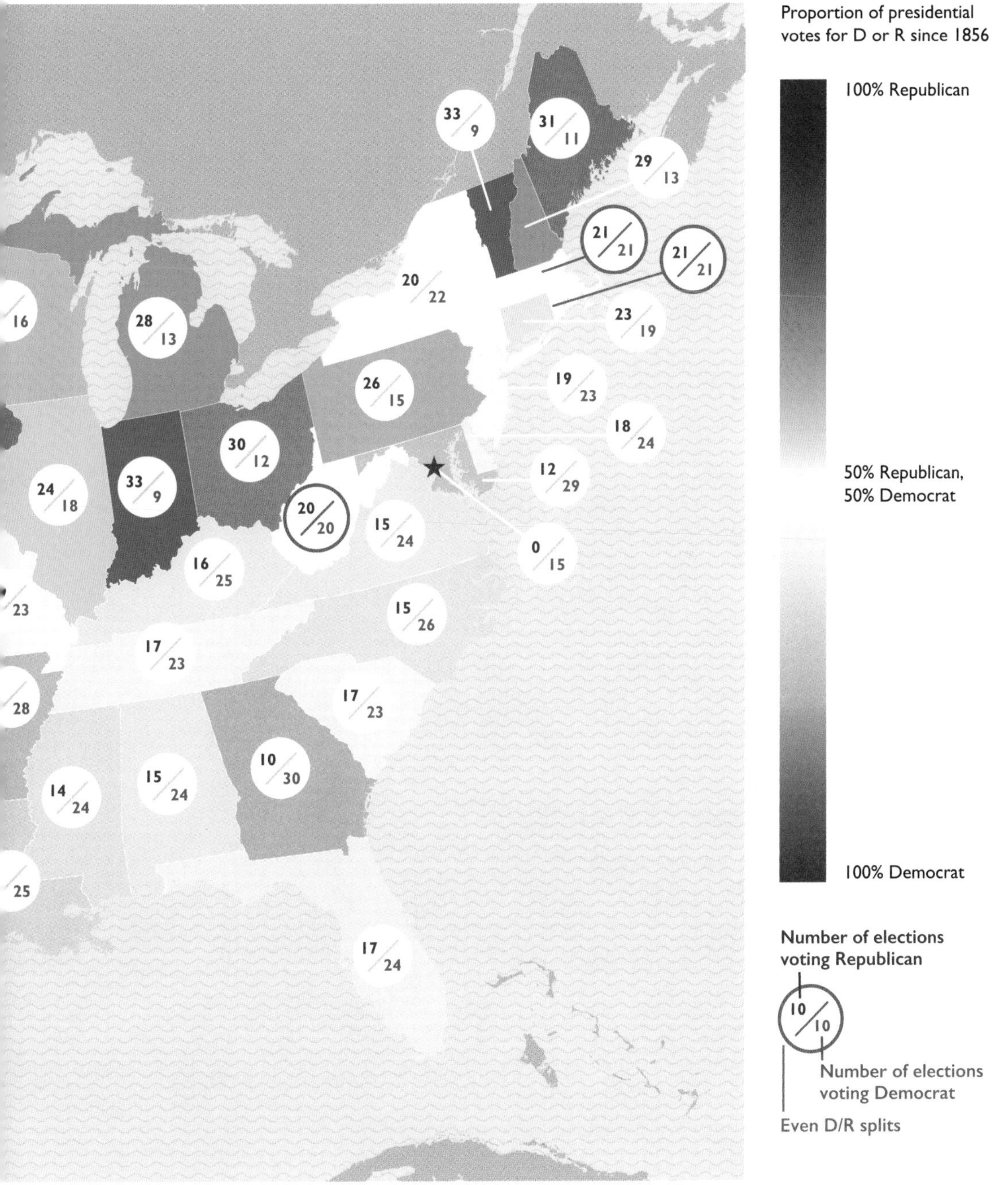

20 **Worth the visit?:** States the candidates made time for during the 2020 general election campaign*

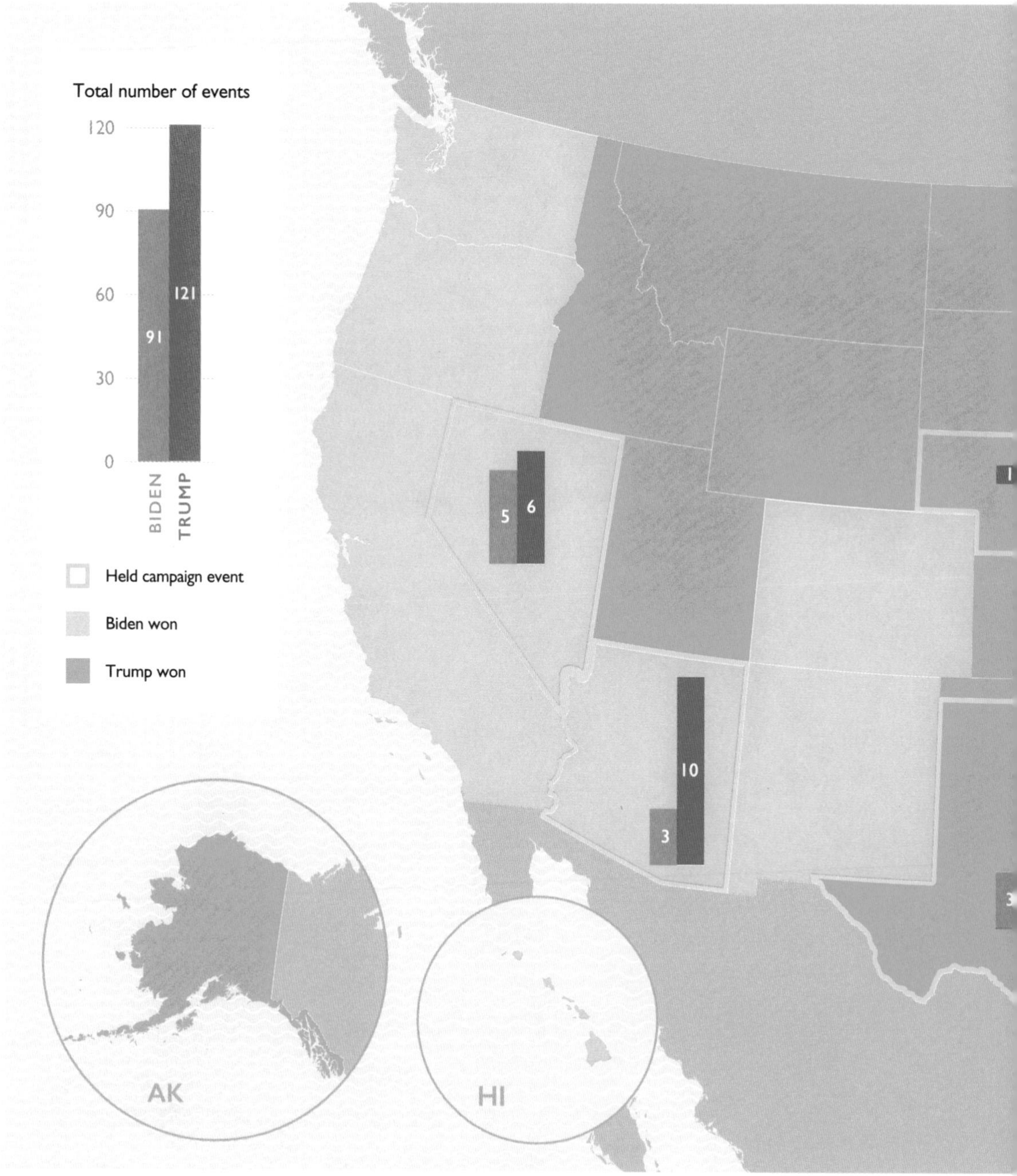

Presidential election outcomes are often determined by a select group of swing states—which is why, from August 28, 2020 (the day after the Republican National Convention ended), through Election Day, the Trump-Pence and Biden-Harris tickets held their 212 campaign events in only 17 states. This is nothing new; since 2008, 22 states haven't had a single general election campaign event from either major party.

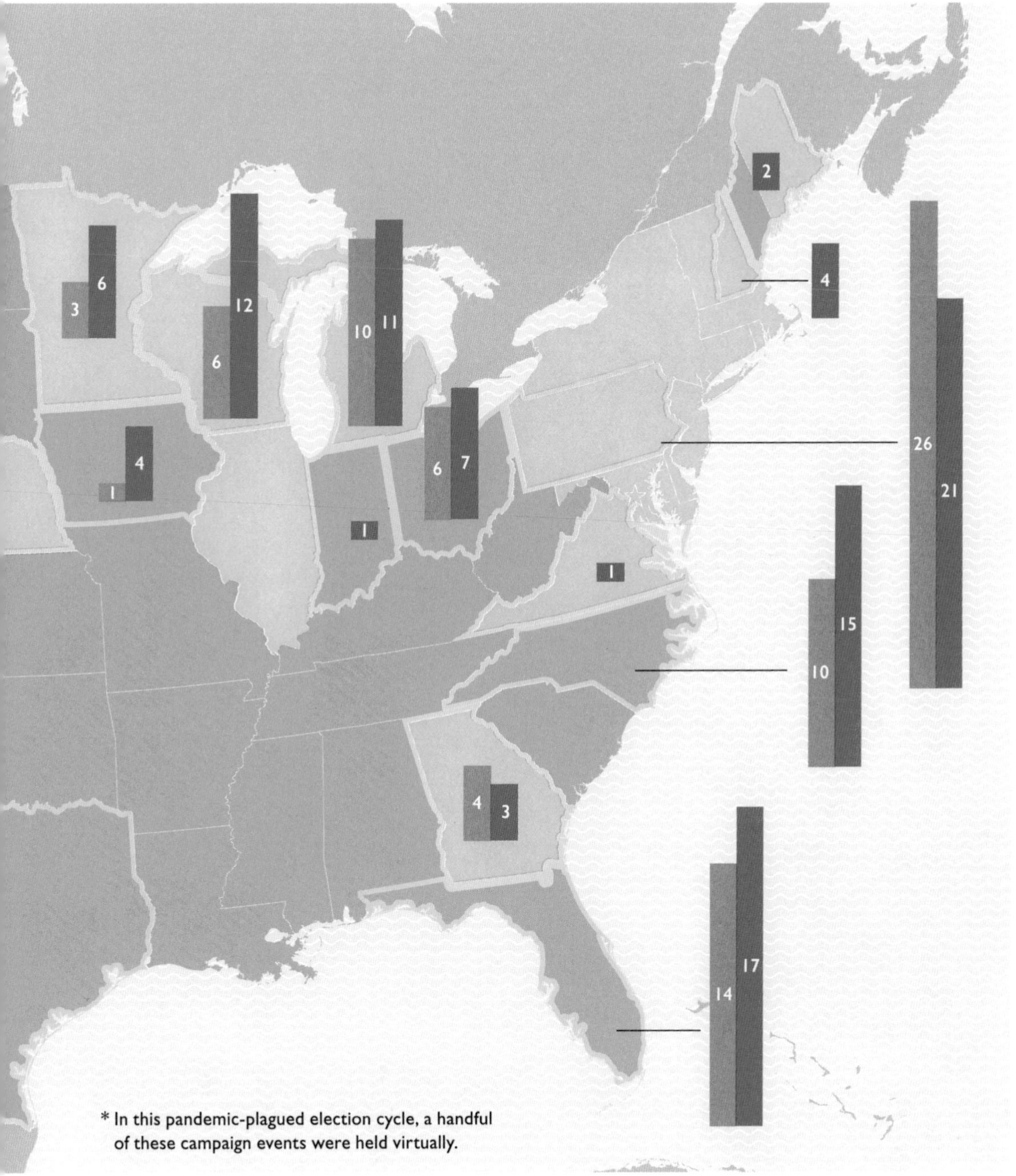

21 "Did Not Vote" ends winning streak:
A historic turnout finally beats the lack of turnout

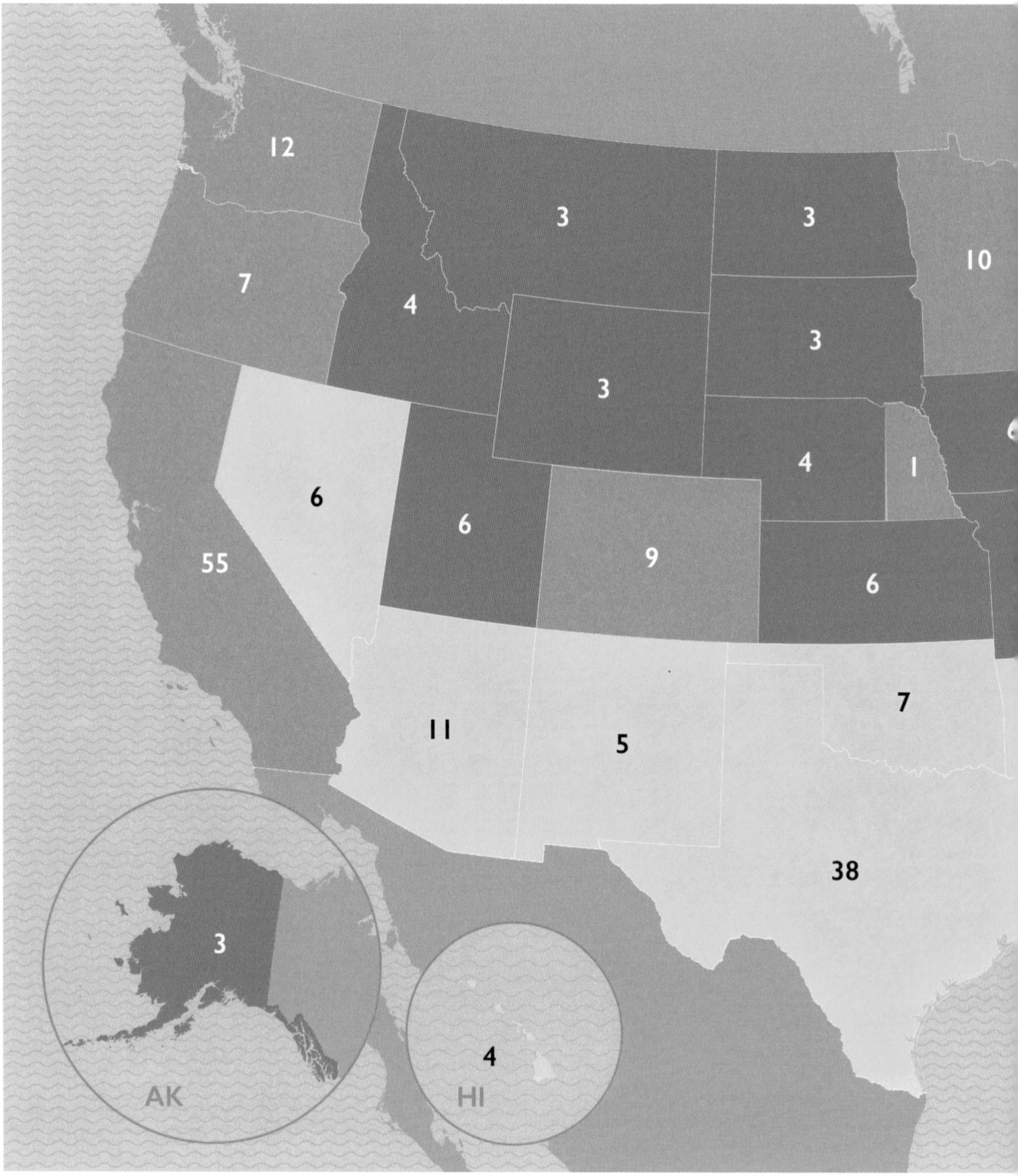

With 67 percent of all voting-eligible citizens participating in the 2020 US election—the highest voter turnout this century—a long-running streak came to an end: victory for a hypothetical "Did Not Vote" candidate. In 2016, "Did Not Vote" would have defeated both Donald Trump and Hillary Clinton in a landslide (see inset below). But in 2020, an election that saw the largest-ever uptick in voter turnout between two presidential elections (17 million more people voted in 2020 than in 2016), the actual leading candidates finally beat out voter apathy. Your vote matters.

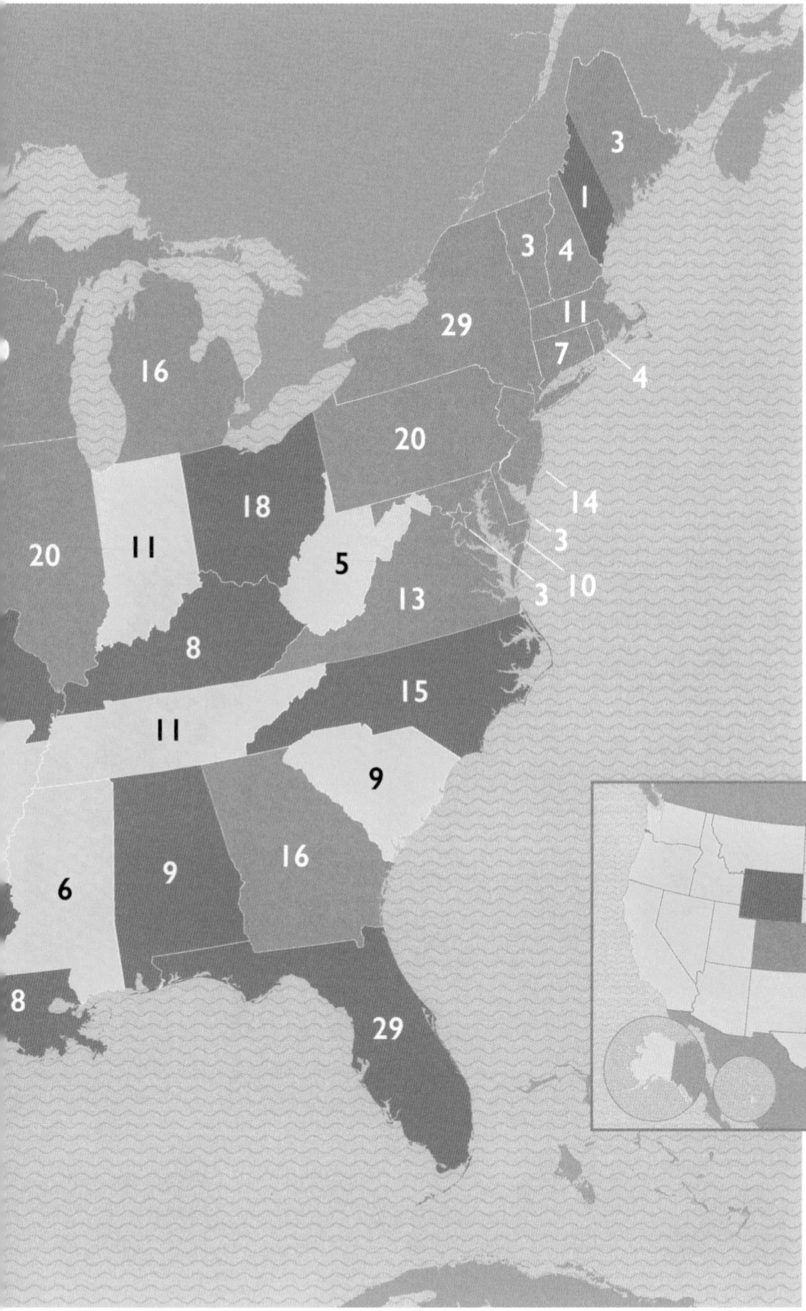

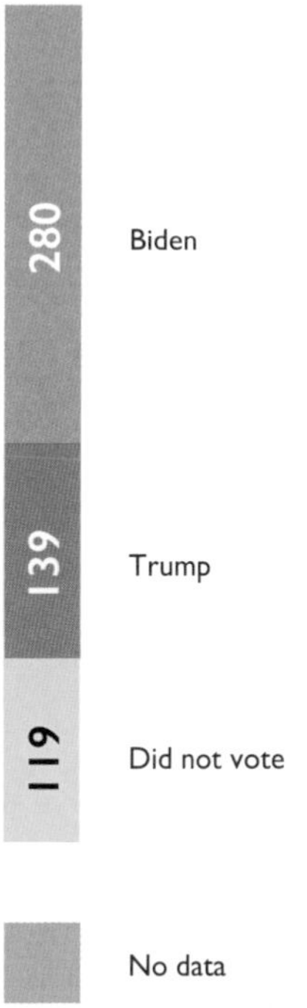

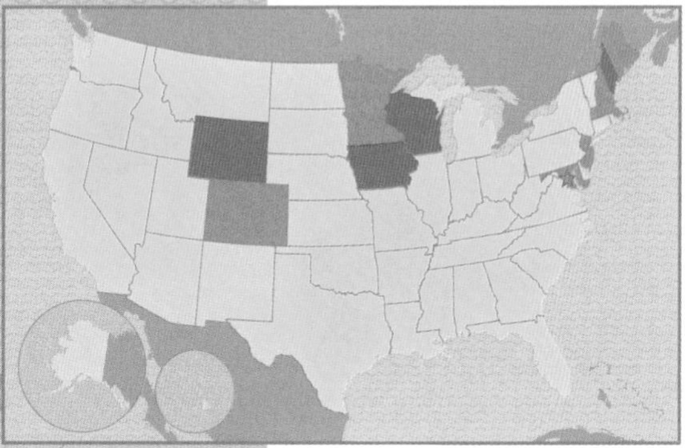

"Did Not Vote" would have won in 2016, with 454 electoral votes to Clinton's 64 and Trump's 20.

22 States that picked the greats:
Predicting the pantheon of best presidents

There were 18 elections for the 10 presidents whom historians consider the all-time greatest, according to a 2017 C-SPAN survey. This map shows which states voted most reliably for them, based on only the number of elections the states actually voted in. For various reasons (e.g., they weren't states yet, or were excluded in 1864 for being Confederate states), not every state voted in all of these elections.

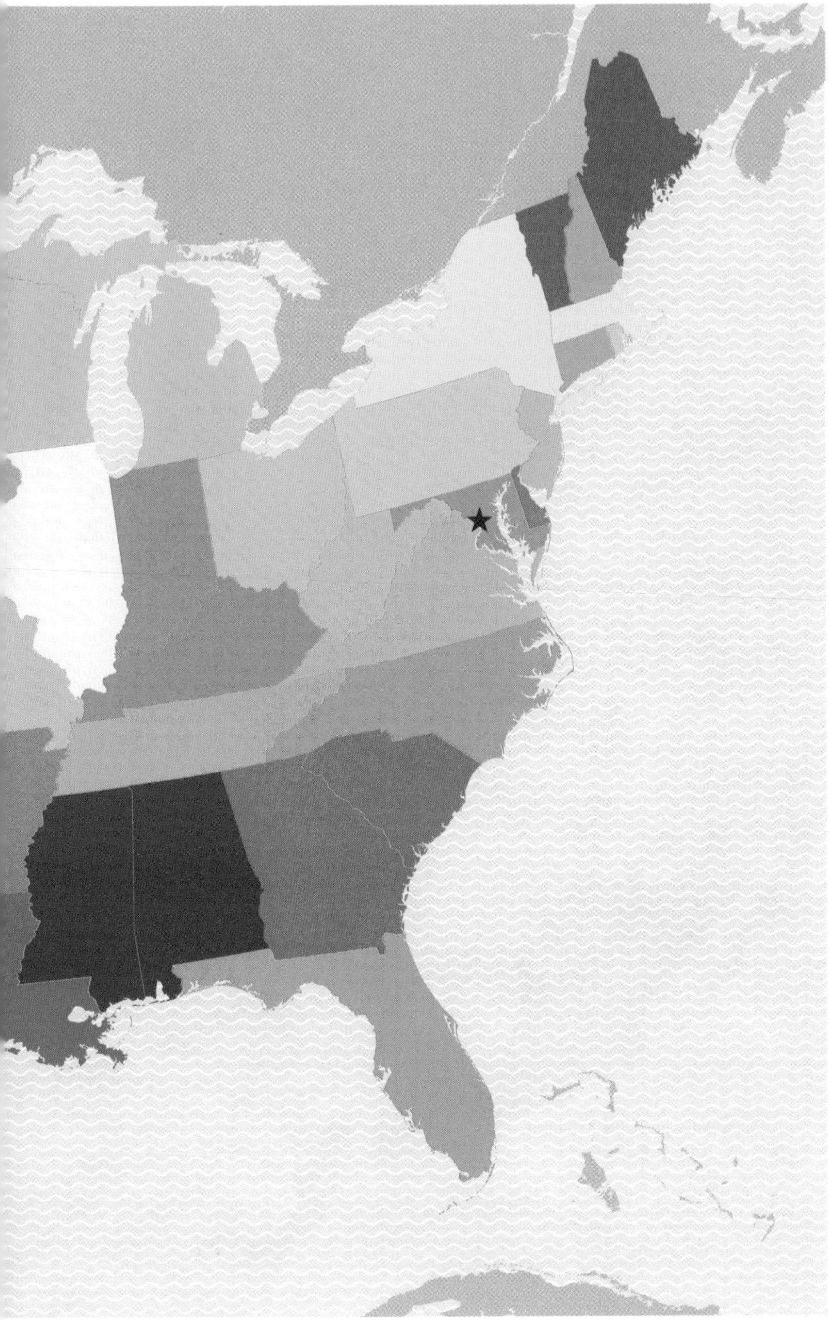

Percentage of times electing a "Top 10" president

100

77

55

33

No data

"Top 10" presidents

1. Abraham Lincoln
2. George Washington
3. Franklin D. Roosevelt
4. Theodore Roosevelt
5. Dwight D. Eisenhower
6. Harry S. Truman
7. Thomas Jefferson
8. John F. Kennedy
9. Ronald Reagan
10. Lyndon B. Johnson

23 **The price of leadership:**
Salaries of North American heads of state

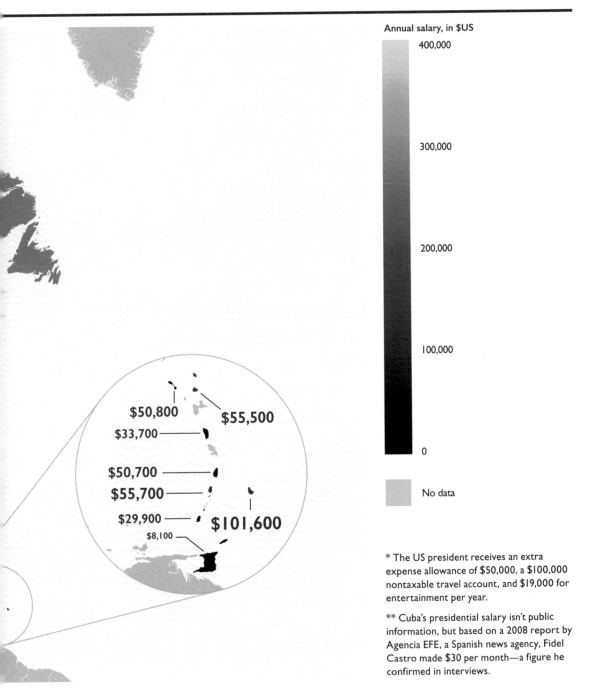

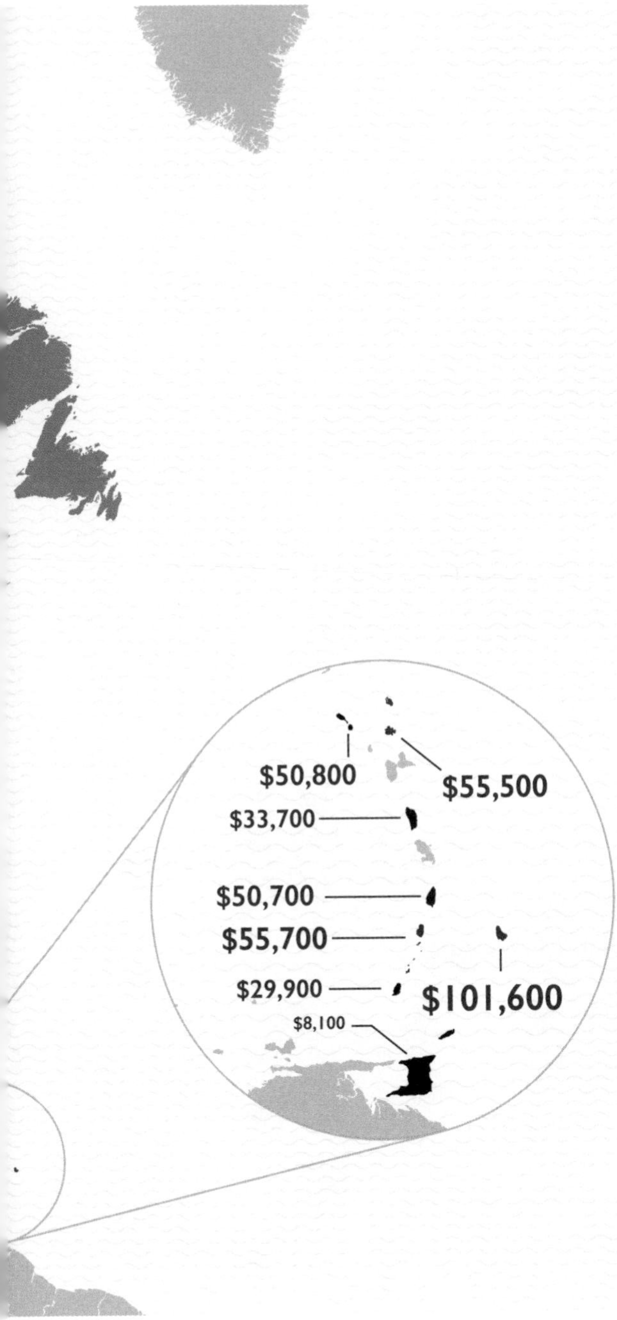

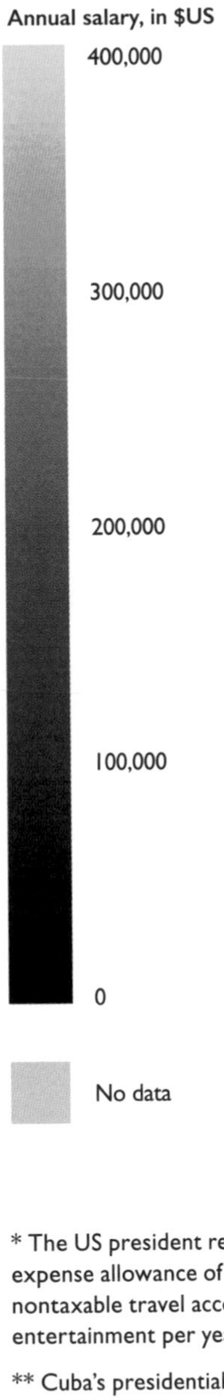

* The US president receives an extra expense allowance of $50,000, a $100,000 nontaxable travel account, and $19,000 for entertainment per year.

** Cuba's presidential salary isn't public information, but based on a 2008 report by Agencia EFE, a Spanish news agency, Fidel Castro made $30 per month—a figure he confirmed in interviews.

24 **Land of equality?:** If every state's population had Wyoming's representation in the Senate

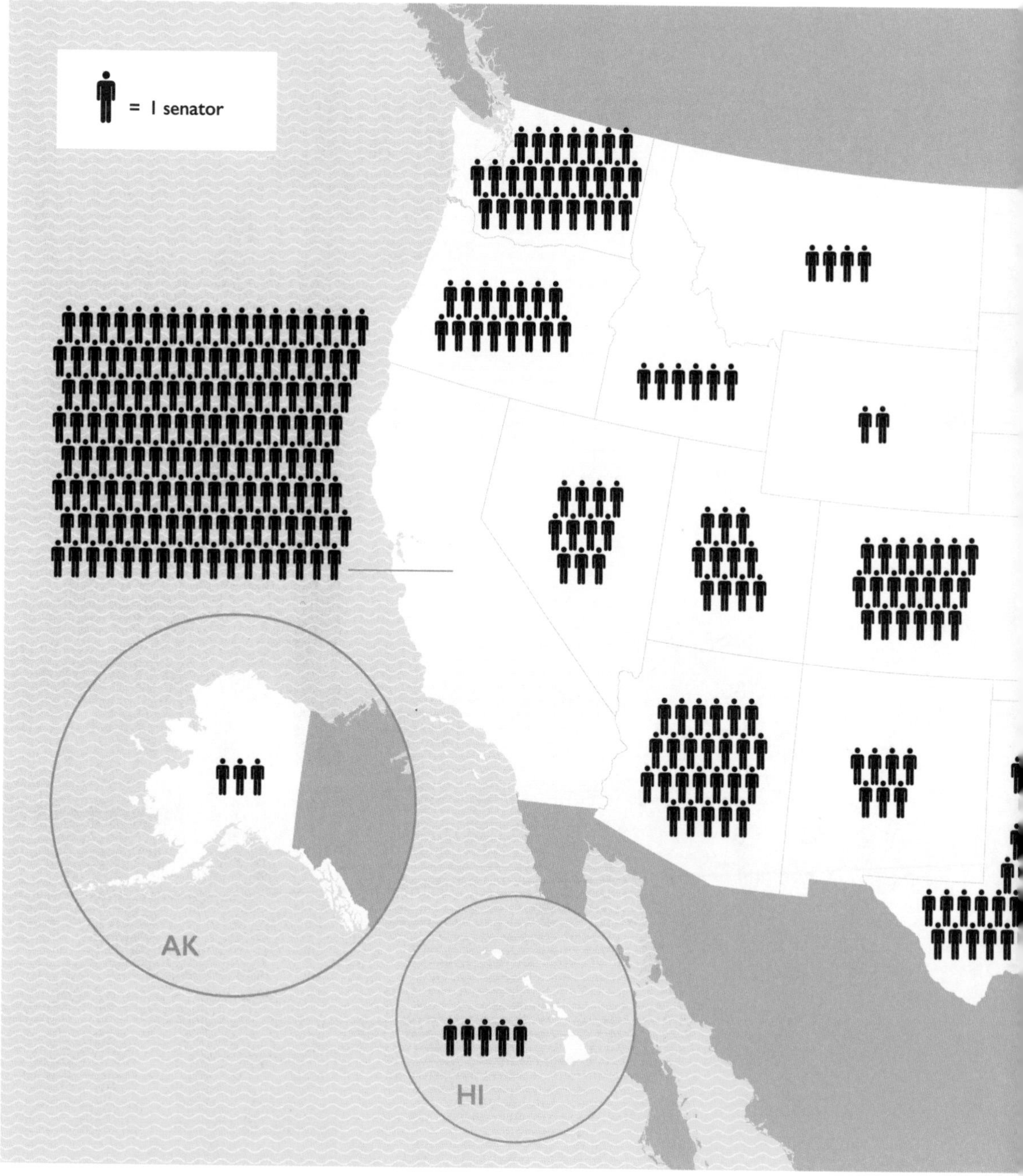

Neither of our great legislative bodies is truly representative. In the House, gerrymandering ensures that the distribution of seats between parties doesn't match the percentage of votes they received. And in the Senate, where every state has two senators, Wyoming—the least populous state, with 578,759 people—has one senator for every 289,380 Wyomingites. If California, the most populous state, enjoyed the same representation—one senator for every 289,380 citizens—it would have 137 senators. This map imagines what equal population-based Senate representation across all 50 states would look like.

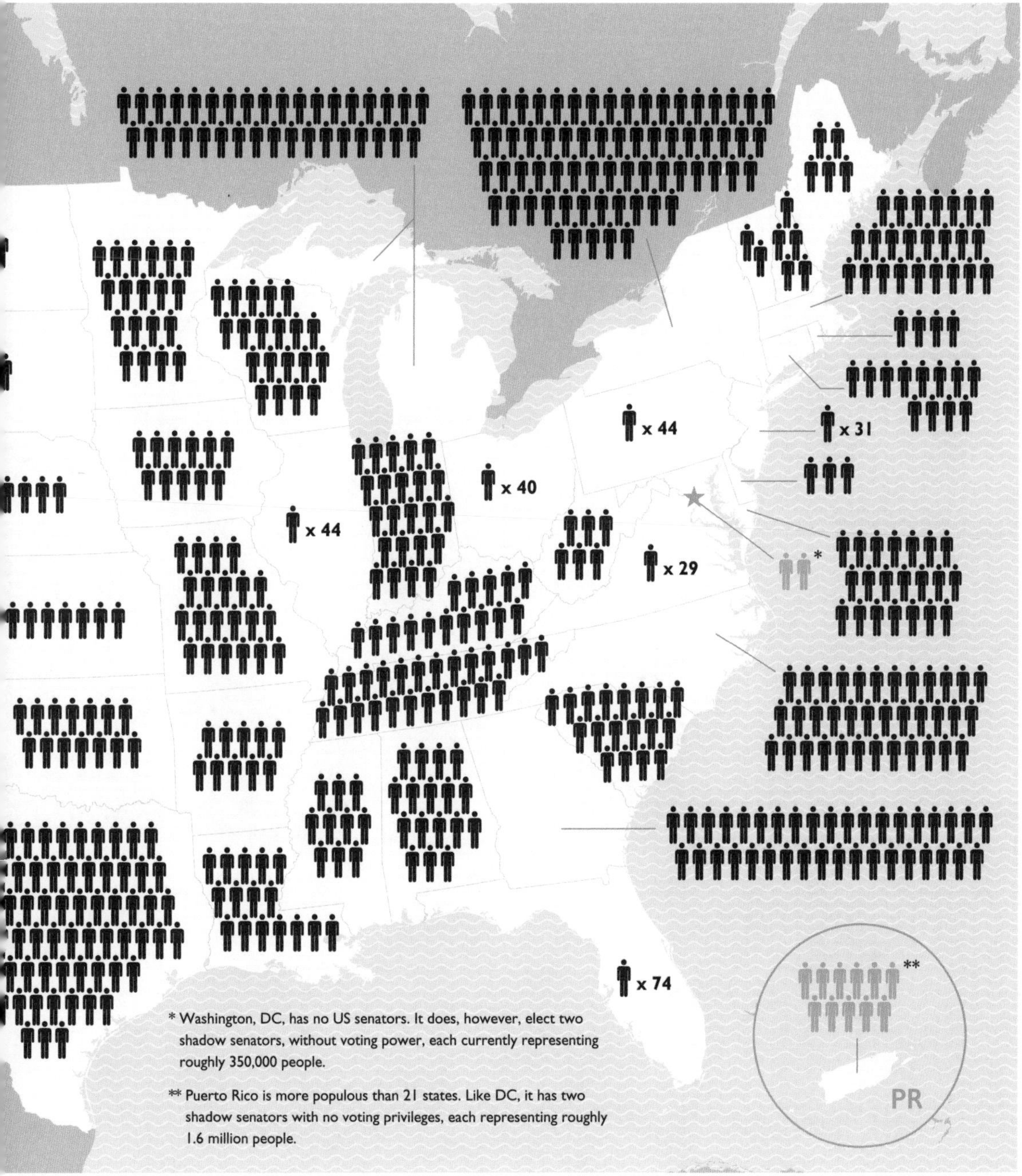

* Washington, DC, has no US senators. It does, however, elect two shadow senators, without voting power, each currently representing roughly 350,000 people.

** Puerto Rico is more populous than 21 states. Like DC, it has two shadow senators with no voting privileges, each representing roughly 1.6 million people.

25 Coaching all the way to the bank:
The highest-paid public employee by state

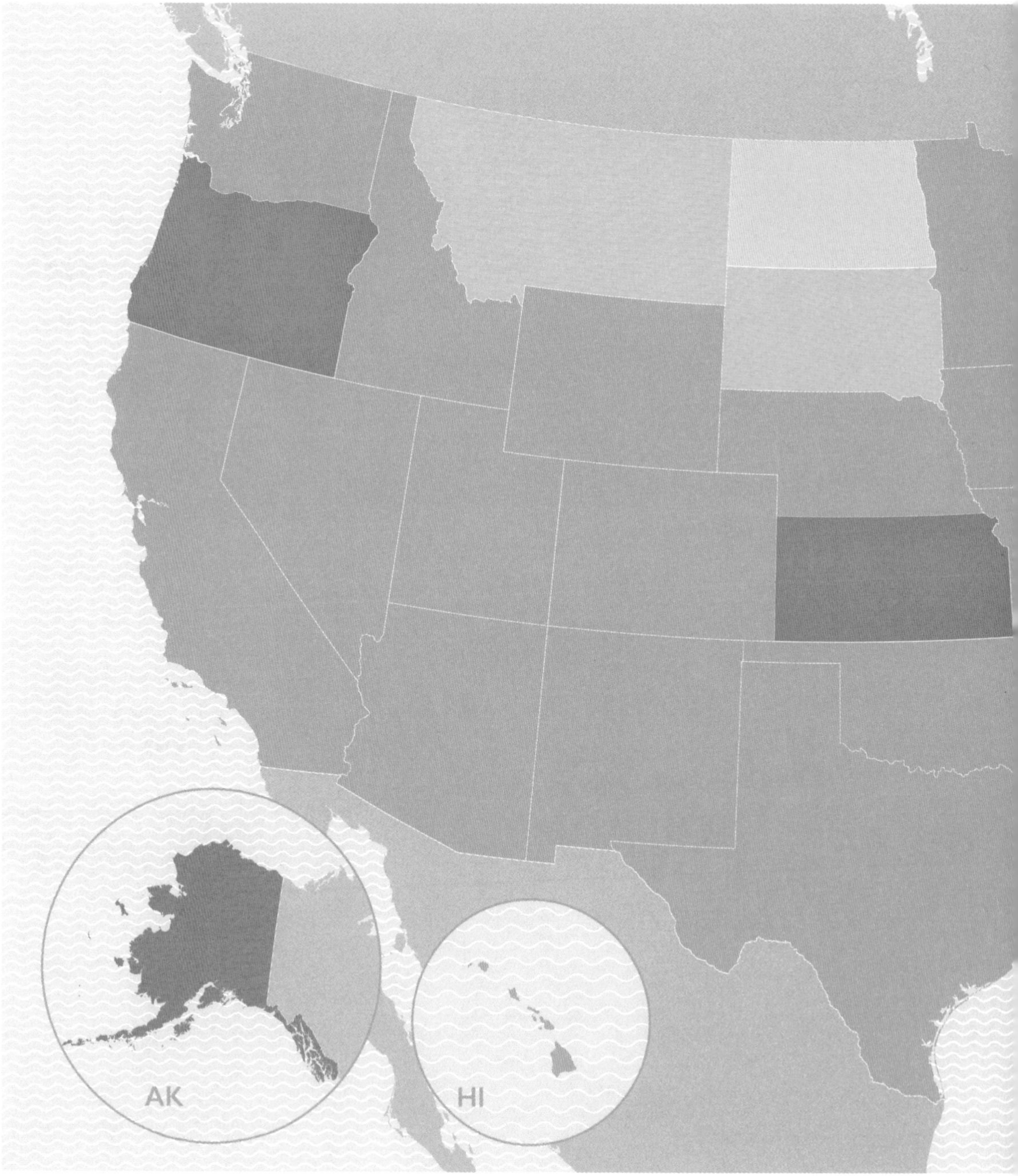

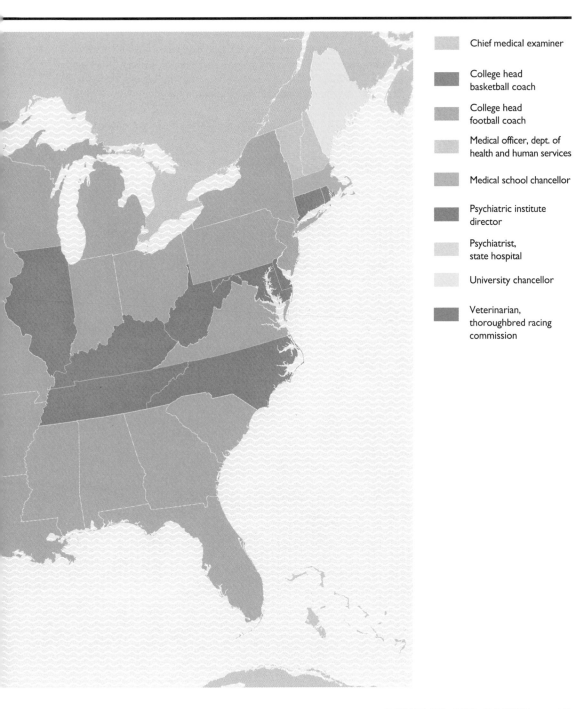

Chief medical examiner

College head basketball coach

College head football coach

Medical officer, dept. of health and human services

Medical school chancellor

Psychiatric institute director

Psychiatrist, state hospital

University chancellor

Veterinarian, thoroughbred racing commission

26 **Monroe's lament:** The dependencies of North America

In an address to Congress in 1823, President James Monroe issued a warning to enterprising countries outside the Western Hemisphere: Stay out. This major and enduring declaration of US foreign policy—which would come to be known as the Monroe Doctrine—drew a line in the sand, making clear that the western half of the world was the US's domain, and that European meddling and further colonization was unacceptable. Here are the dependencies that have persisted despite all that, preserving some foreign influence in North American waters.

Colombia | Denmark | France | Netherlands | United Kingdom

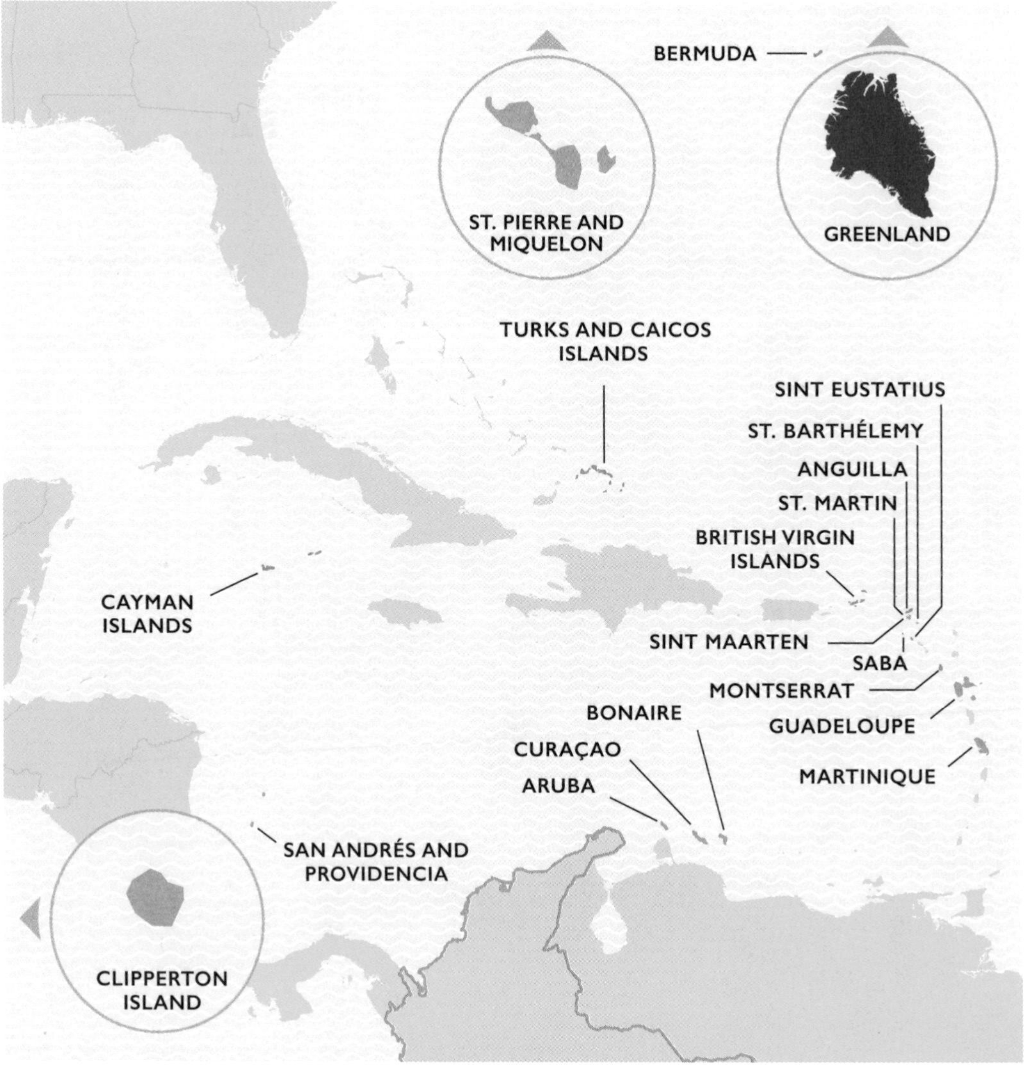

27 **Bang for the buck:**
Who pays the most for their military?

Per capita, the US spends about four times as much on its military as the rest of the continent combined. Costa Rica and Panama formally abolished their militaries in 1948 and 1990, respectively. Haiti's armed forces disbanded in 1995, though recently there's been a movement to remobilize.

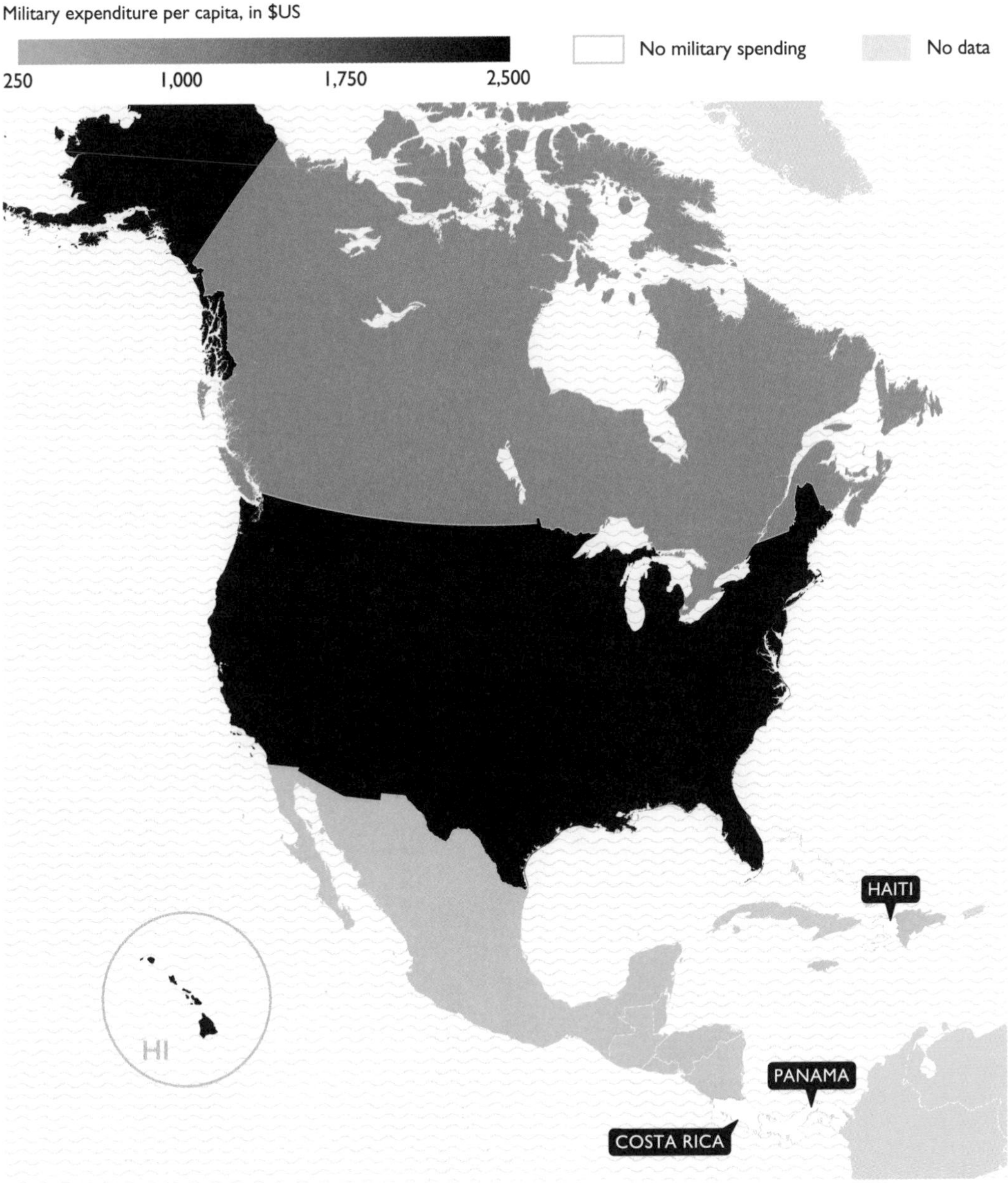

28 **Living on the edge:** 2 out of 3 Americans live within 100 miles of the border

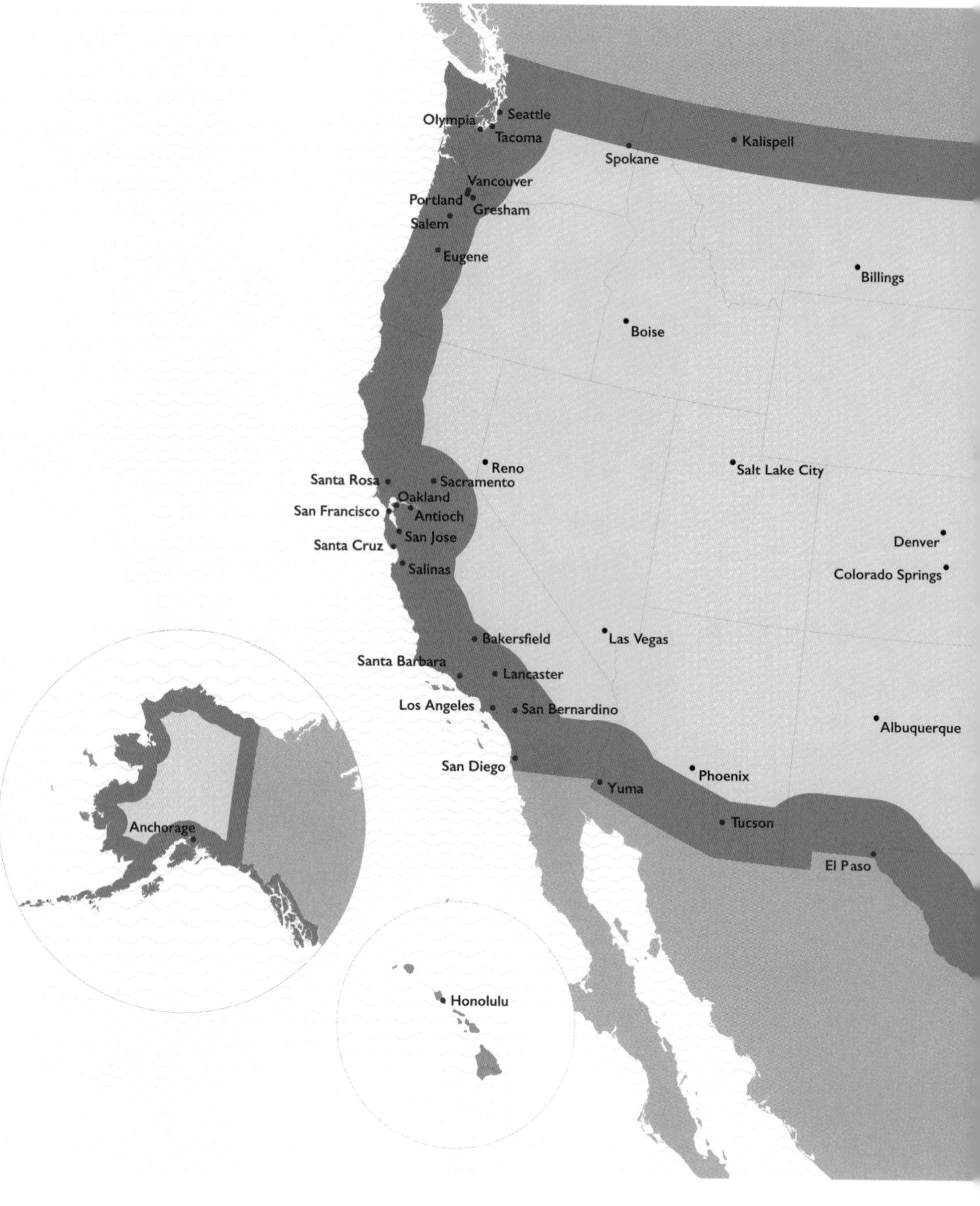

This is trivia with real consequence. At the time of writing, the Department of Homeland Security has the power to search without a warrant for undocumented immigrants in any vehicle within this 100-mile (160 km) zone.

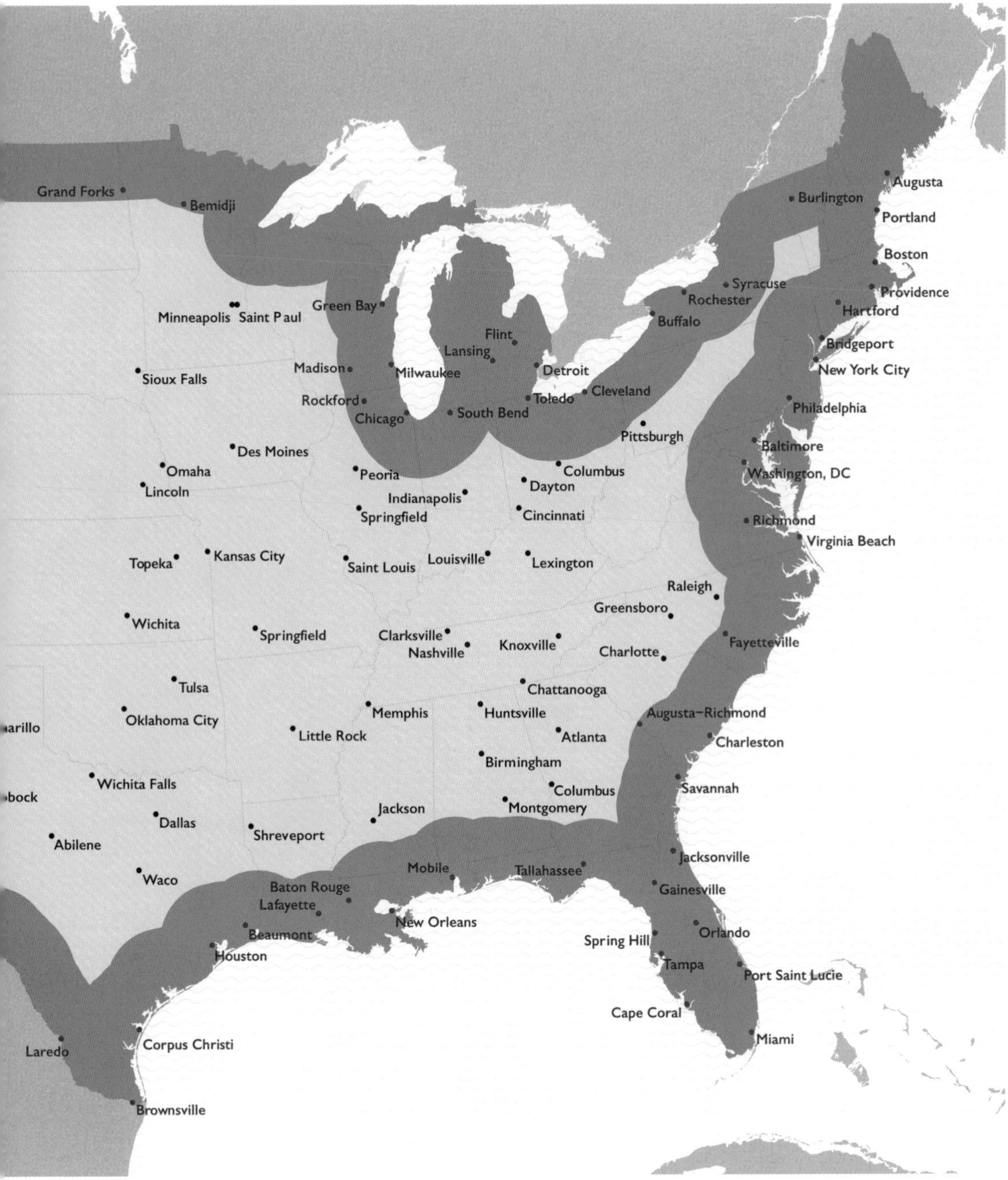

29 **Where service comes first:** Proportion of veterans by state

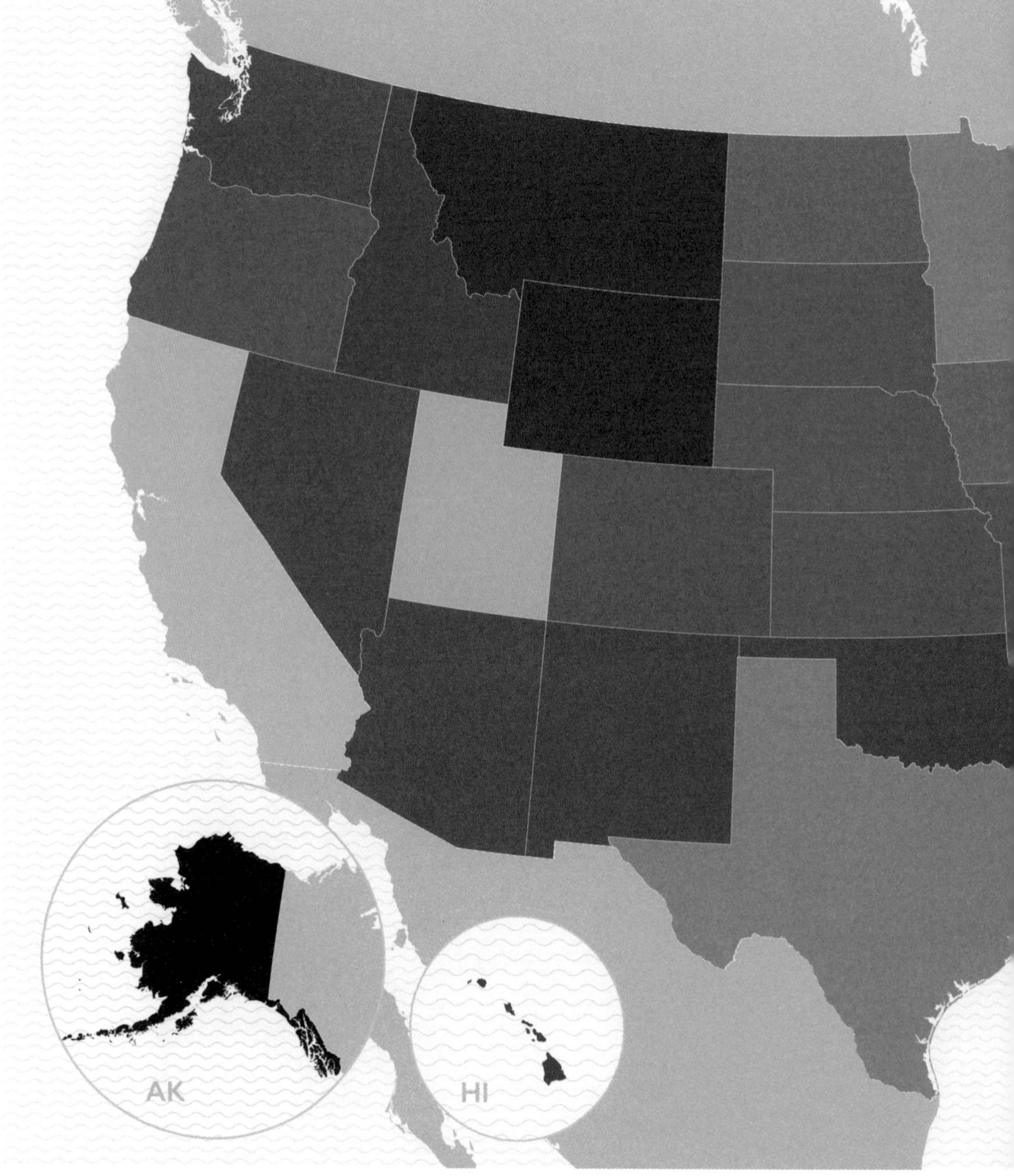

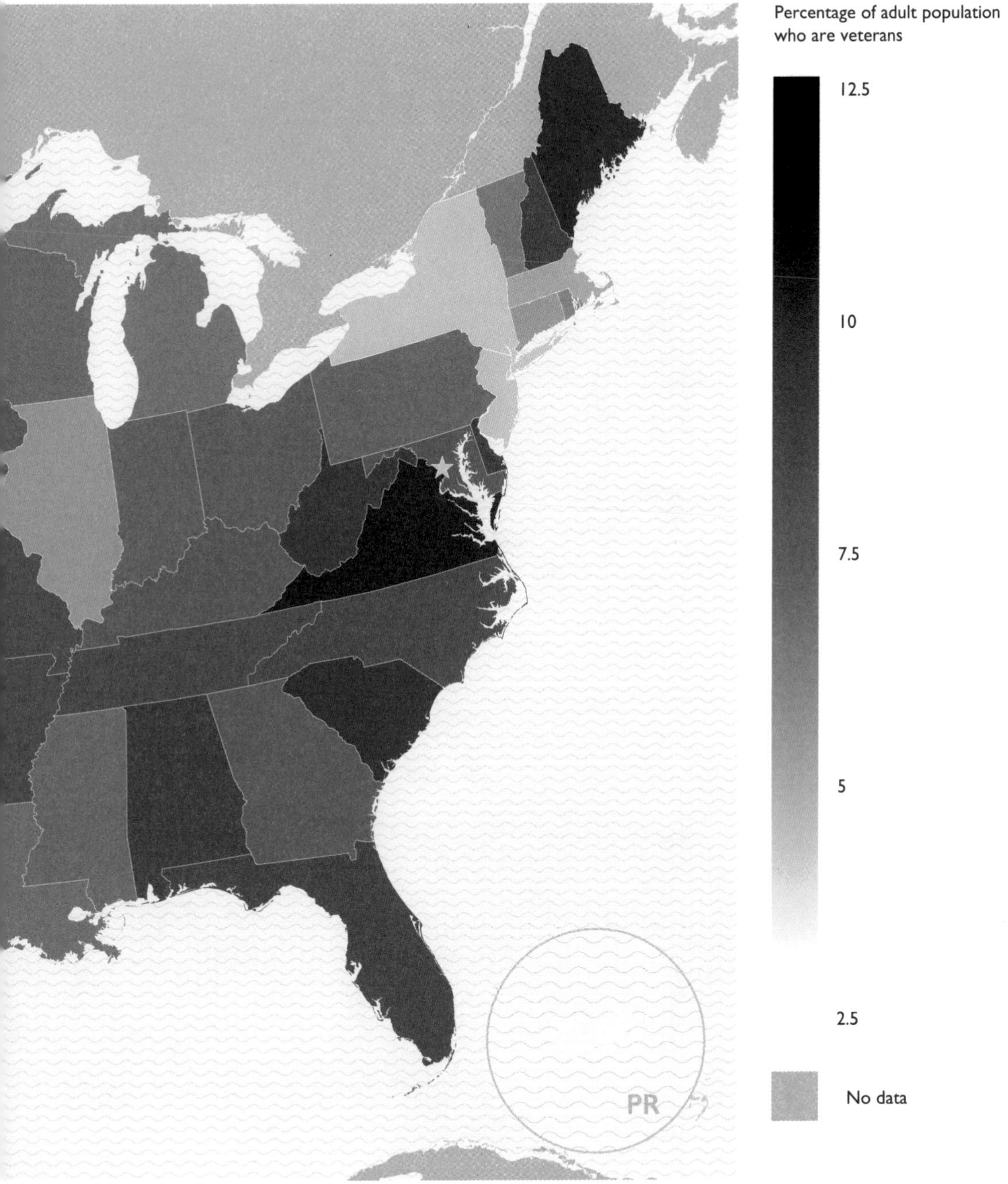
PR
Percentage of adult population who are veterans
12.5
10
7.5
5
2.5
No data

3
NATURE

Missing the forest for the trees: The can't-miss trees of North America

30

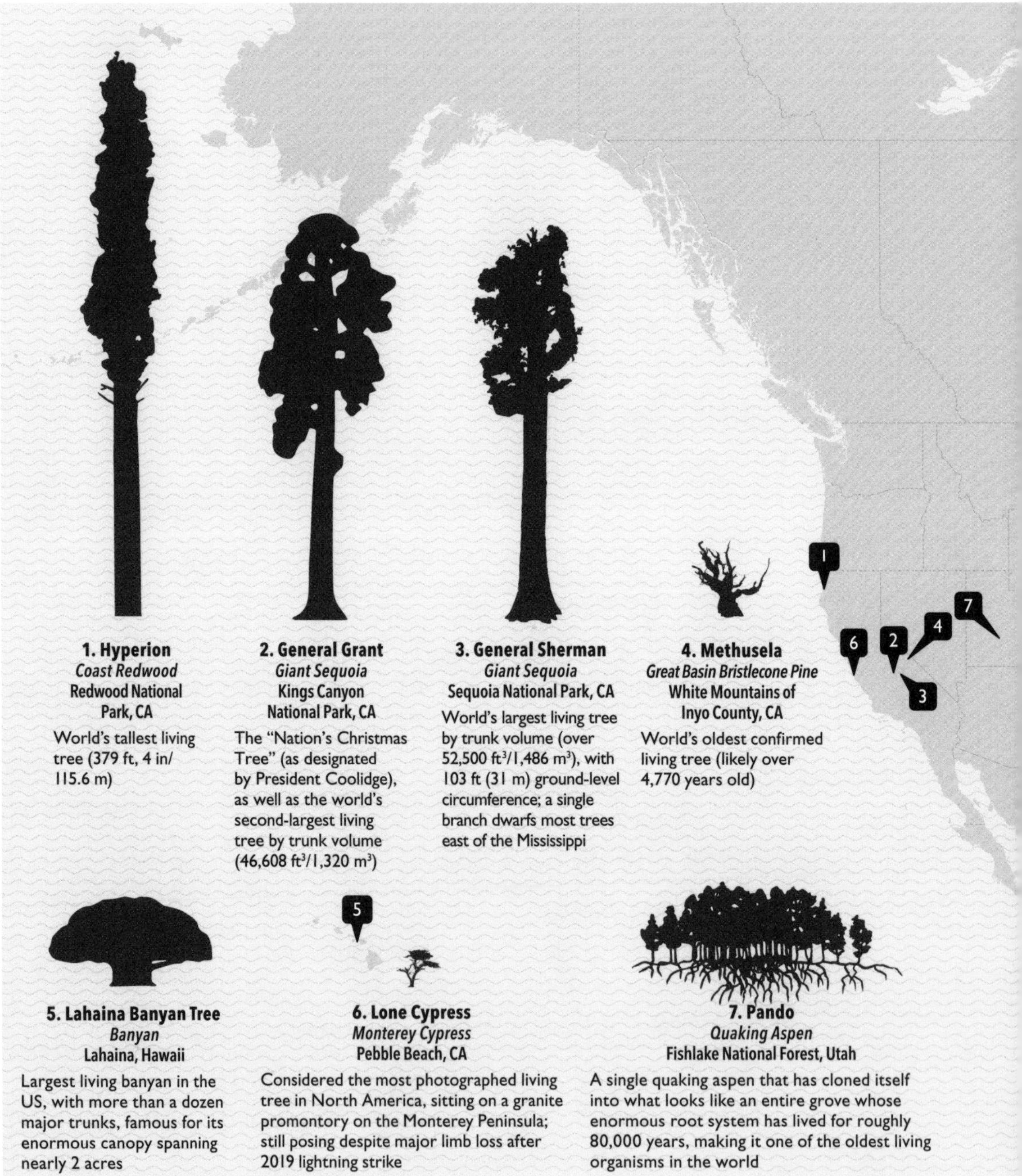

1. Hyperion
Coast Redwood
Redwood National Park, CA

World's tallest living tree (379 ft, 4 in/115.6 m)

2. General Grant
Giant Sequoia
Kings Canyon National Park, CA

The "Nation's Christmas Tree" (as designated by President Coolidge), as well as the world's second-largest living tree by trunk volume (46,608 ft^3/1,320 m^3)

3. General Sherman
Giant Sequoia
Sequoia National Park, CA

World's largest living tree by trunk volume (over 52,500 ft^3/1,486 m^3), with 103 ft (31 m) ground-level circumference; a single branch dwarfs most trees east of the Mississippi

4. Methusela
Great Basin Bristlecone Pine
White Mountains of Inyo County, CA

World's oldest confirmed living tree (likely over 4,770 years old)

5. Lahaina Banyan Tree
Banyan
Lahaina, Hawaii

Largest living banyan in the US, with more than a dozen major trunks, famous for its enormous canopy spanning nearly 2 acres

6. Lone Cypress
Monterey Cypress
Pebble Beach, CA

Considered the most photographed living tree in North America, sitting on a granite promontory on the Monterey Peninsula; still posing despite major limb loss after 2019 lightning strike

7. Pando
Quaking Aspen
Fishlake National Forest, Utah

A single quaking aspen that has cloned itself into what looks like an entire grove whose enormous root system has lived for roughly 80,000 years, making it one of the oldest living organisms in the world

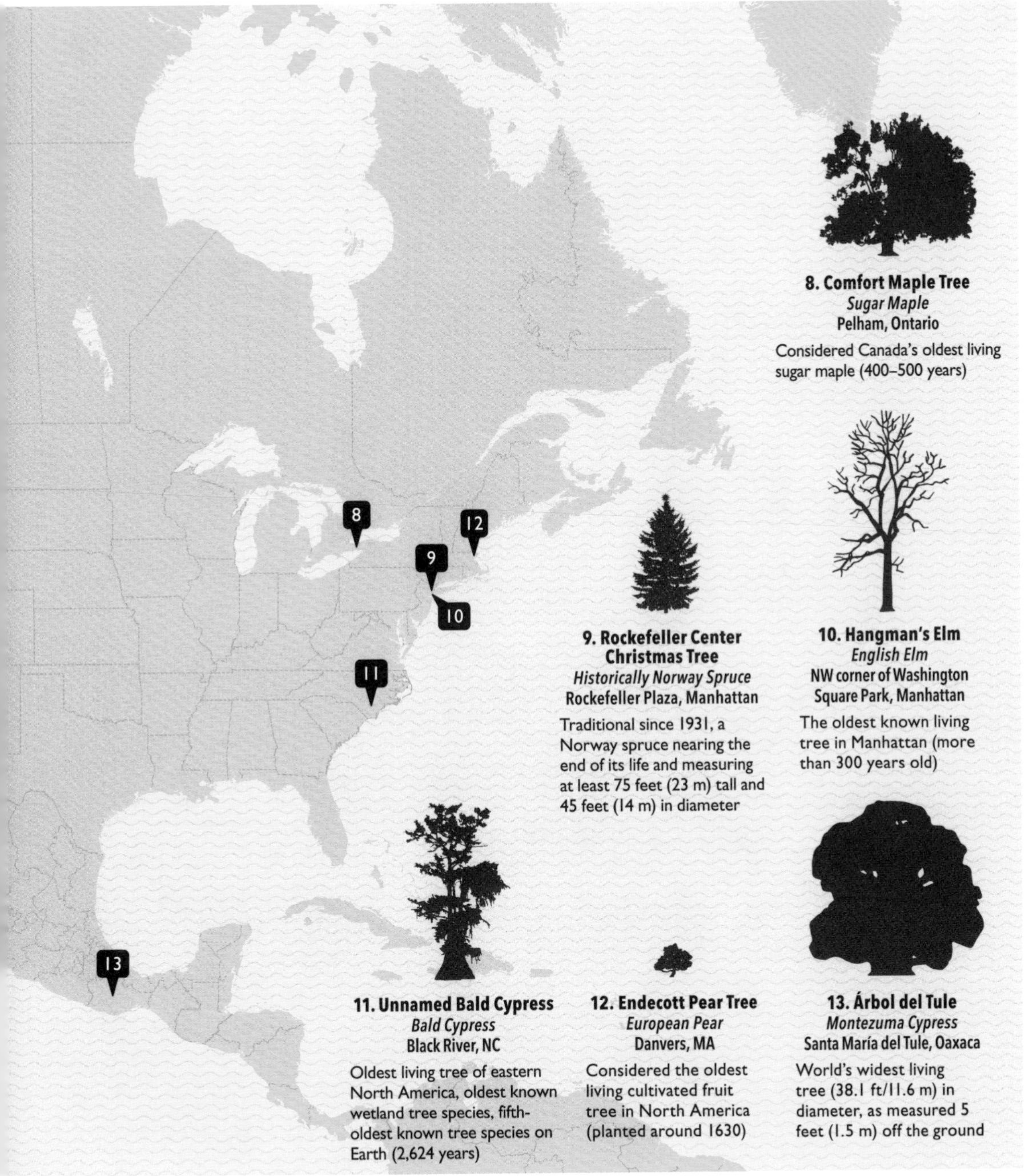

8. Comfort Maple Tree
Sugar Maple
Pelham, Ontario

Considered Canada's oldest living sugar maple (400–500 years)

9. Rockefeller Center Christmas Tree
Historically Norway Spruce
Rockefeller Plaza, Manhattan

Traditional since 1931, a Norway spruce nearing the end of its life and measuring at least 75 feet (23 m) tall and 45 feet (14 m) in diameter

10. Hangman's Elm
English Elm
NW corner of Washington Square Park, Manhattan

The oldest known living tree in Manhattan (more than 300 years old)

11. Unnamed Bald Cypress
Bald Cypress
Black River, NC

Oldest living tree of eastern North America, oldest known wetland tree species, fifth-oldest known tree species on Earth (2,624 years)

12. Endecott Pear Tree
European Pear
Danvers, MA

Considered the oldest living cultivated fruit tree in North America (planted around 1630)

13. Árbol del Tule
Montezuma Cypress
Santa María del Tule, Oaxaca

World's widest living tree (38.1 ft/11.6 m) in diameter, as measured 5 feet (1.5 m) off the ground

31 **Dog days:**
When does the hottest day of the year fall?

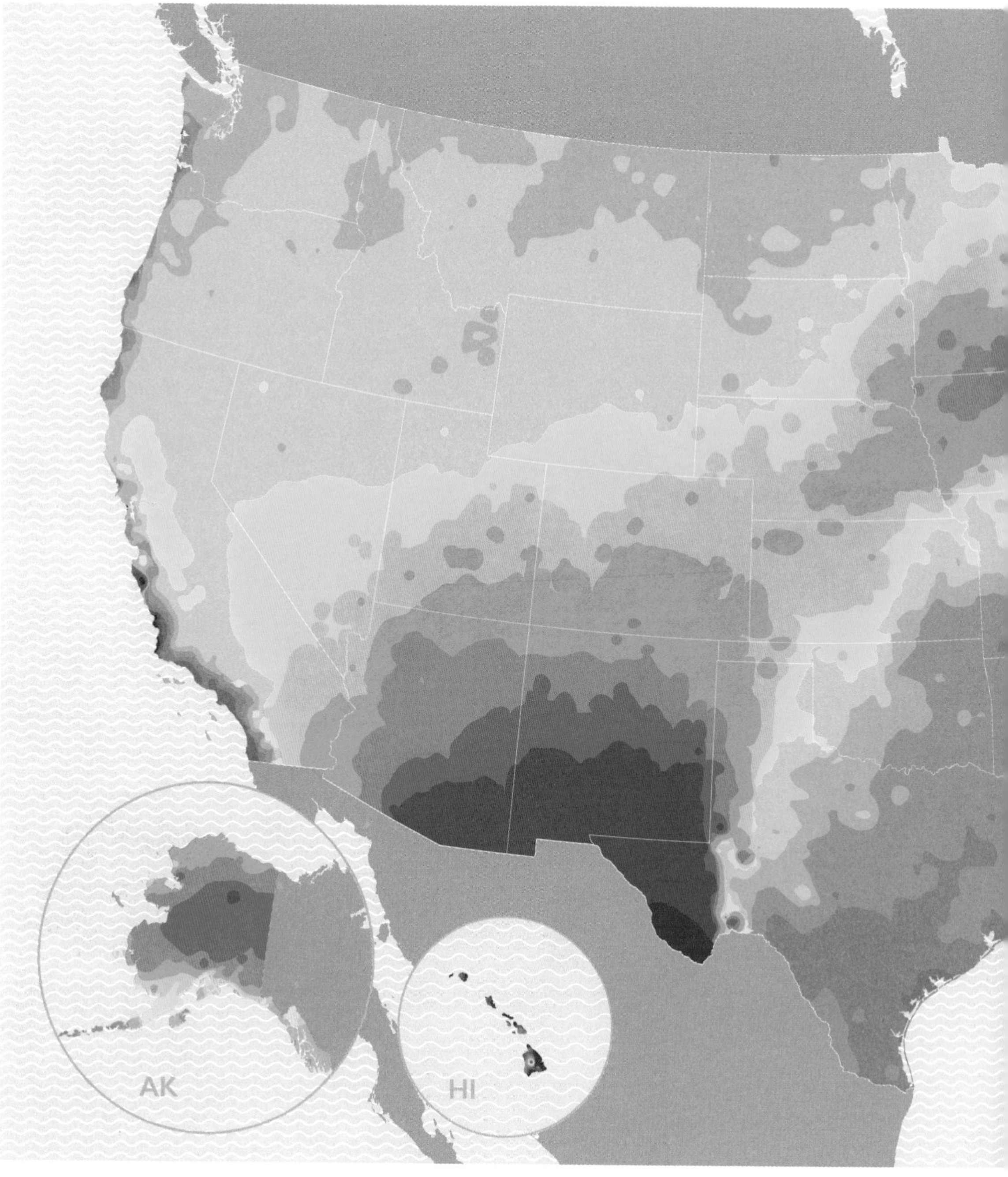

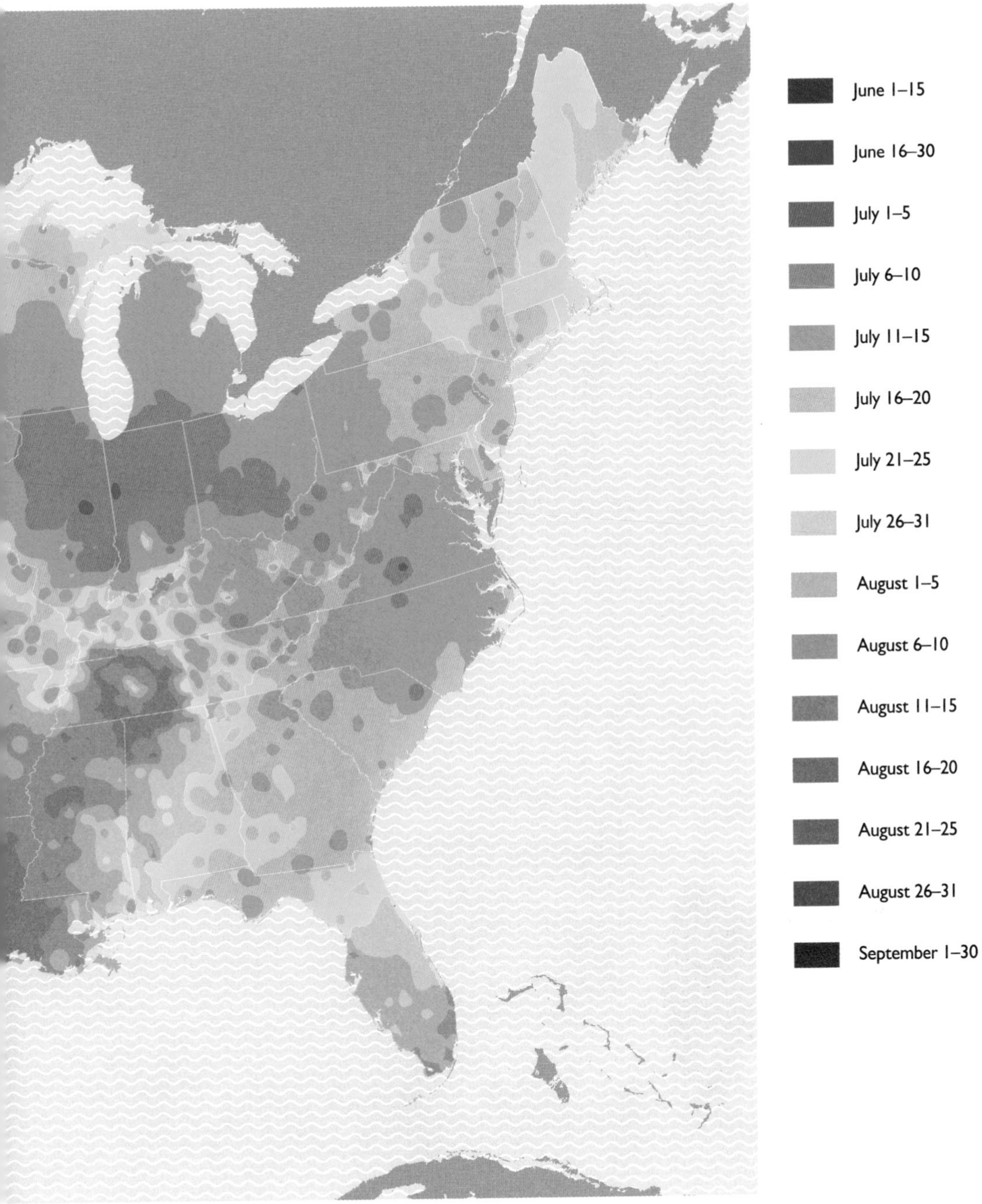
June 1–15
June 16–30
July 1–5
July 6–10
July 11–15
July 16–20
July 21–25
July 26–31
August 1–5
August 6–10
August 11–15
August 16–20
August 21–25
August 26–31
September 1–30

32

Spring hasn't sprung:
Punxsutawney Phil predicts late blooms

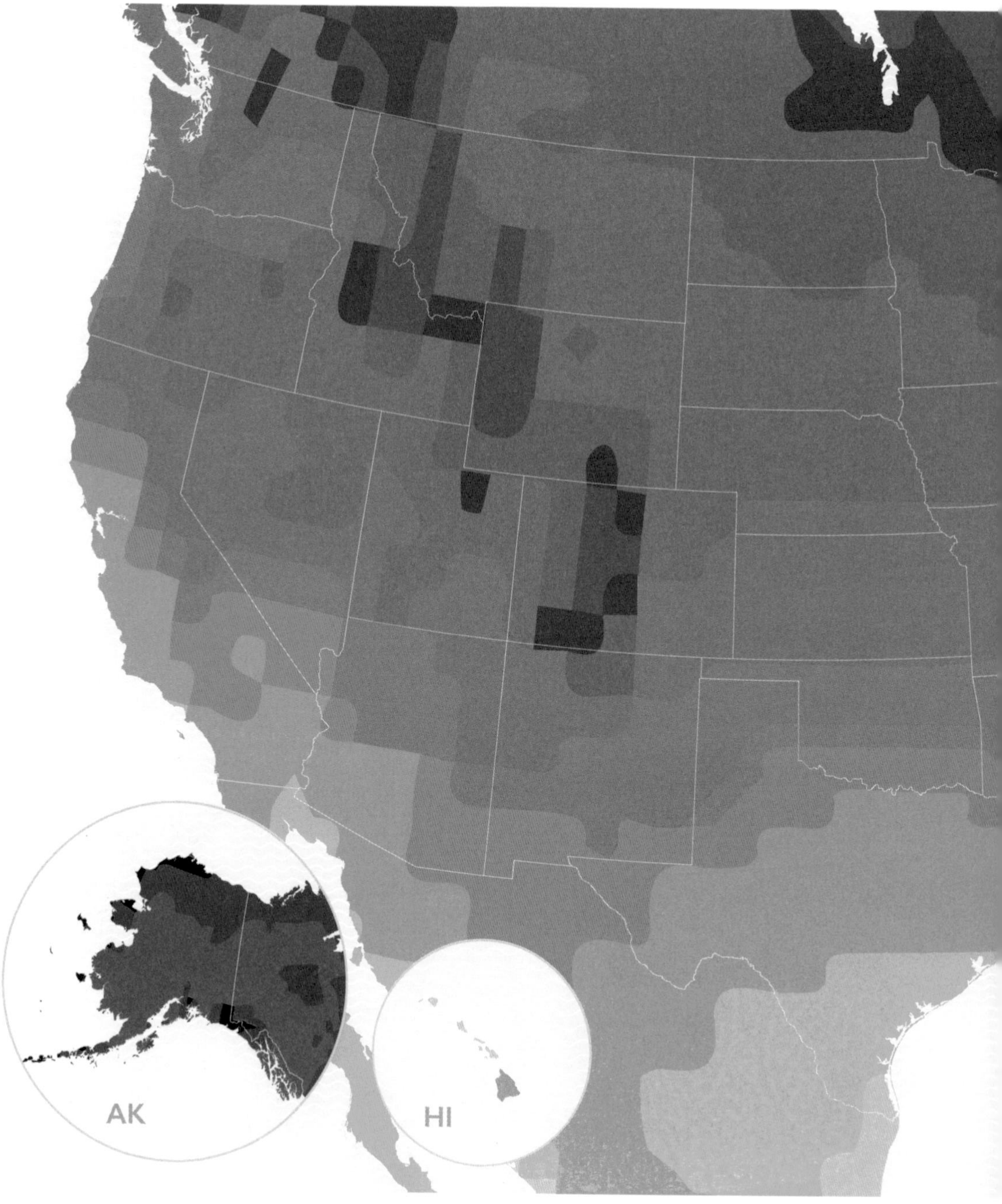

Every February 2, Groundhog Day, Punxsutawney Phil makes his own forecast at Gobbler's Knob in Punxsutawney, PA. Since 1886, the resident groundhog has seen his shadow (predicting a long winter) 84 percent of the time. Phil's on to something: In Punxsutawney, PA, the first blooms don't usually appear until April 23 or so—certainly no early spring, the first day of which is officially March 19, 20, or 21.

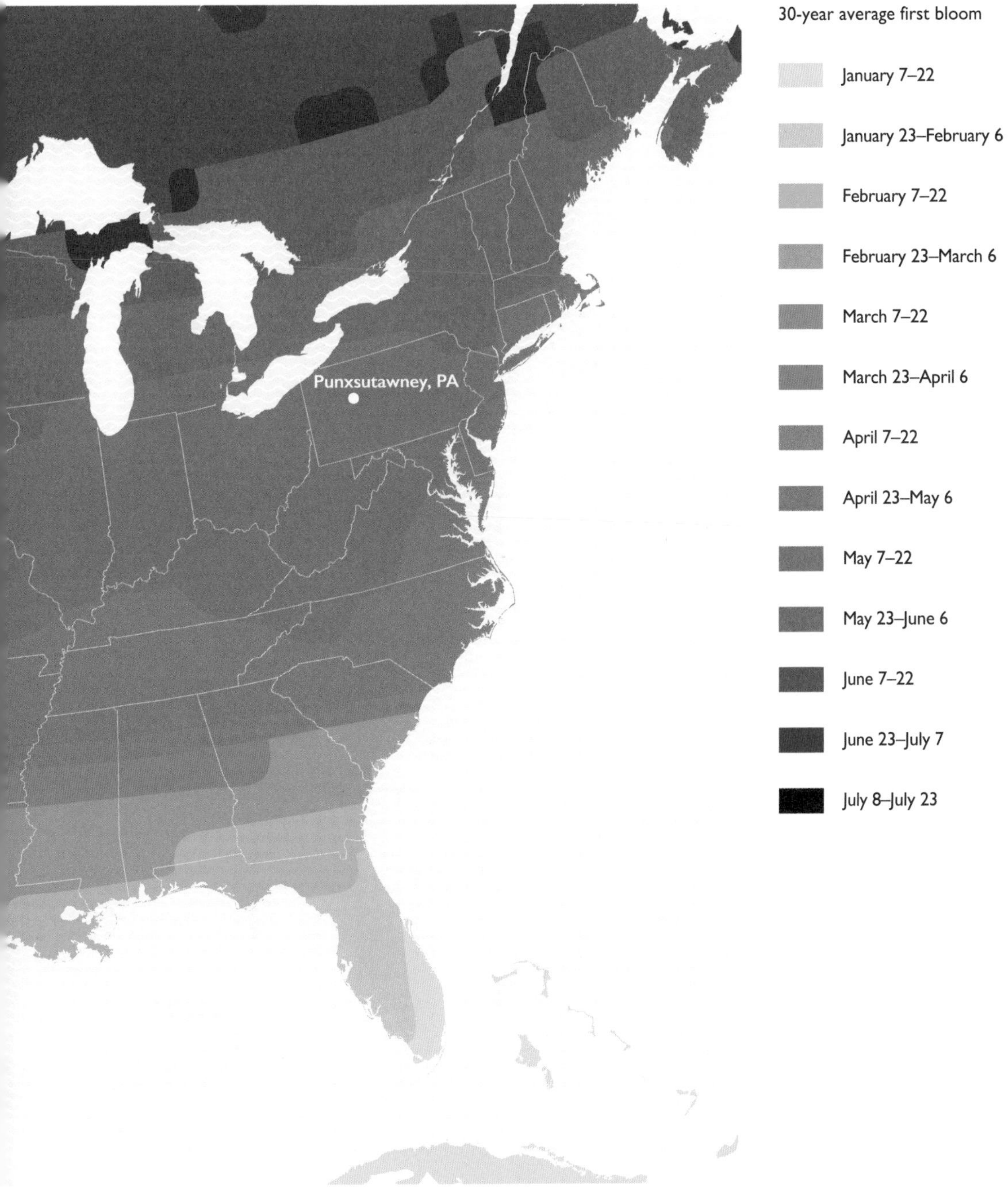

33 **Land before time:**
The greatest dinosaur finds in North America

1. Dinosaur Provincial Park, AB | 1912
A UNESCO World Heritage Site representing the most complete fossil record of the late Cretaceous, including famous discoveries of *Corythosaurus*, *Albertosaurus*, and *Struthiomimus*

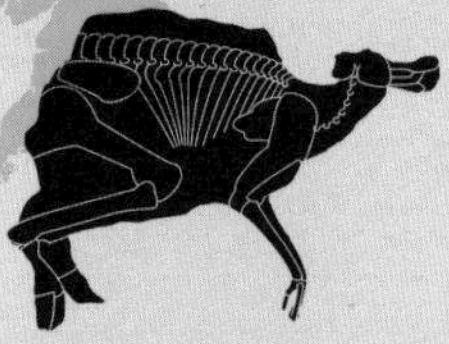

2. Malta, MT | 2000
Guinness World Record holder for best-preserved dinosaur, a juvenile *Brachylophosaurus* "mummy" (only the fourth dino discovery to earn that designation), so well preserved that even its last meal is intact in its stomach

3. Hell Creek Formation, MT | 1903
First *Tyrannosaurus rex* discovered

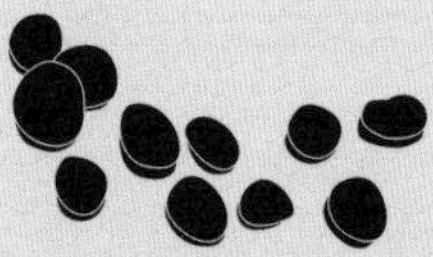

4. Devil's Coulee, AB | 2020
Analysis of fossilized eggshells of *Troodon*, *Maiasaura*, and *Megaloolithus* show that they were warmer than their environment, and possibly warm-blooded—evolutionarily, somewhere between birds and reptiles

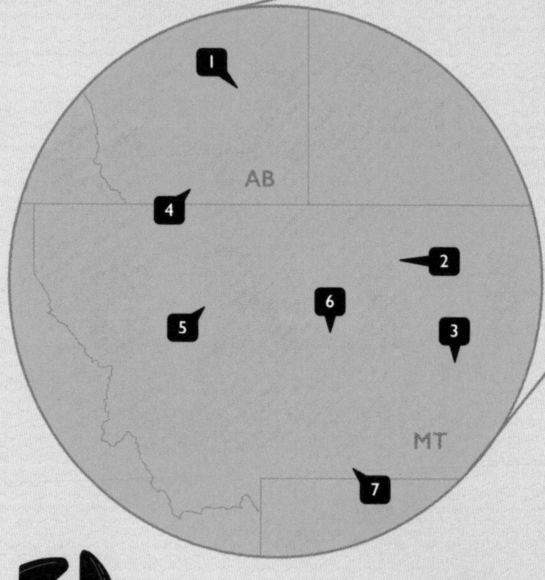

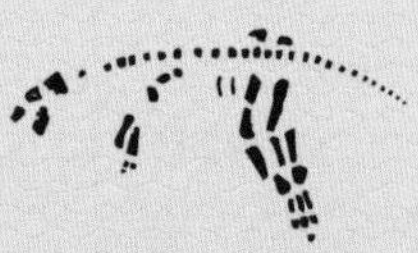

5. Egg Mountain, MT | 1978
First baby dino bones discovered in a nest, revealing how some dinos looked after their young

6. Judith Landing, MT | 1854
First dinosaur remains found in North America (teeth of a duckbilled hadrosaur then called *Trachodon*)

7. Bridger, MT | 1964
Revolutionary fossil discovery of fast, ferocious predator *Deinonychus*, which showed some dinos may be warm-blooded and ancestors of birds

North America is a treasure trove of dinosaur remains. In the US, a greater variety of dinosaurs have been found than in any other country. Paleontologists have discovered 75 different species of dinosaurs in Montana, more than in any other state. And the Cleveland-Lloyd Quarry at Jurassic National Monument in Utah, with more than 12,000 bones, contains the highest concentration of Jurassic-aged dinosaur bones in the world. In Canada, Dinosaur Provincial Park in Alberta is especially rich with fossils, yielding more than 150 complete skeletons and over 50 species.

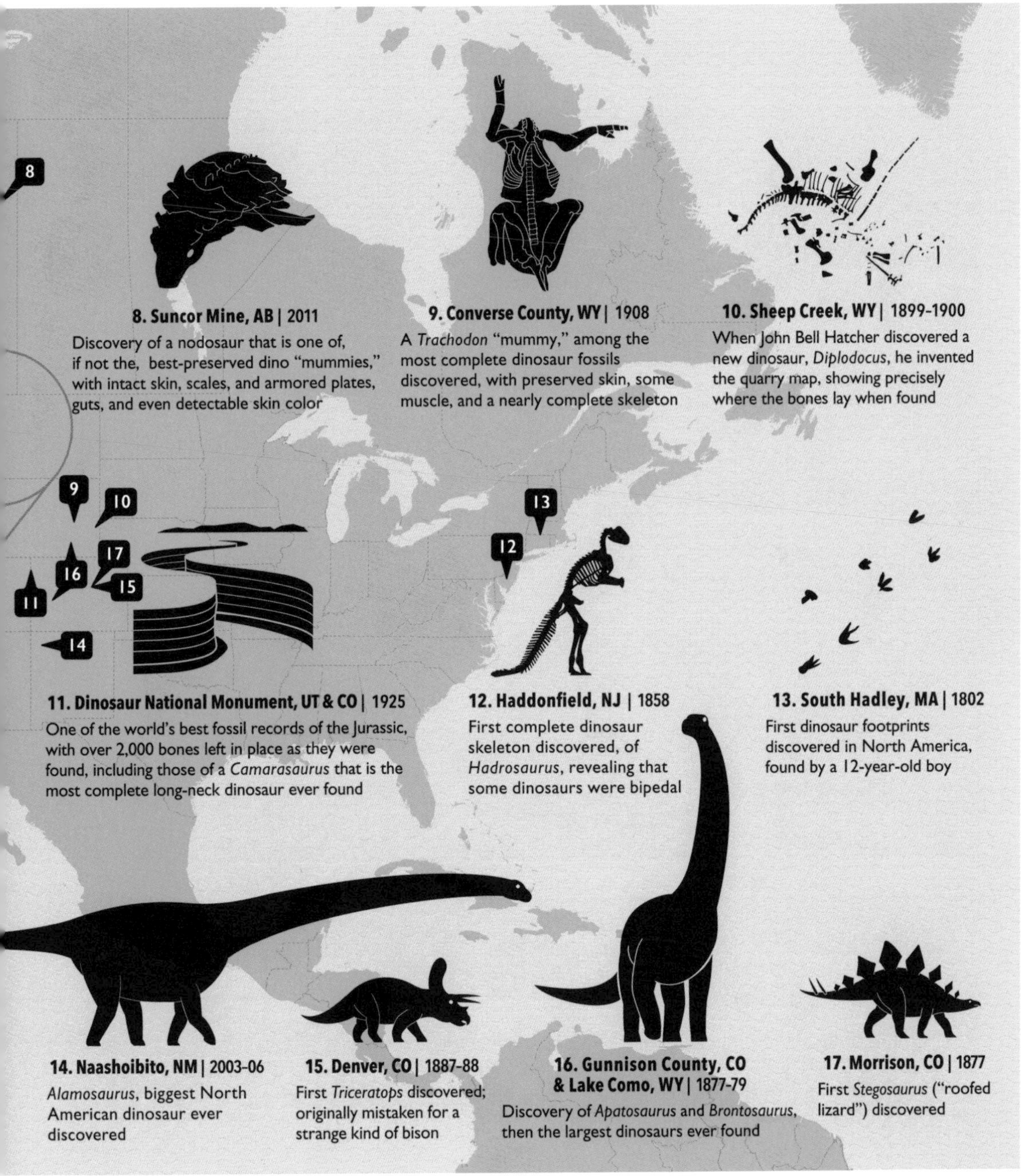

8. Suncor Mine, AB | 2011
Discovery of a nodosaur that is one of, if not the, best-preserved dino "mummies," with intact skin, scales, and armored plates, guts, and even detectable skin color

9. Converse County, WY | 1908
A *Trachodon* "mummy," among the most complete dinosaur fossils discovered, with preserved skin, some muscle, and a nearly complete skeleton

10. Sheep Creek, WY | 1899-1900
When John Bell Hatcher discovered a new dinosaur, *Diplodocus*, he invented the quarry map, showing precisely where the bones lay when found

11. Dinosaur National Monument, UT & CO | 1925
One of the world's best fossil records of the Jurassic, with over 2,000 bones left in place as they were found, including those of a *Camarasaurus* that is the most complete long-neck dinosaur ever found

12. Haddonfield, NJ | 1858
First complete dinosaur skeleton discovered, of *Hadrosaurus*, revealing that some dinosaurs were bipedal

13. South Hadley, MA | 1802
First dinosaur footprints discovered in North America, found by a 12-year-old boy

14. Naashoibito, NM | 2003-06
Alamosaurus, biggest North American dinosaur ever discovered

15. Denver, CO | 1887-88
First *Triceratops* discovered; originally mistaken for a strange kind of bison

16. Gunnison County, CO & Lake Como, WY | 1877-79
Discovery of *Apatosaurus* and *Brontosaurus*, then the largest dinosaurs ever found

17. Morrison, CO | 1877
First *Stegosaurus* ("roofed lizard") discovered

34 Our biggest hits:
North America's largest impact craters

OROSIRIAN
2,050–1,800 MYA
1.Sudbury
STATHERIAN
1,800–1,600 MYA
ECTASIAN
1,400–1,200 MYA
STENIAN
1,200–1,000 MYA
2. Santa Fe
TONIAN
1,000–850 MYA
CRYOGENIAN
850–635 MYA
EDIACARAN
635–541 MYA
ORDOVICIAN
485–443 MYA
9. Ames
10. Calvin
11. Slate Islands
12. Pilot
SILURIAN
443–419 MYA
13. Glasford
14. Couture
DEVONIAN
419–359 MYA
15. Nicholson
16. La Moinerie
17. Brent
18. Elbow
19. Flynn Creek
CARBONIFEROUS
359–299 MYA
20. West Hawk
21. Charlevoix
22. Serpent Mound
23. Crooked Creek
24. Middlesboro
25. Decaturville
26. Île Rouleau
PERMIAN
299–252 MYA
27. Lac à l'Eau Claire Ouest (Clearwater West)
28. Lac à l'Eau Claire Est (Clearwater East)
29. Des Plaines
TRIASSIC
252–201 MYA
30. Gow
31. Tunnunik
32. Saint Martin
33. Manicouagan
JURASSIC
201–146 MYA
34. Red Wing
35. Wells Creek
36. Cloud Creek
37. Viewfield
38. Upheaval Dome
CRETACEOUS
146–66 MYA
39. Carswell
40. Sierra Madera
41. Deep Bay
42. Kentland
43. Avak
44. Steen River

Earth's surface is constantly being recycled and resurfaced through plate tectonics and erosion, so older impact craters are all but erased. Highlights below include: Sudbury (#1), in Ontario, the third-largest impact crater in the world after Vredefort in South Africa and, in Mexico's Yucatán Peninsula, Chicxulub (#49), itself famous for likely wiping out the dinosaurs, along with roughly two thirds of all other species on Earth; Chesapeake Bay (#55), the largest in the US; and Barringer (#57) in Arizona, extremely well preserved because of its relative youth.

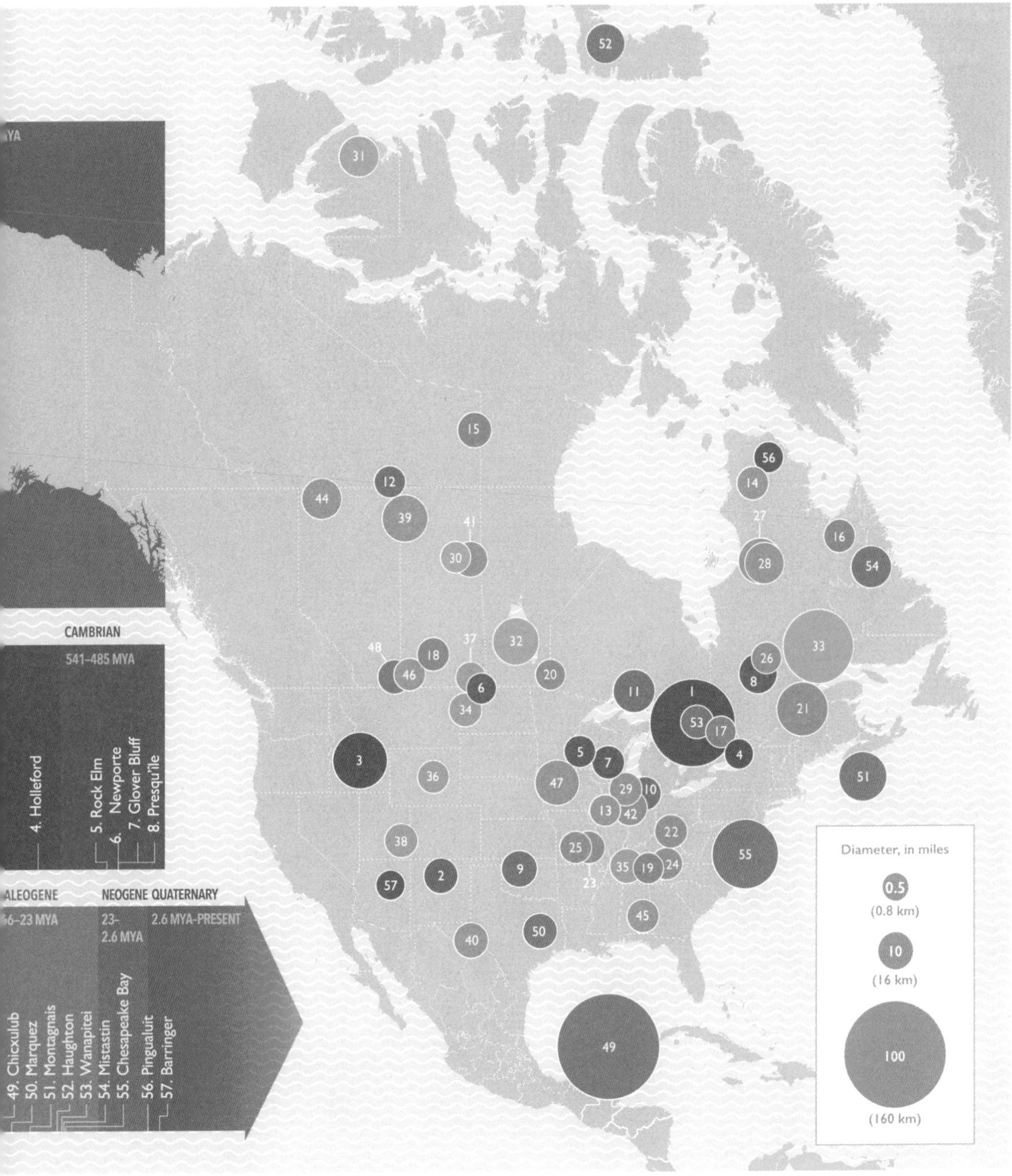

35 Above the fruited plain:
The highest elevation in each state

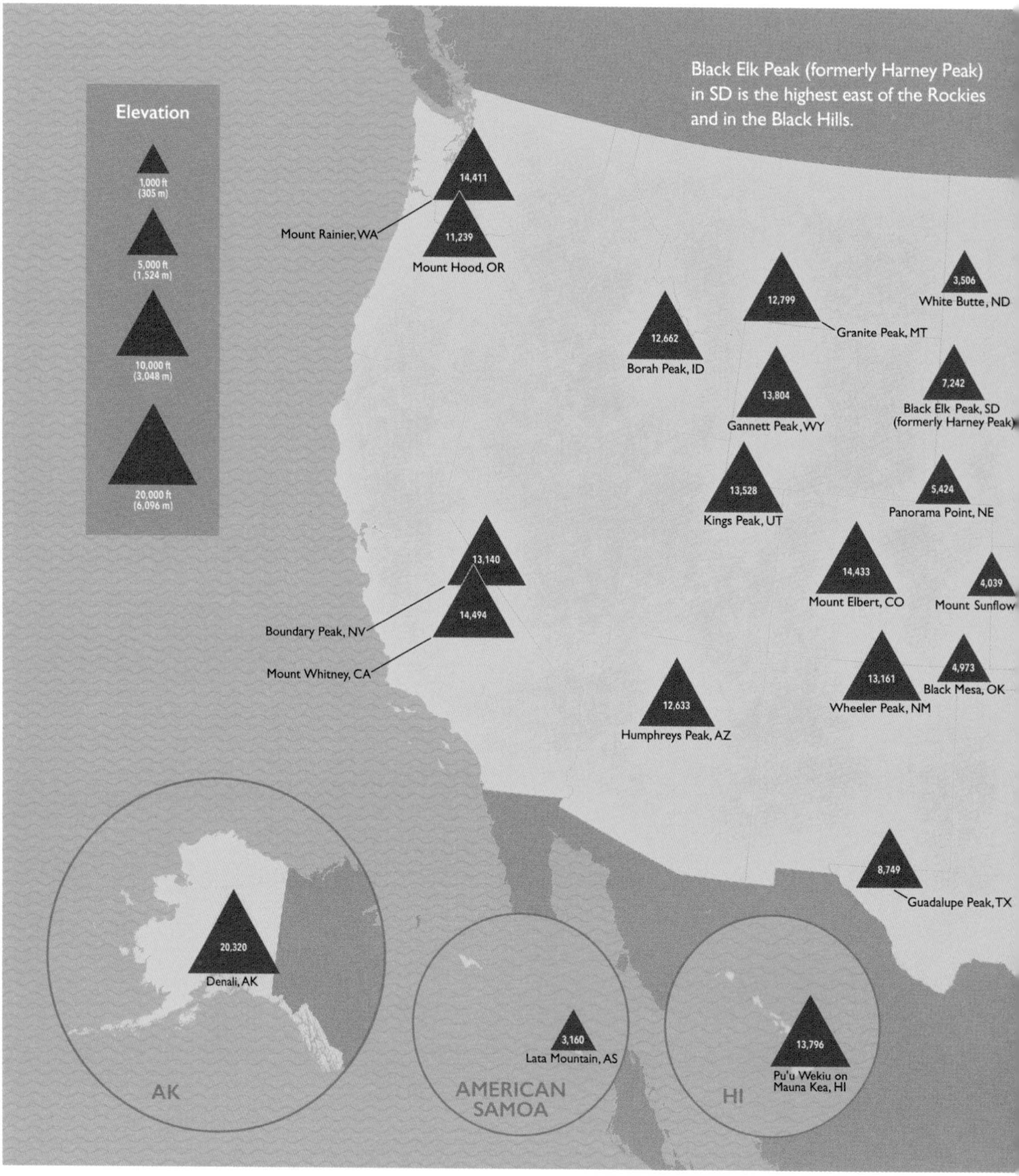

Black Elk Peak (formerly Harney Peak) in SD is the highest east of the Rockies and in the Black Hills.
Elevation
1,000 ft (305 m)
5,000 ft (1,524 m)
10,000 ft (3,048 m)
20,000 ft (6,096 m)
14,411
Mount Rainier, WA
11,239
Mount Hood, OR
12,662
Borah Peak, ID
12,799
Granite Peak, MT
3,506
White Butte, ND
13,804
Gannett Peak, WY
7,242
Black Elk Peak, SD (formerly Harney Peak)
13,528
Kings Peak, UT
5,424
Panorama Point, NE
13,140
Boundary Peak, NV
14,494
Mount Whitney, CA
14,433
Mount Elbert, CO
4,039
Mount Sunflow
13,161
Wheeler Peak, NM
4,973
Black Mesa, OK
12,633
Humphreys Peak, AZ
8,749
Guadalupe Peak, TX
20,320
Denali, AK
AK
3,160
Lata Mountain, AS
AMERICAN SAMOA
13,796
Pu'u Wekiu on Mauna Kea, HI
HI

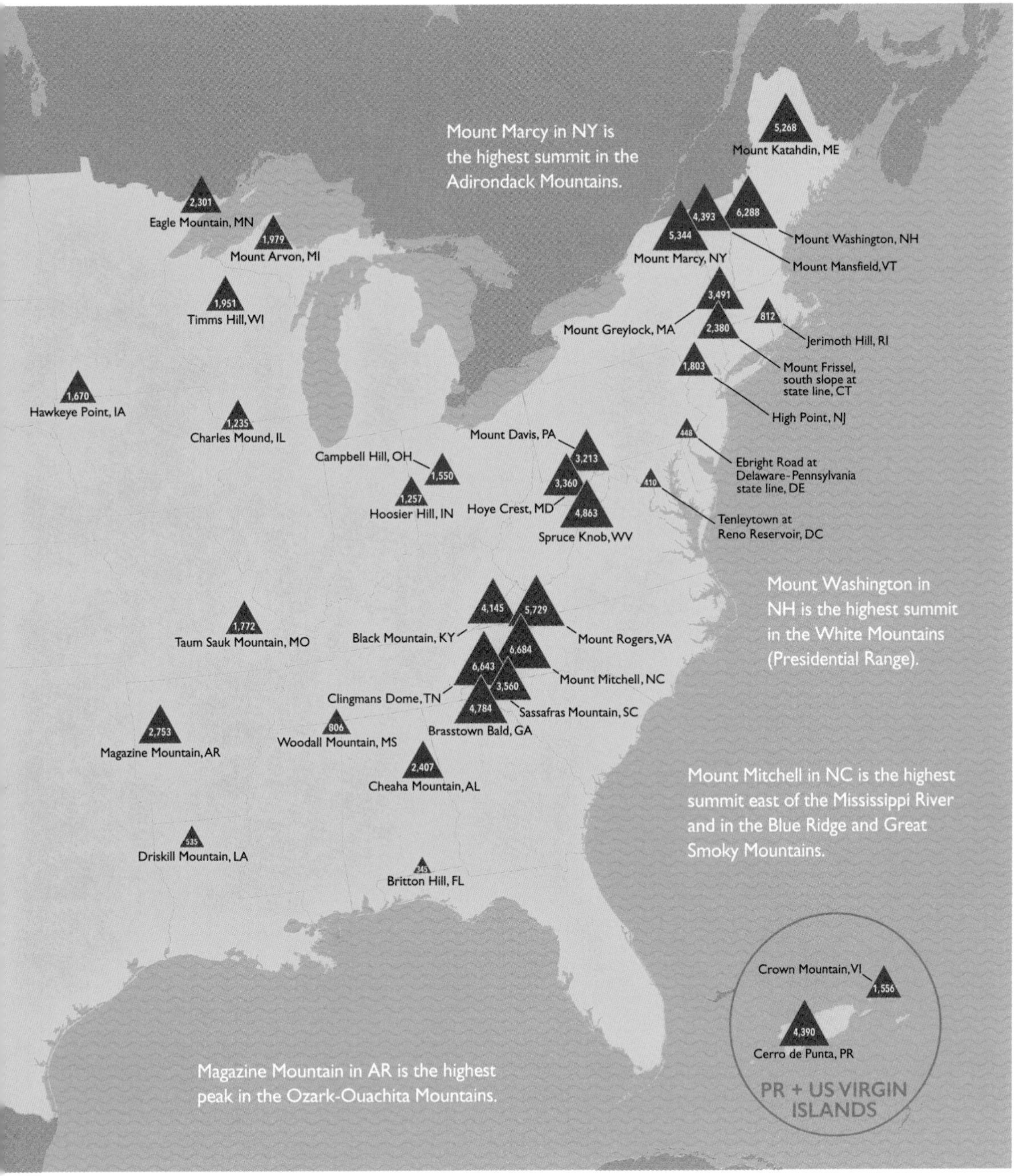
Mount Marcy in NY is the highest summit in the Adirondack Mountains.
5,268
Mount Katahdin, ME
2,301
Eagle Mountain, MN
1,979
Mount Arvon, MI
4,393
6,288
5,344
Mount Washington, NH
Mount Marcy, NY
Mount Mansfield, VT
1,951
Timms Hill, WI
3,491
812
Mount Greylock, MA
2,380
Jerimoth Hill, RI
1,803
Mount Frissel, south slope at state line, CT
1,670
Hawkeye Point, IA
1,235
Charles Mound, IL
High Point, NJ
Mount Davis, PA
448
3,213
Ebright Road at Delaware-Pennsylvania state line, DE
Campbell Hill, OH
1,550
3,360
410
1,257
Hoosier Hill, IN
Hoye Crest, MD
4,863
Tenleytown at Reno Reservoir, DC
Spruce Knob, WV
Mount Washington in NH is the highest summit in the White Mountains (Presidential Range).
4,145
5,729
1,772
Taum Sauk Mountain, MO
Black Mountain, KY
Mount Rogers, VA
6,684
6,643
Mount Mitchell, NC
3,560
Clingmans Dome, TN
4,784
Sassafras Mountain, SC
2,753
806
Brasstown Bald, GA
Magazine Mountain, AR
Woodall Mountain, MS
2,407
Cheaha Mountain, AL
Mount Mitchell in NC is the highest summit east of the Mississippi River and in the Blue Ridge and Great Smoky Mountains.
535
Driskill Mountain, LA
345
Britton Hill, FL
Crown Mountain, VI
1,556
4,390
Cerro de Punta, PR
Magazine Mountain in AR is the highest peak in the Ozark-Ouachita Mountains.
PR + US VIRGIN ISLANDS

36 And now, the weather:
North American weather extremes

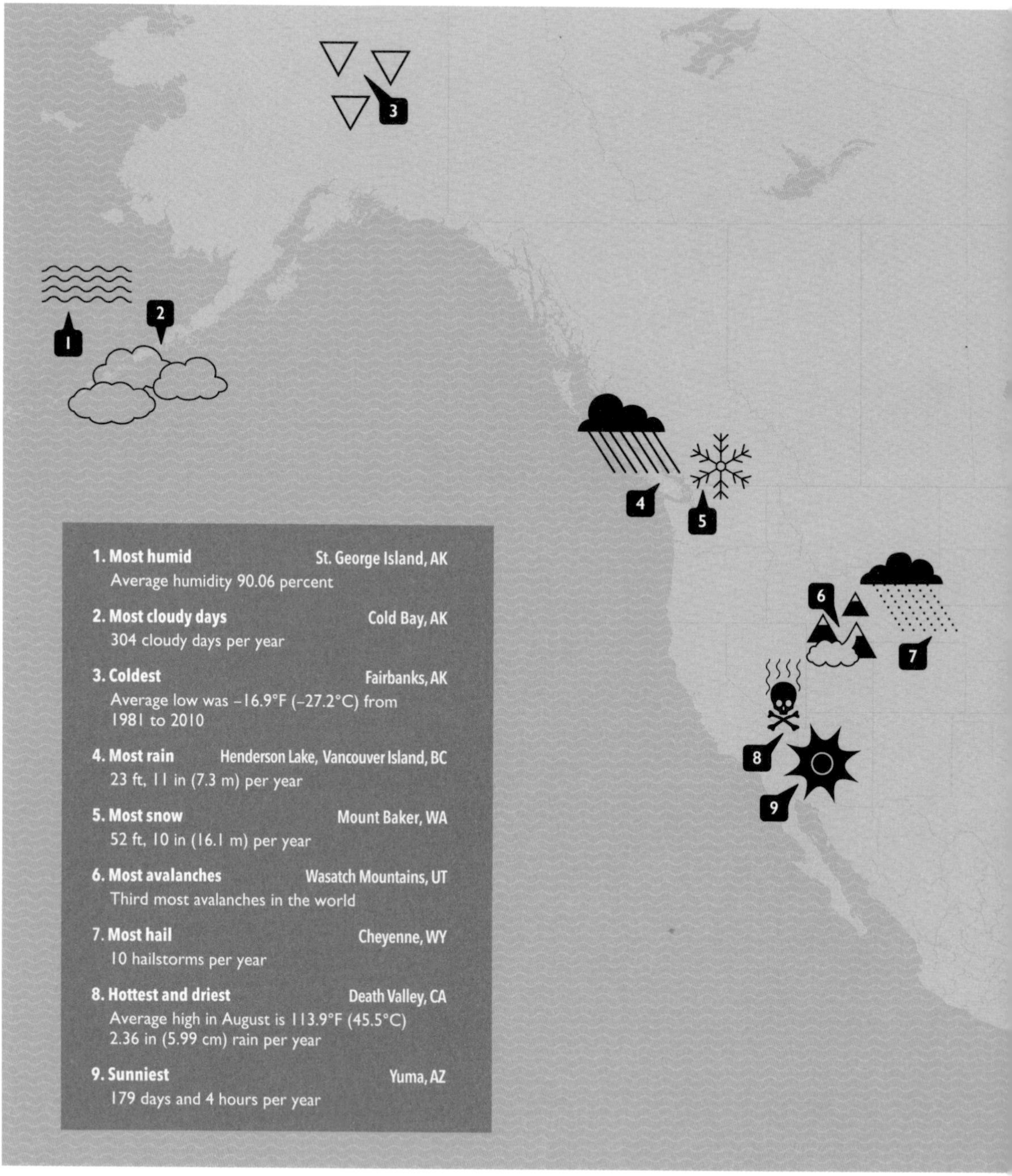

1. Most humid — St. George Island, AK
Average humidity 90.06 percent

2. Most cloudy days — Cold Bay, AK
304 cloudy days per year

3. Coldest — Fairbanks, AK
Average low was −16.9°F (−27.2°C) from 1981 to 2010

4. Most rain — Henderson Lake, Vancouver Island, BC
23 ft, 11 in (7.3 m) per year

5. Most snow — Mount Baker, WA
52 ft, 10 in (16.1 m) per year

6. Most avalanches — Wasatch Mountains, UT
Third most avalanches in the world

7. Most hail — Cheyenne, WY
10 hailstorms per year

8. Hottest and driest — Death Valley, CA
Average high in August is 113.9°F (45.5°C)
2.36 in (5.99 cm) rain per year

9. Sunniest — Yuma, AZ
179 days and 4 hours per year

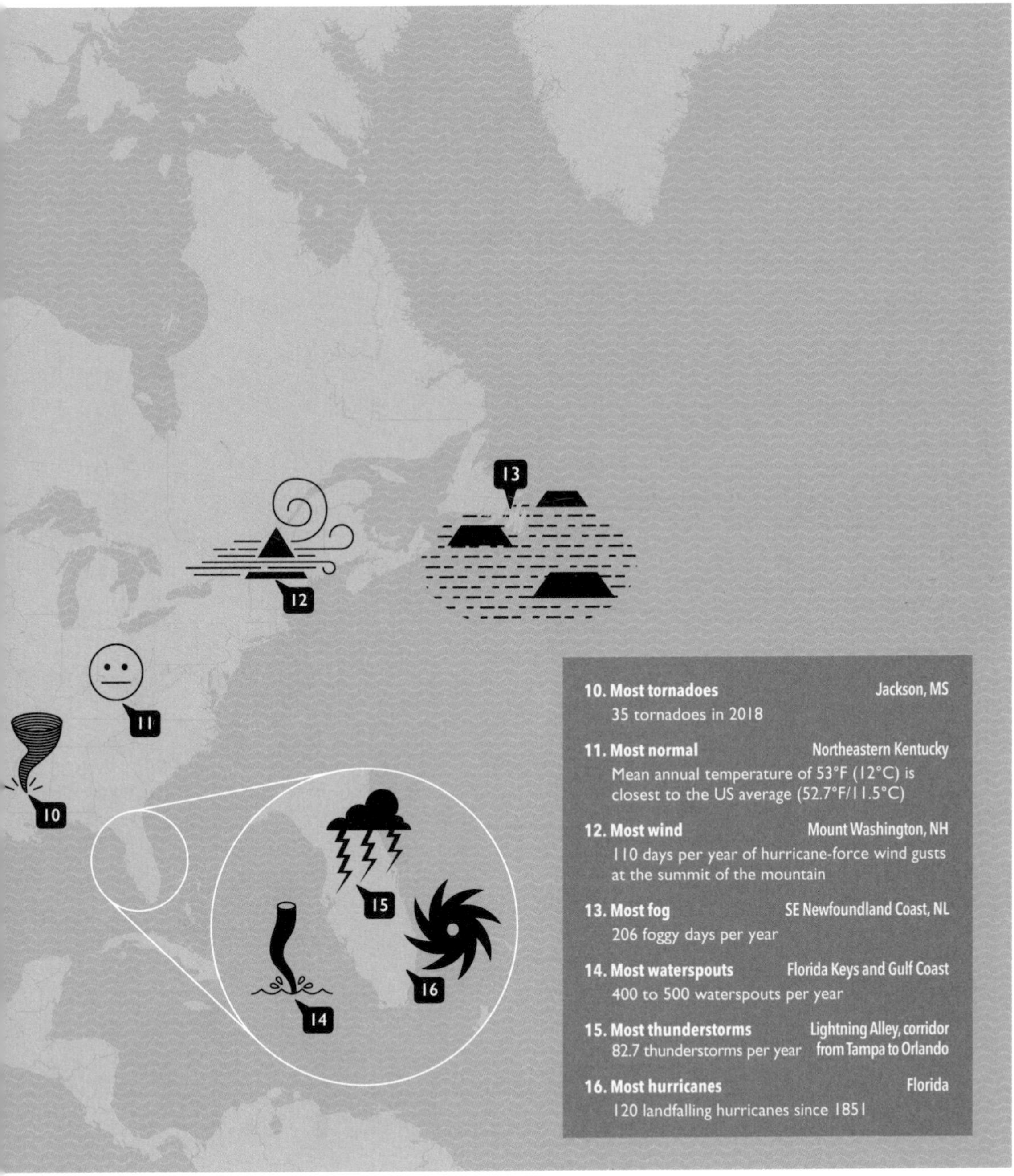

10. Most tornadoes — Jackson, MS
35 tornadoes in 2018

11. Most normal — Northeastern Kentucky
Mean annual temperature of 53°F (12°C) is closest to the US average (52.7°F/11.5°C)

12. Most wind — Mount Washington, NH
110 days per year of hurricane-force wind gusts at the summit of the mountain

13. Most fog — SE Newfoundland Coast, NL
206 foggy days per year

14. Most waterspouts — Florida Keys and Gulf Coast
400 to 500 waterspouts per year

15. Most thunderstorms — Lightning Alley, corridor from Tampa to Orlando
82.7 thunderstorms per year

16. Most hurricanes — Florida
120 landfalling hurricanes since 1851

37 Caving the day:
Cavernous wonders across North America

1. Jewel Cave National Monument
Second-longest cave in the US and third-longest in the world

2. Wind Cave National Park
One of the longest and most complex caves in the world, famous for its boxwork—a calcite formation so rare that the majority worldwide is found here

3. Mammoth Cave National Park
World's longest cave (more than 412 miles/663 km long)

4. Crystal Cave
Home to the world's largest geode

5. Craighead Caverns
Home to the world's second-largest underground lake, nicknamed the Lost Sea

6. Luray Caverns
Home of the world's largest musical instrument, the Great Stalacpipe Organ, whose rubber mallets tap ancient stalactites to produce notes

7 . Cave of the Crystals
Milky-white gypsum crystals nearly 40 feet (12 m) long fill this horseshoe-shaped cave

8. Carlsbad Caverns National Park
120 known caves; the second-largest is Carlsbad Cavern, whose Big Room is 8.2 acres—the biggest easily accessible cave chamber in North America

9. Ellison's Cave
Home to the harrowing pit known as Fantastic, the deepest cave plunge in the continental US (586 feet/179 m deep)

10. Kazumura Cave
Longest lava tube cave in the world

11. Sistema Huautla
Deepest cave in the Western Hemisphere (5,118 feet/1,560 km deep)

12. Sistema Sac Actun
World's longest underwater cave system (231 miles/372 km long)

13. Cueva Martín Infierno
World's tallest standalone stalagmite (220 feet/67 m tall)

14. Parque Nacional de las Cavernas del Río Camuy
The Rio Camuy, the third-largest underground river in the world, runs through this cave

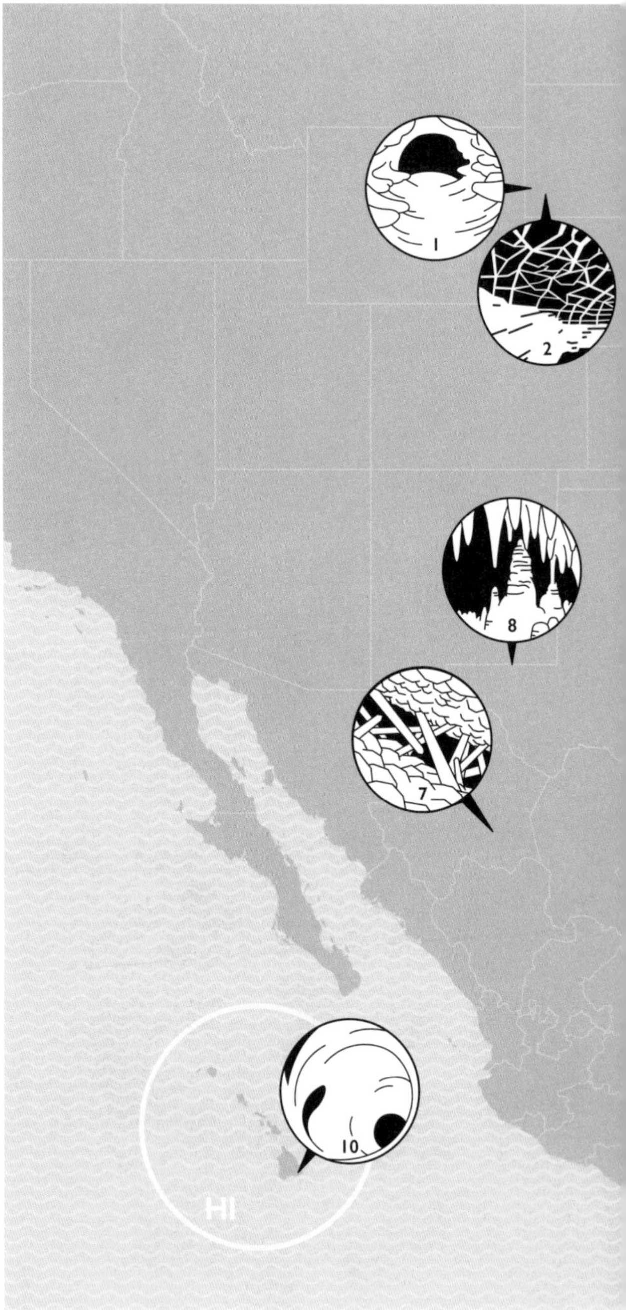

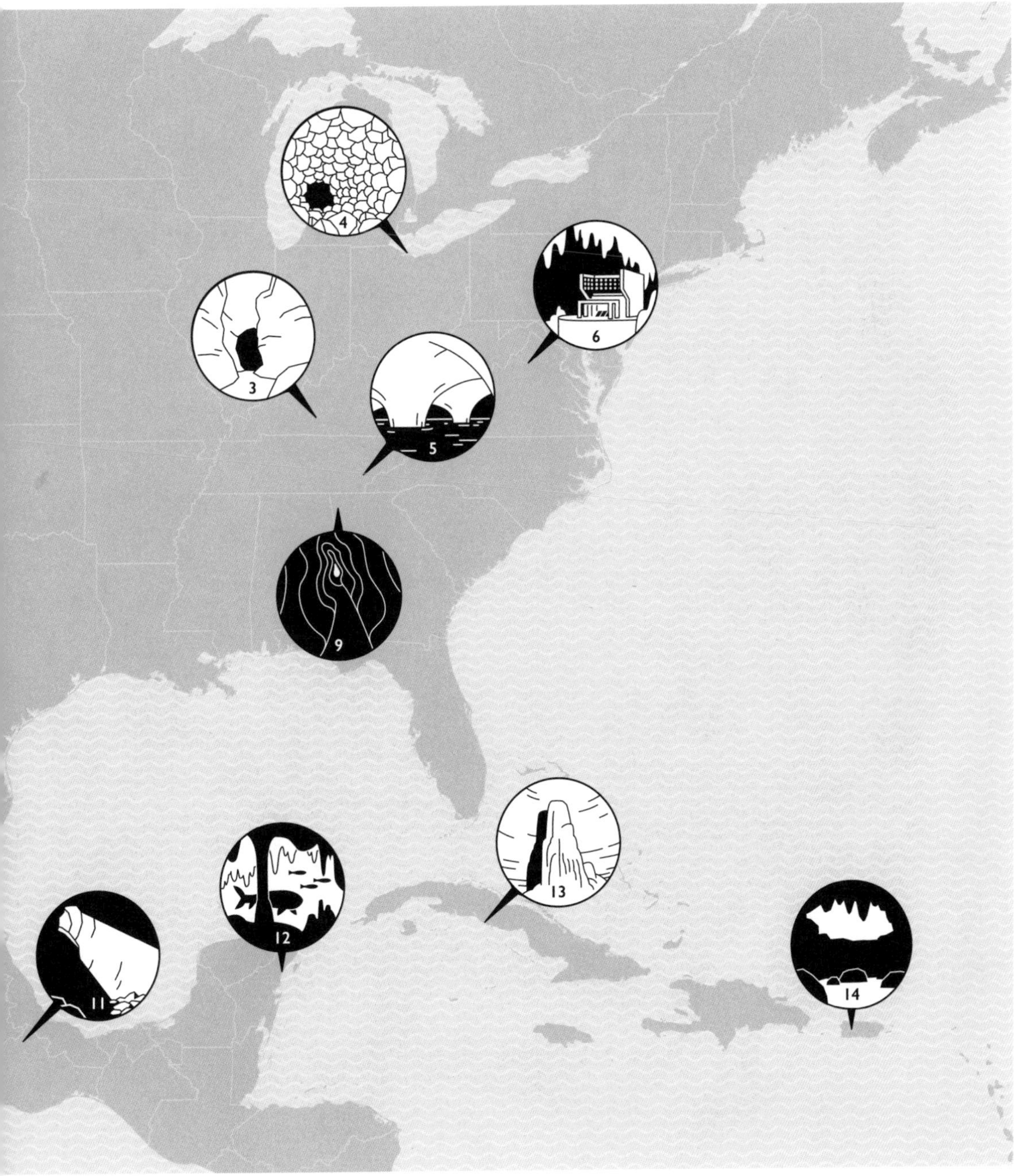
4
6
3
5
9
13
12
11
14

38 **Supervolcano:** When Yellowstone blanketed the US and Canada in ash

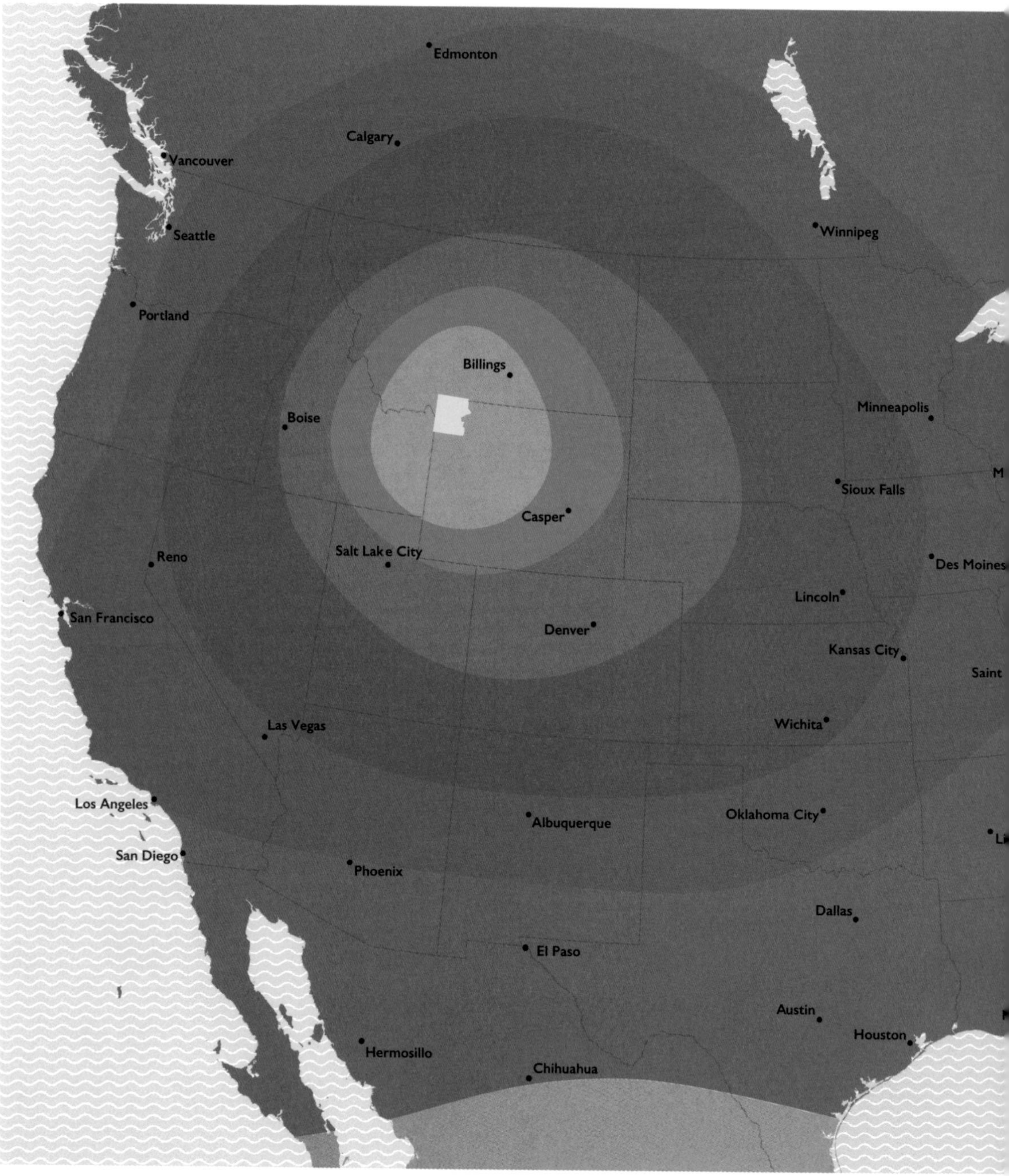

Yellowstone National Park

Depth of volcanic ash

- 39.37 in (More than 1,000 mm)
- 11.81–39.37 in (300–1,000 mm)
- 3.94–11.81 in (100–300 mm)
- 1.18–3.94 in (30–100mm)
- 0.4–1.18 in (10–30 mm)
- 0.12–0.4 in (3–10 mm)
- 0.04–0.12 in (1–3 mm)
- No data

If you've ever seen Old Faithful, along with all the other gushing geysers and boiling hot springs of Yellowstone National Park, you've witnessed the enormous power that lies underfoot. It's magma that makes Yellowstone bubble and spew—magma from an enormous active volcano that lurks beneath the park. When a volcano belches more than 240 million cubic miles (more than 1 billion km^3) of magma, it becomes a supervolcano. Yellowstone has hit this mark more than once, most recently about 640,000 years ago. It's extremely unlikely it'll erupt anytime soon; in the near term, relatively minor lava flows are far more probable. But Yellowstone is certainly capable of another supereruption. Here, we see a geologist's model of how far the volcanic ash spread, and how deep, during that eruption 640,000 years ago. The wind is believed to have carried lighter particles all the way to the East Coast, blanketing states from Florida to Maine in at least a millimeter or so of ash.

39 Perma-lost:
The Last Glacial Maximum compared to today

Glacial extent during Last Glacial Maximum (19,000 BCE)   Glacial extent today

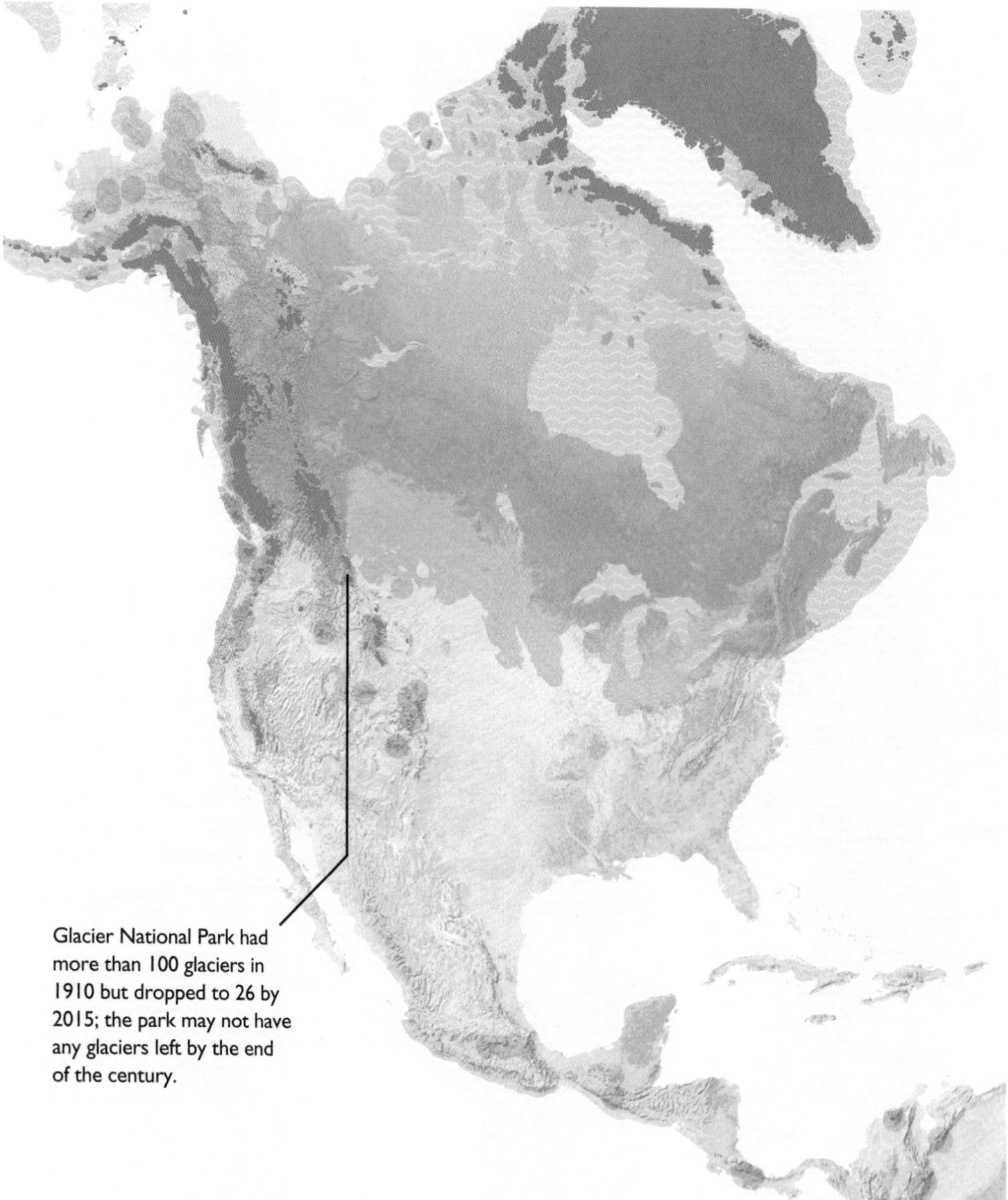
Glacier National Park had more than 100 glaciers in 1910 but dropped to 26 by 2015; the park may not have any glaciers left by the end of the century.

40 **Where darkness reigns:**
The dark skies still left for pristine stargazing

The rise of urbanization and its attendant light pollution has brought with it the loss of truly dark skies. On the Bortle scale, a rating system for sky darkness, a score of 1-2 indicates little or no light. Note that the borders of these regions are, in all likelihood, slightly affected by light, too—mainly along the horizon. North America is running out of land that lies beneath its remaining pristine night skies.

Pristine skies Reduced visibility due to light pollution

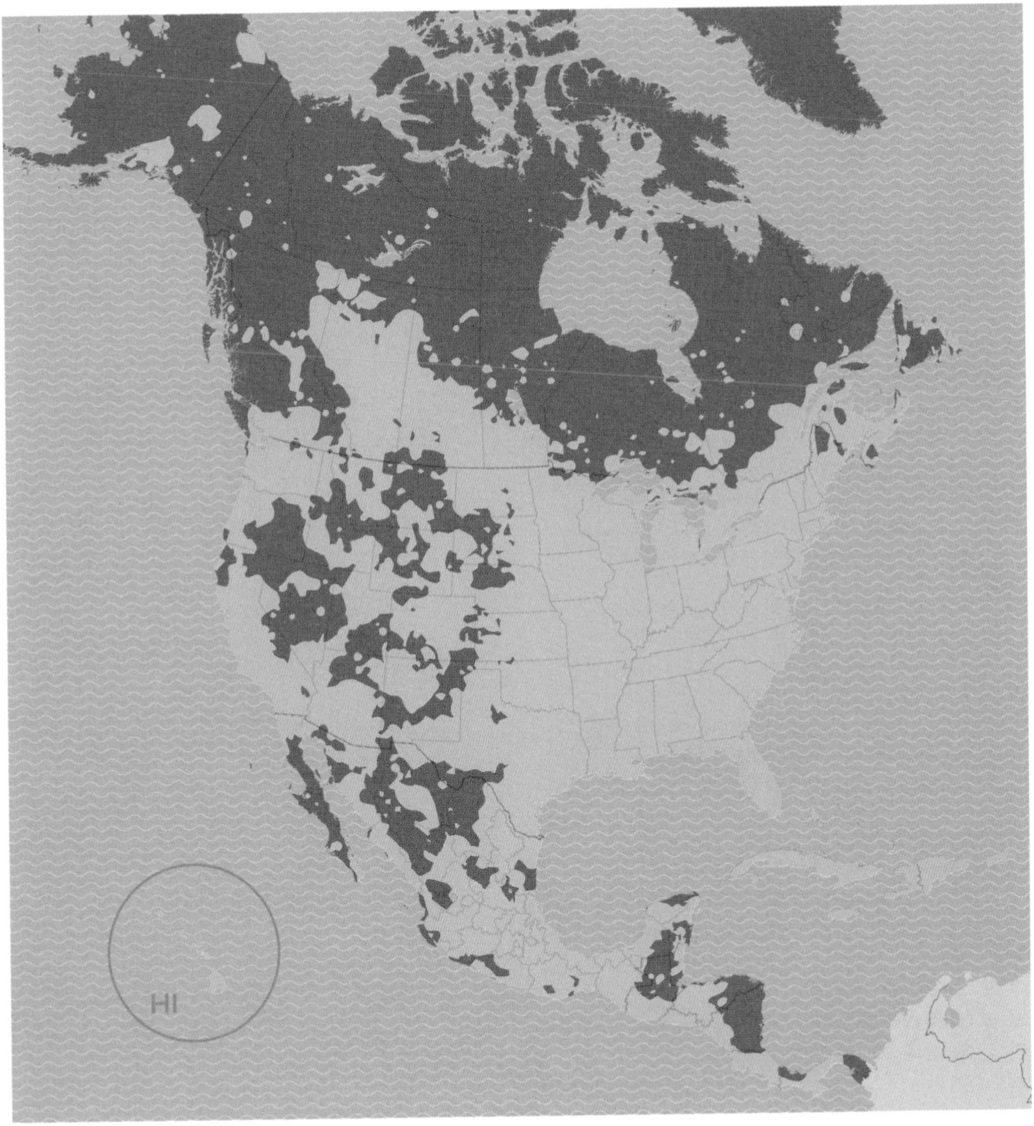

41

Singled out: Endangered or threatened species that live entirely within one state

	Animals	Plants
AK		1. Aleutian shield fern
AL	2. Alabama beach mouse	3. Alabama streak-sorus fern
AR	4. Hell Creek Cave crayfish	
AZ	5. Mount Graham red squirrel	6. Sentry milk-vetch
CA	7. Palos Verdes blue butterfly	8. Santa Barbara Island liveforever
CO	9. Pawnee montane skipper	10. Penland beardtongue
FL	11. Miami tiger beetle	12. Garber's spurge
GA	13. Etowah darter	14. Mat-forming quillwort
HI	15. Hawaiian yellow-faced bee	16. Hoawa
ID	17. Bruneau hot springsnail	18. Slickspot peppergrass
IL	19. Illinois cave amphipod	
KY	20. Relict darter	21. Kentucky glade cress
MA	22. Plymouth red-bellied turtle	
MD	23. Maryland darter	
MI		24. Michigan monkey-flower
MN		25. Minnesota dwarf trout lily
MO	26. Tumbling Creek cavesnail	
MS	27. Pearl darter	
MT	28. Meltwater lednian stonefly	
NC	29. Noonday snail	30. Mountain golden heather
NE	31. Salt Creek tiger beetle	
NJ		32. Knieskern's beaked-rush
NM	33. Koster's springsnail 35. Roswell springsnail 37. Noel's amphipod	34. Gypsum wild buckwheat 36. Lee pincushion cactus
NV	38. Devils Hole pupfish	39. Ash Meadows gumplant 40. Ash Meadows blazingstar
NY	41. Chittenango ovate amber snail	
OH	42. Scioto madtom	
OR	43. Hutton tui chub	44. Malheur wire-lettuce
TN	45. Nashville crayfish	46. Ruth's golden aster
TX	47. Diminutive amphipod 49. Phantom springsnail 50. Phantom tryonia	48. Little Aguja pondweed
UT	51. June sucker	52. Autumn buttercup
VA	53. Shenandoah salamander	54. Virginia round-leaf birch
WA	55. Roy Prairie pocket gopher	56. Umtanum Desert buckwheat
WI		57. Fassett's locoweed
WV	58. Guyandotte River crayfish	
WY	59. Kendall Warm Springs dace	60. Desert yellowhead
PR	61. Golden coquí	62. *Leptocereus grantianus*

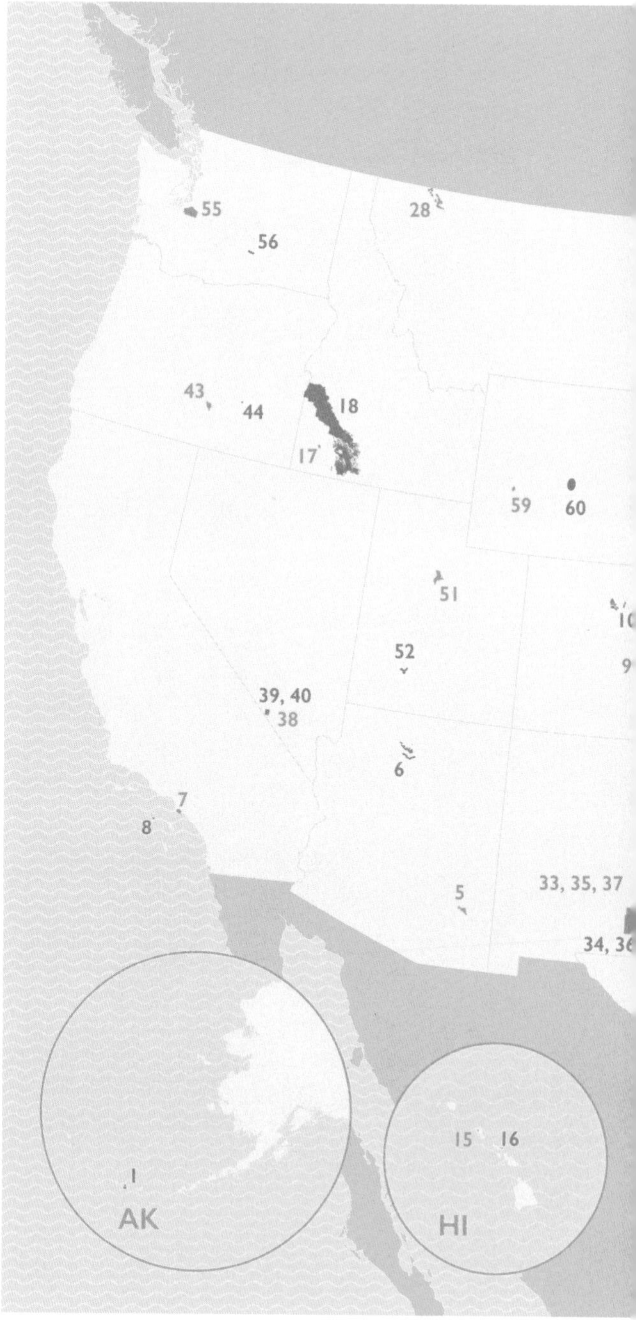

Some threatened and endangered species are so few in number that their current range is confined to one state—and nowhere else on the planet. Where there is more than one such plant or animal within a state, we've included the one that inhabits the smallest area. Some states have none, and some have only a plant or animal, not both. Also, due to a handful of species sharing the same ranges, there are ties in Texas, New Mexico, and Nevada.

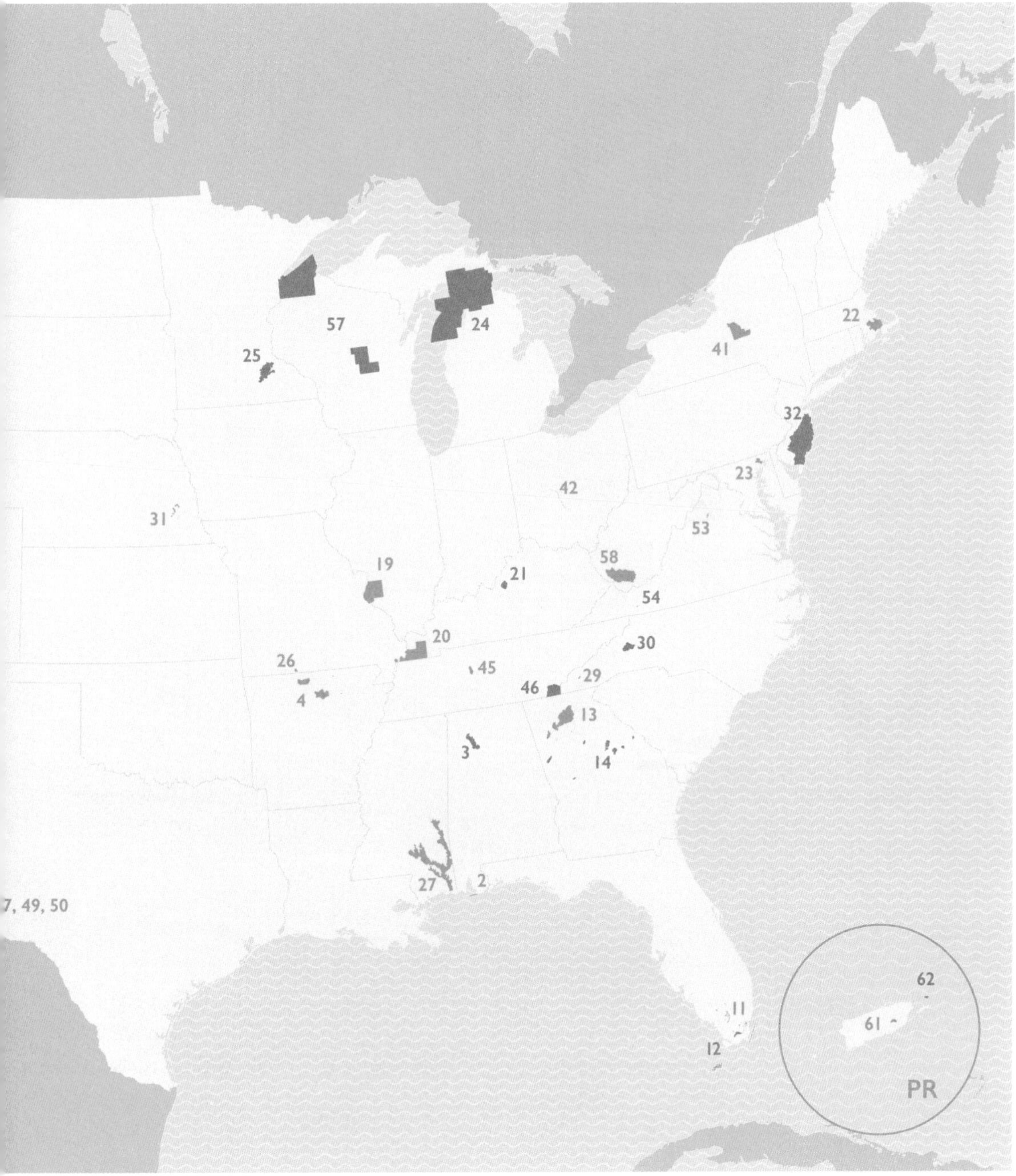

42 **A second act:** North American species that have staged comebacks

Bald Eagle*

1963	416 pairs in lower 48
2007	11,052 pairs in lower 48

American Peregrine Falcon*

1975	39 pairs in lower 48
1999	1,650 pairs in lower 48

* The peregrine falcon and bald eagle are widespread throughout North America.

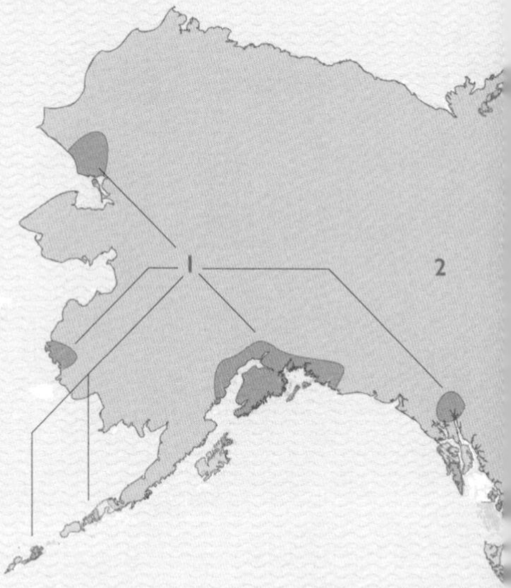

1. Aleutian Cackling Goose

1975	790
2021	200,000

2. Grizzly Bear

1975	700–800 in lower 48
2019	1,913 in lower 48

4. California Condor

1985	9 in the wild
2012	386 (213 wild)

5. Black-Footed Ferret

1991	0 in the wild
2010	1,410 in the wild

6. Swift Fox

1938	0 in the wild in CAN
2006	647 in the wild in CAN

7. Whooping Crane

1941	16
2011	599

10. Lesser Long-Nosed Bat

1988	< 1,000
2018	200,000

11. Louisiana Black Bear

1950s	80–120 in Louisiana
2016	1,000 in Louisiana

12. Okaloosa Darter

1994	1,500
2011	> 500,000

13. Tennessee Purple Coneflower

1979	3 colonies
2011	35 colonies

The full North American ranges of some notable survivors in the history of endangered species protection crisscross the continent. Some plants and animals were fortunate enough to be taken off the endangered species list due to recovery. Others, for all their progress, are still threatened or even endangered in all or part of their range, including a few members of the inaugural endangered species list of 1967: the whooping crane, California condor, and black-footed ferret.

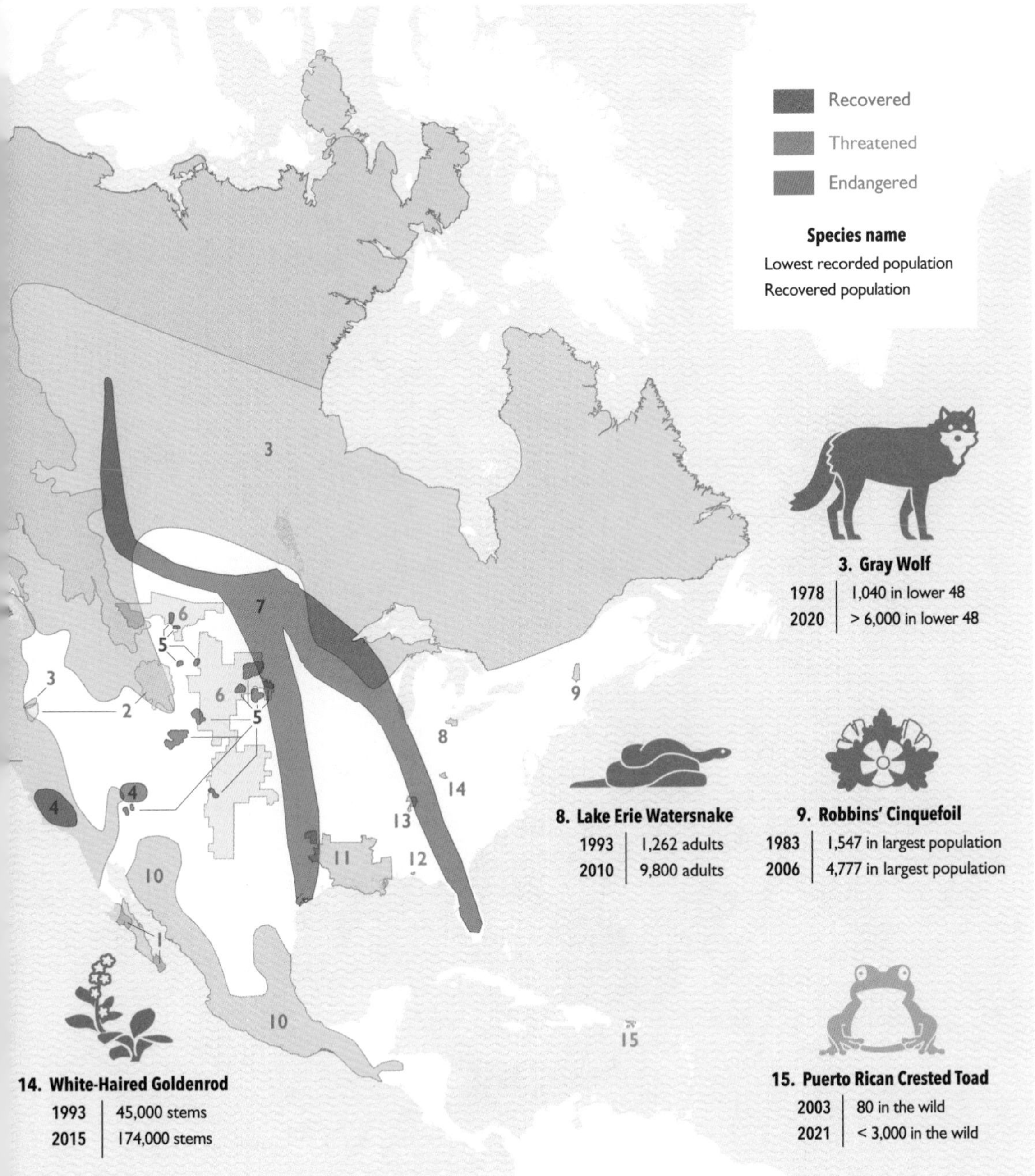

43 Let there be light:
Local time of sunset on the summer solstice

Before 8:00 PM	8:00–8:30 PM	8:30–9:00 PM	9:00–9:30 PM
9:30–10:00 PM	10:00–10:30 PM	10:30–11:00 PM	11:00–11:30 PM
11:30 PM to midnight	After midnight	24 hours of daylight	No data

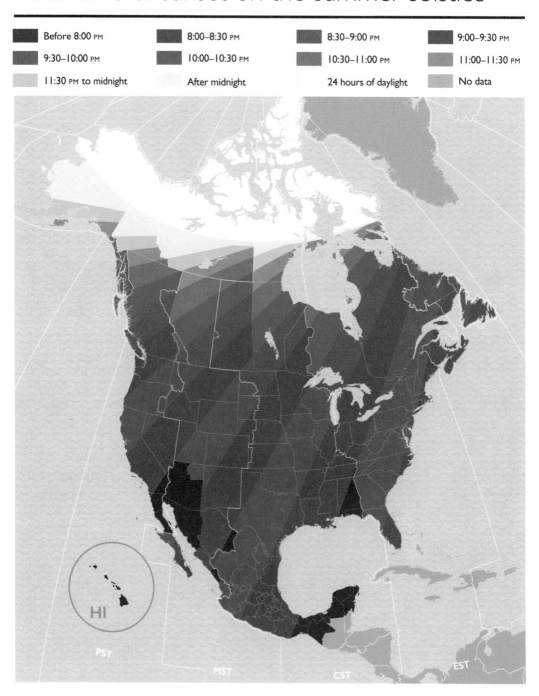

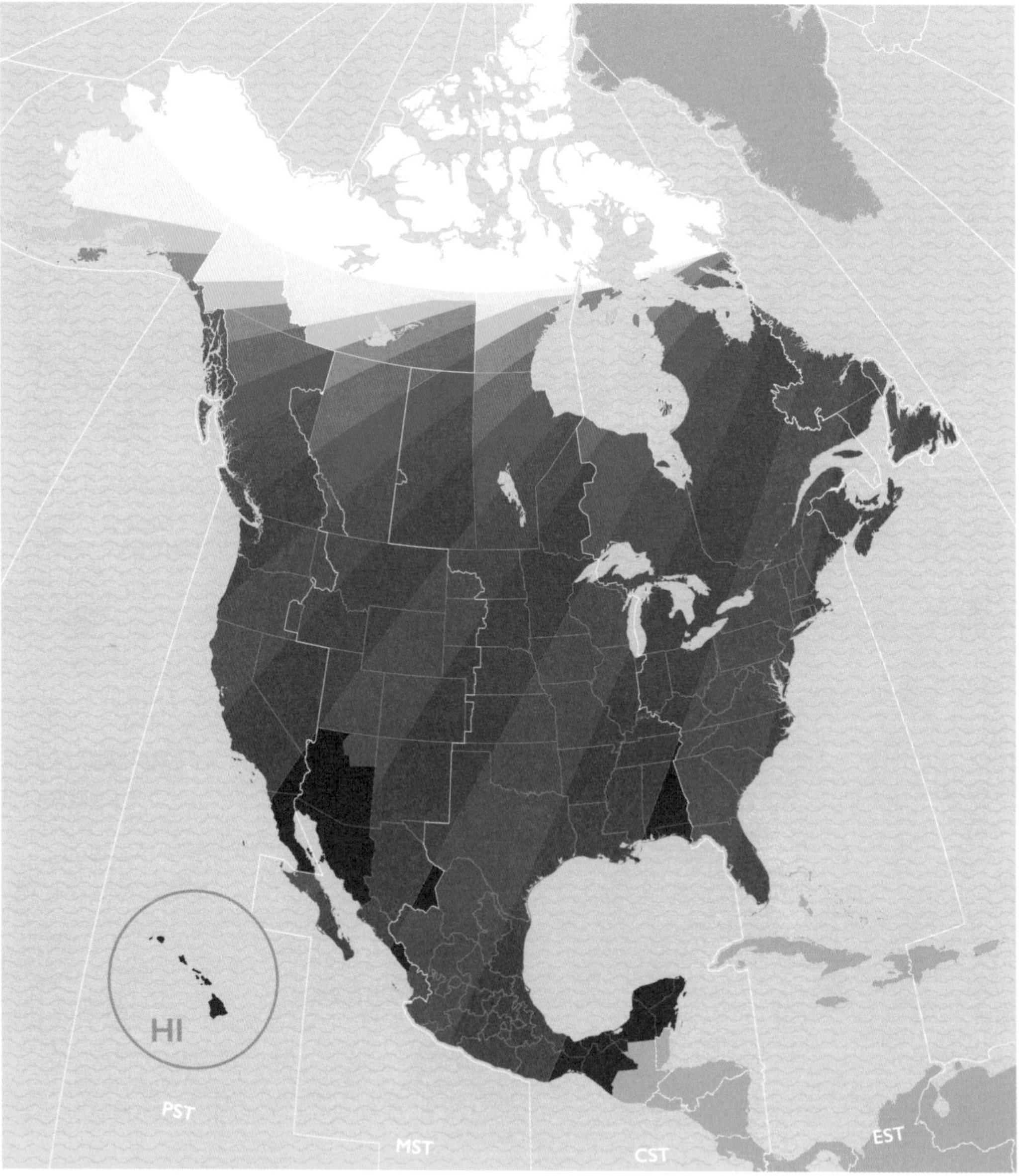

44 **Don't let the sun go down on me:** Local time of sunset on the winter solstice

- 6:00–6:30 PM
- 5:30–6:00 PM
- 5:00–5:30 PM
- 4:30–5:00 PM
- 4:00–4:30 PM
- 3:30–4:00 PM
- 3:00–3:30 PM
- Before 3:00 PM
- At least one day with 24 hours of darkness
- No data

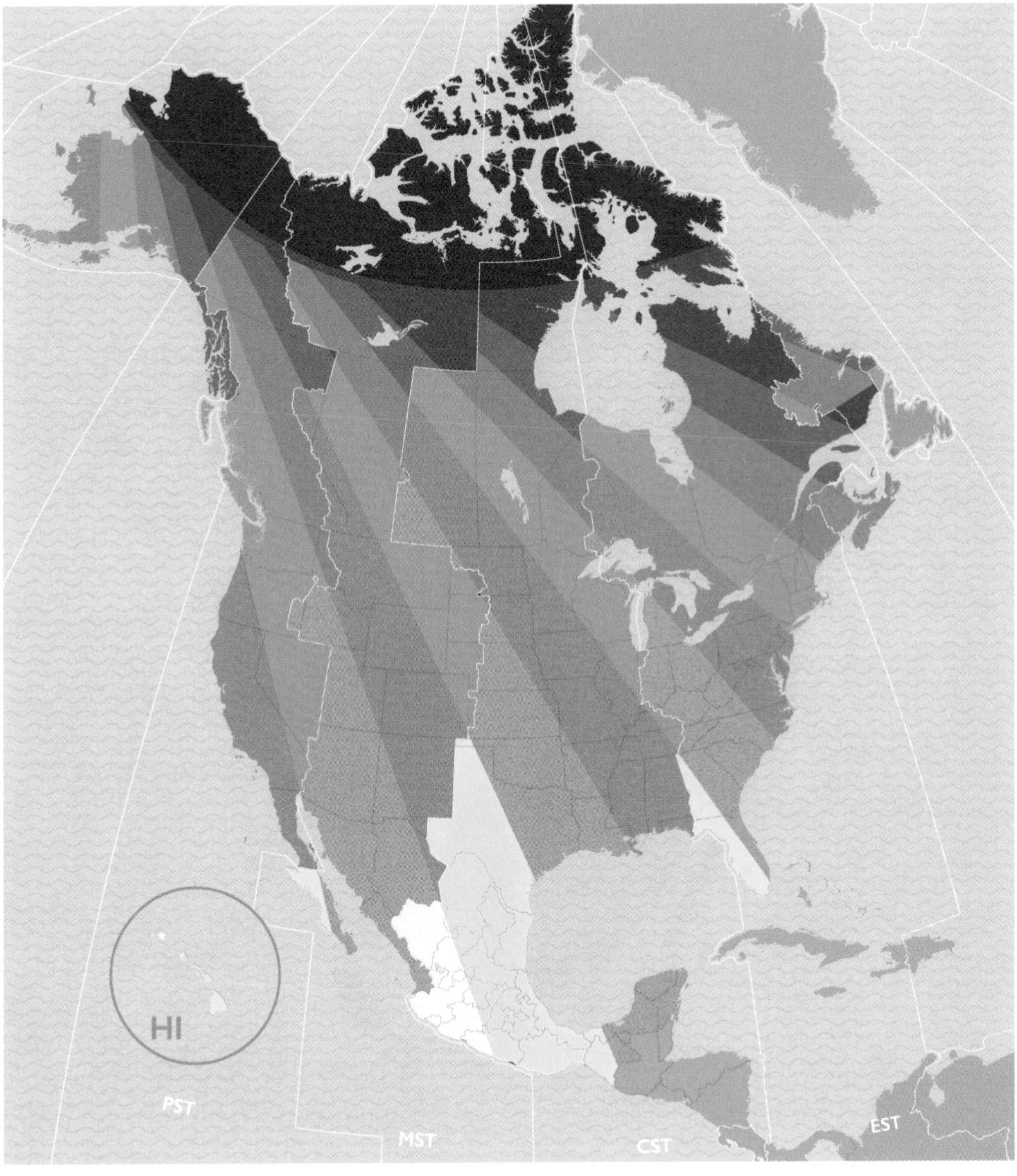

45 Too much time on my hands:
Hours of daylight on the summer solstice

24 | 20–24 | 18–20 | 17–18 | 16–17 | 15.5–16 | 15–15.5 | 14.5–15 | 14–14.5 | Under 14 | No data

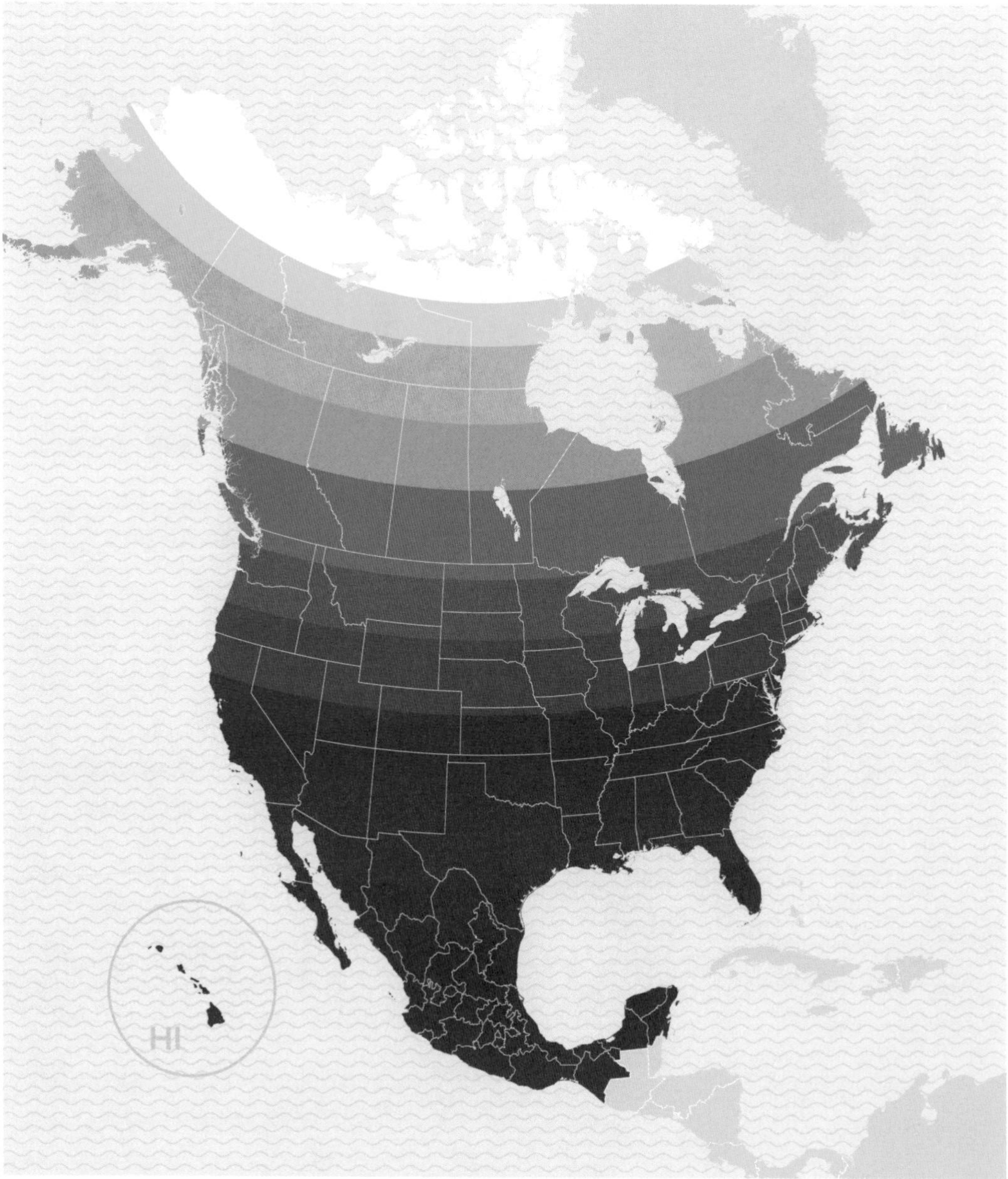

46 **Not enough time in the day:**
Hours of daylight on the winter solstice

Under 3	3–4	4–5	5–6	6–6.5
6.5–7	7–7.5	7.5–8	8–8.5	8.5–9
9–9.5	9.5–10	10–10.5	Over 10.5	No data

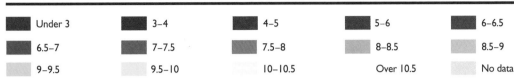

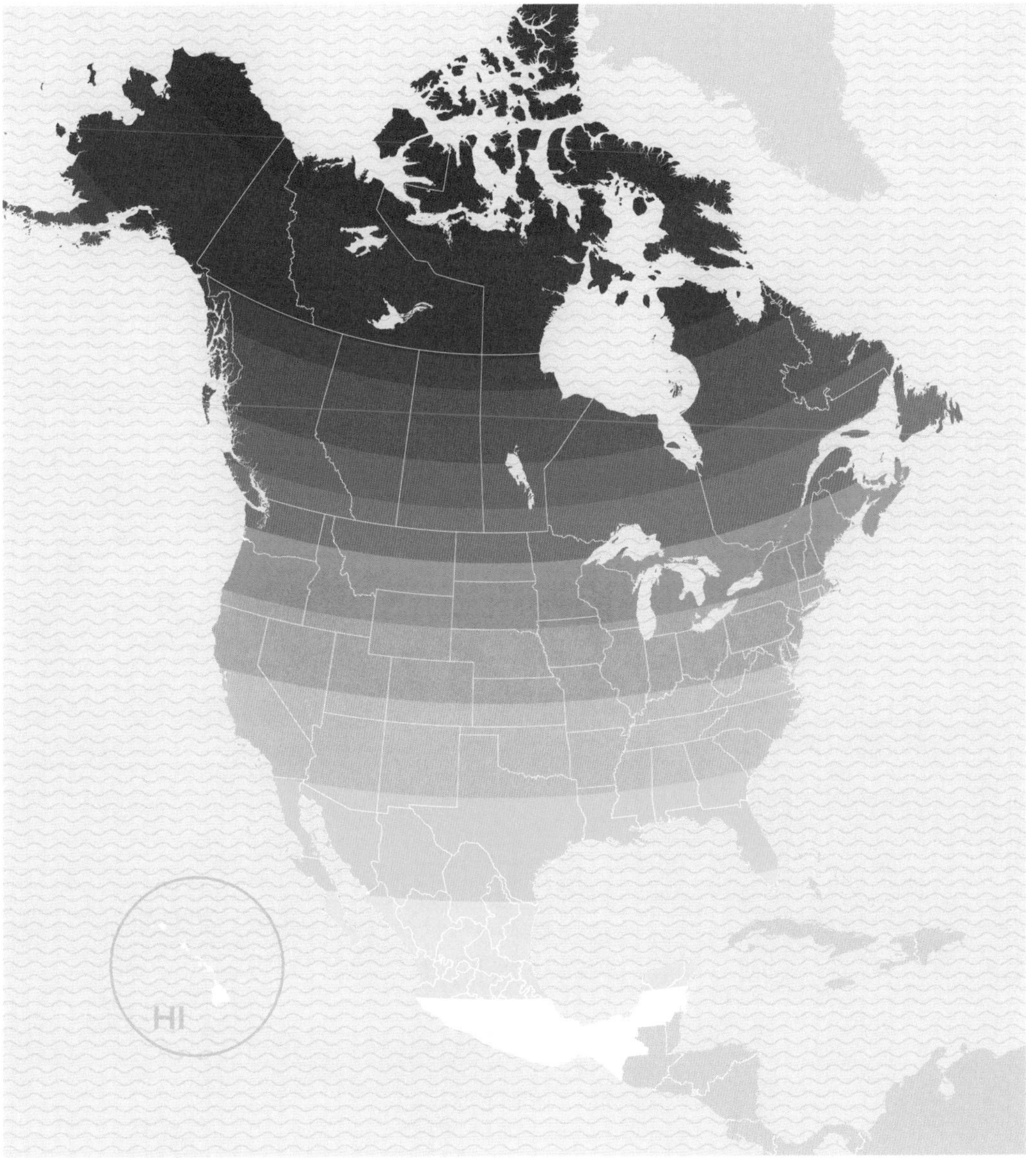

4

CULTURE AND SPORTS

47 **Tallest, fastest, steepest:** Record-setting amusement parks and rides of North America

1. Full Throttle — **Six Flags Magic Mountain, CA**
Largest loop on a roller coaster in North America (loop diameter of 127 ft, 1 in/38.7 m)

2. High Roller — **Las Vegas, NV**
North America's tallest Ferris/observation wheel (550 ft/168 m)

3. Goliath — **Six Flags Great America, IL**
Largest drop for a wood-tracked roller coaster in the world (180 ft/55 m)

4. Cedar Point — **Sandusky, OH**
Most rides at an amusement park, with 71

5. Leap-the-Dips — **Lakemont Park, PA**
Oldest operating roller coaster in the world

6. TMNT Shellraiser — **Nickelodeon Universe American Dream, NJ**
Steepest roller coaster drop in the world at 121.5° (beyond vertical)

7. Outlaw Run — **Silver Dollar City, MO**
Most inversions for a wood-tracked roller coaster in the world with three (tied for first)

8. The Beast — **King's Island, OH**
Longest wood-tracked roller coaster in the world at 7,359 ft (2,243 m)

9. Steel Curtain — **Kennywood, PA**
Most inversions for a roller coaster in North America with eight (tied with Avalancha for third in the world)

10. Kingda Ka — **Six Flags Great Adventure, NJ**
1) Fastest roller coaster in North America at 128 mph (206 km/h) and second-fastest in the world
2) Tallest roller coaster in the world at 456 ft (139 m)
3) Largest roller coaster drop in the world at 418 ft (127 m)

11. Switchback Railway — **Coney Island, NYC**
First US roller coaster built as an amusement park ride

12. Shock Wave — **Six Flags Over Texas, TX**
Highest g-force in North America (second in the world) at 5.9g

13. Fury 325 — **Carowinds, NC**
Longest steel roller coaster in North America (third in the world) at 6,602 ft (2,012 m)

14. Magic Kingdom Park — **Walt Disney World, FL**
Most visited amusement park in the world, with 20,963,000 visitors in 2019

15. Avalancha — **Xetulul, Guatemala**
Most inversions for a roller coaster in North America with eight (tied with Steel Curtain for third in the world)

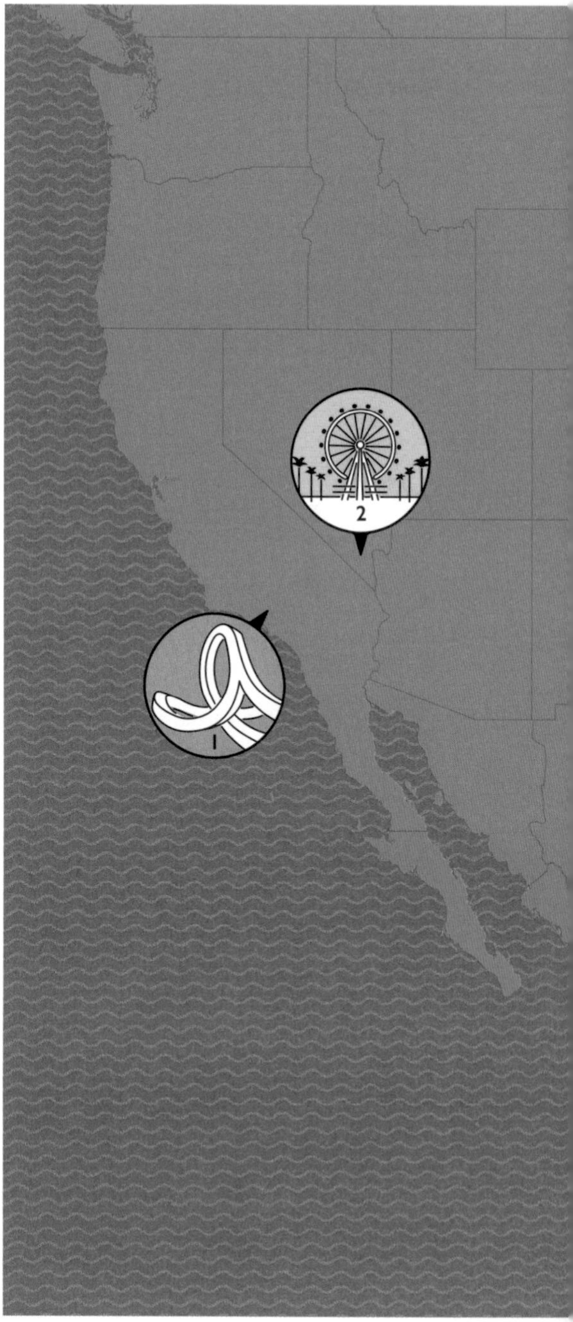

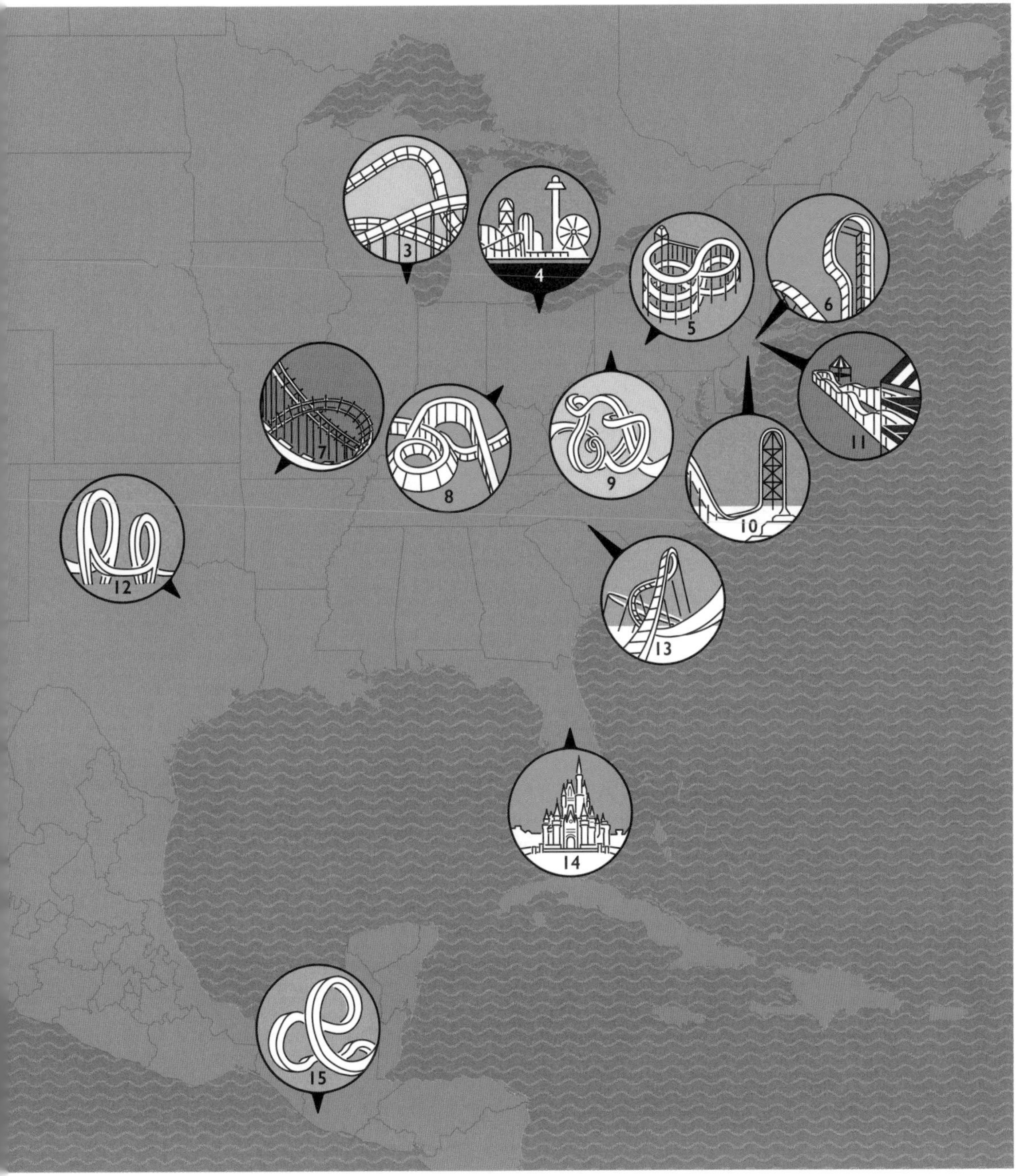
3
4
5
6
7
8
9
10
11
12
13
14
15

48 Ones for the books:
Iconic moments in North American sports

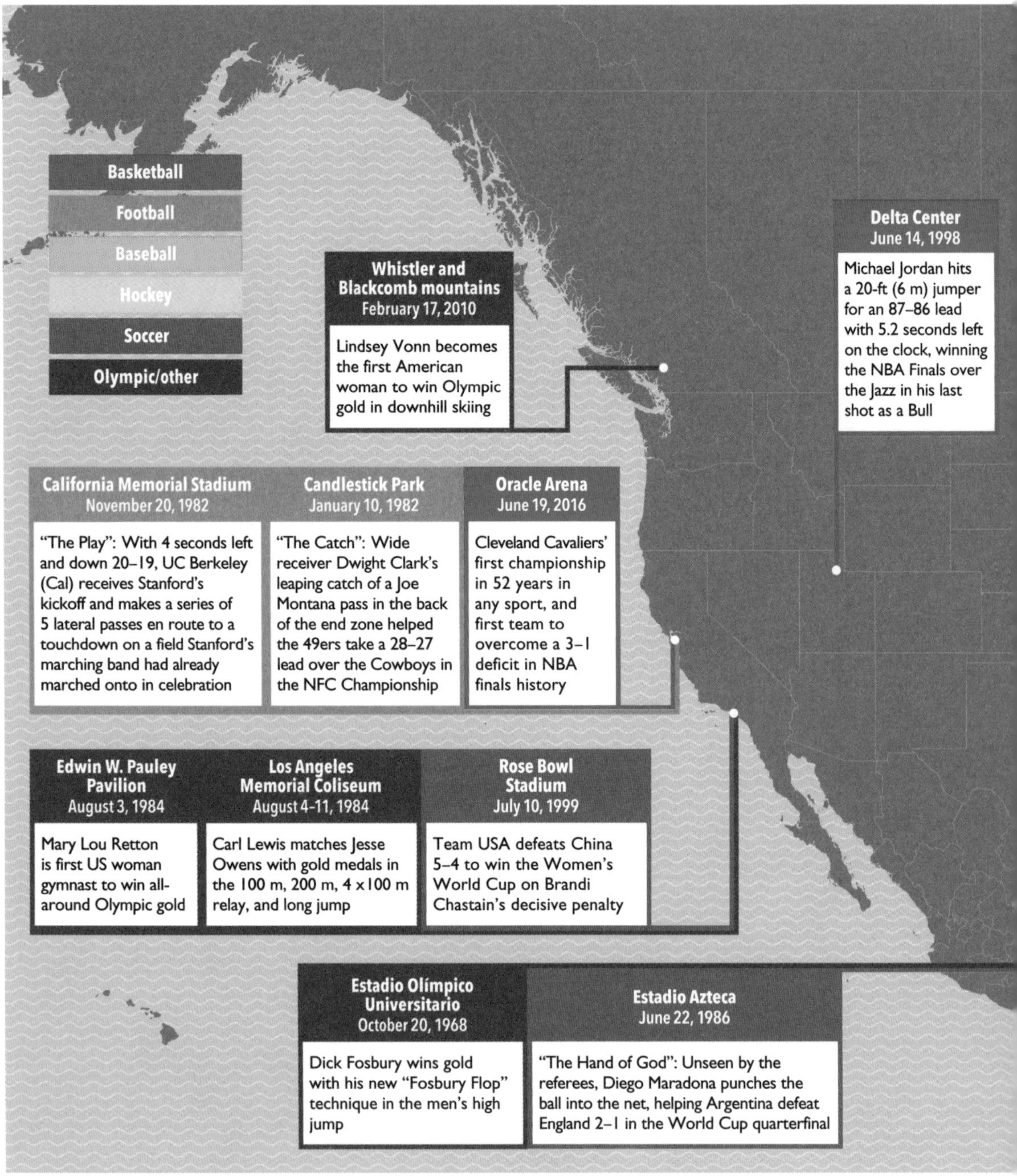
Basketball
Football
Baseball
Hockey
Soccer
Olympic/other
Whistler and Blackcomb mountains
February 17, 2010
Lindsey Vonn becomes the first American woman to win Olympic gold in downhill skiing
Delta Center
June 14, 1998
Michael Jordan hits a 20-ft (6 m) jumper for an 87–86 lead with 5.2 seconds left on the clock, winning the NBA Finals over the Jazz in his last shot as a Bull
California Memorial Stadium
November 20, 1982
"The Play": With 4 seconds left and down 20–19, UC Berkeley (Cal) receives Stanford's kickoff and makes a series of 5 lateral passes en route to a touchdown on a field Stanford's marching band had already marched onto in celebration
Candlestick Park
January 10, 1982
"The Catch": Wide receiver Dwight Clark's leaping catch of a Joe Montana pass in the back of the end zone helped the 49ers take a 28–27 lead over the Cowboys in the NFC Championship
Oracle Arena
June 19, 2016
Cleveland Cavaliers' first championship in 52 years in any sport, and first team to overcome a 3–1 deficit in NBA finals history
Edwin W. Pauley Pavilion
August 3, 1984
Mary Lou Retton is first US woman gymnast to win all-around Olympic gold
Los Angeles Memorial Coliseum
August 4-11, 1984
Carl Lewis matches Jesse Owens with gold medals in the 100 m, 200 m, 4 x 100 m relay, and long jump
Rose Bowl Stadium
July 10, 1999
Team USA defeats China 5–4 to win the Women's World Cup on Brandi Chastain's decisive penalty
Estadio Olímpico Universitario
October 20, 1968
Dick Fosbury wins gold with his new "Fosbury Flop" technique in the men's high jump
Estadio Azteca
June 22, 1986
"The Hand of God": Unseen by the referees, Diego Maradona punches the ball into the net, helping Argentina defeat England 2–1 in the World Cup quarterfinal

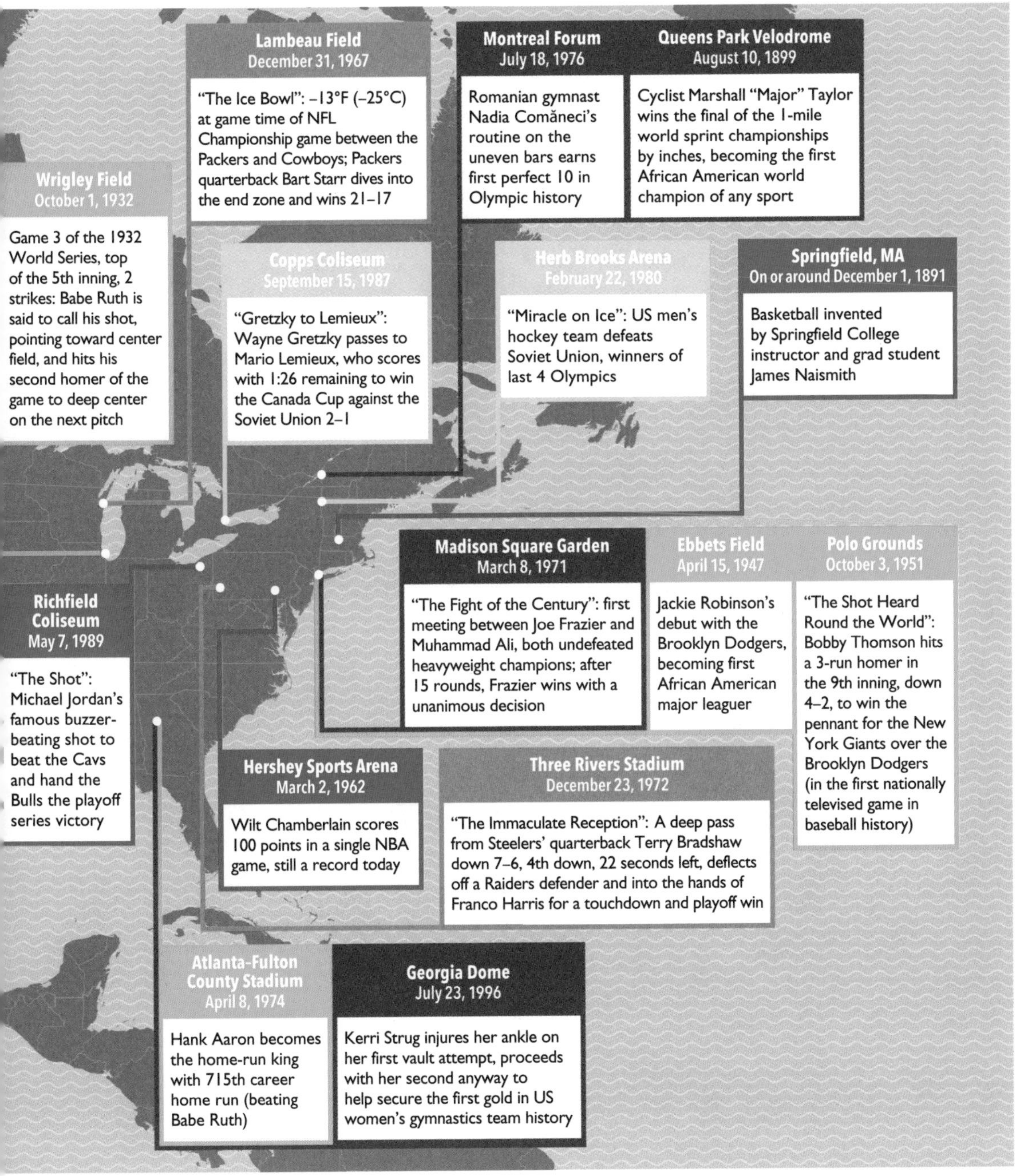
Lambeau Field
December 31, 1967
"The Ice Bowl": –13°F (–25°C) at game time of NFL Championship game between the Packers and Cowboys; Packers quarterback Bart Starr dives into the end zone and wins 21–17
Montreal Forum
July 18, 1976
Romanian gymnast Nadia Comăneci's routine on the uneven bars earns first perfect 10 in Olympic history
Queens Park Velodrome
August 10, 1899
Cyclist Marshall "Major" Taylor wins the final of the 1-mile world sprint championships by inches, becoming the first African American world champion of any sport
Wrigley Field
October 1, 1932
Game 3 of the 1932 World Series, top of the 5th inning, 2 strikes: Babe Ruth is said to call his shot, pointing toward center field, and hits his second homer of the game to deep center on the next pitch
Copps Coliseum
September 15, 1987
"Gretzky to Lemieux": Wayne Gretzky passes to Mario Lemieux, who scores with 1:26 remaining to win the Canada Cup against the Soviet Union 2–1
Herb Brooks Arena
February 22, 1980
"Miracle on Ice": US men's hockey team defeats Soviet Union, winners of last 4 Olympics
Springfield, MA
On or around December 1, 1891
Basketball invented by Springfield College instructor and grad student James Naismith
Madison Square Garden
March 8, 1971
"The Fight of the Century": first meeting between Joe Frazier and Muhammad Ali, both undefeated heavyweight champions; after 15 rounds, Frazier wins with a unanimous decision
Ebbets Field
April 15, 1947
Jackie Robinson's debut with the Brooklyn Dodgers, becoming first African American major leaguer
Polo Grounds
October 3, 1951
"The Shot Heard Round the World": Bobby Thomson hits a 3-run homer in the 9th inning, down 4–2, to win the pennant for the New York Giants over the Brooklyn Dodgers (in the first nationally televised game in baseball history)
Richfield Coliseum
May 7, 1989
"The Shot": Michael Jordan's famous buzzer-beating shot to beat the Cavs and hand the Bulls the playoff series victory
Hershey Sports Arena
March 2, 1962
Wilt Chamberlain scores 100 points in a single NBA game, still a record today
Three Rivers Stadium
December 23, 1972
"The Immaculate Reception": A deep pass from Steelers' quarterback Terry Bradshaw down 7–6, 4th down, 22 seconds left, deflects off a Raiders defender and into the hands of Franco Harris for a touchdown and playoff win
Atlanta-Fulton County Stadium
April 8, 1974
Hank Aaron becomes the home-run king with 715th career home run (beating Babe Ruth)
Georgia Dome
July 23, 1996
Kerri Strug injures her ankle on her first vault attempt, proceeds with her second anyway to help secure the first gold in US women's gymnastics team history

49 **Take me out to the crowd:** The 20 largest North American stadiums by seating capacity

RANK	NAME	HOME TEAM	CAPACITY
1.	Michigan Stadium	Michigan Wolverines	107,601
2.	Beaver Stadium	Penn State Nittany Lions	106,572
3.	Ohio Stadium	Ohio State Buckeyes	102,780
4.	Kyle Field	Texas A&M Aggies	102,733
5.	Tiger Stadium	LSU Tigers	102,321
6.	Neyland Stadium	Tennessee Volunteers	102,037
7.	Bryant–Denny Stadium	Alabama Crimson Tide	101,821
8.	Darrell K Royal–Texas Memorial Stadium	Texas Longhorns	100,119
9.	Sanford Stadium	Georgia Bulldogs	92,746
10.	Cotton Bowl	SMU Mustangs	92,100
11.	Rose Bowl	UCLA Bruins	91,136
12.	Ben Hill Griffin Stadium	Florida Gators	88,548
13.	Jordan-Hare Stadium	Auburn Tigers	87,451
14.	Estadio Azteca	Club América and Cruz Azul	87,000
15.	Memorial Stadium	Nebraska Cornhuskers	86,047
16.	Memorial Stadium	Clemson Tigers	83,301
17.	Gaylord Family Oklahoma Memorial Stadium	Oklahoma Sooners	83,000+
18.	MetLife Stadium	NY Giants and NY Jets	82,500
19.	Lambeau Field	Green Bay Packers	81,441
20.	FedExField	Washington Football Team	80,301
*	Commonwealth Stadium The largest stadium in Canada	Edmonton Elks and FC Edmonton	56,400
**	Estadio Latinoamericano The largest stadium in Cuba	Industriales	55,000

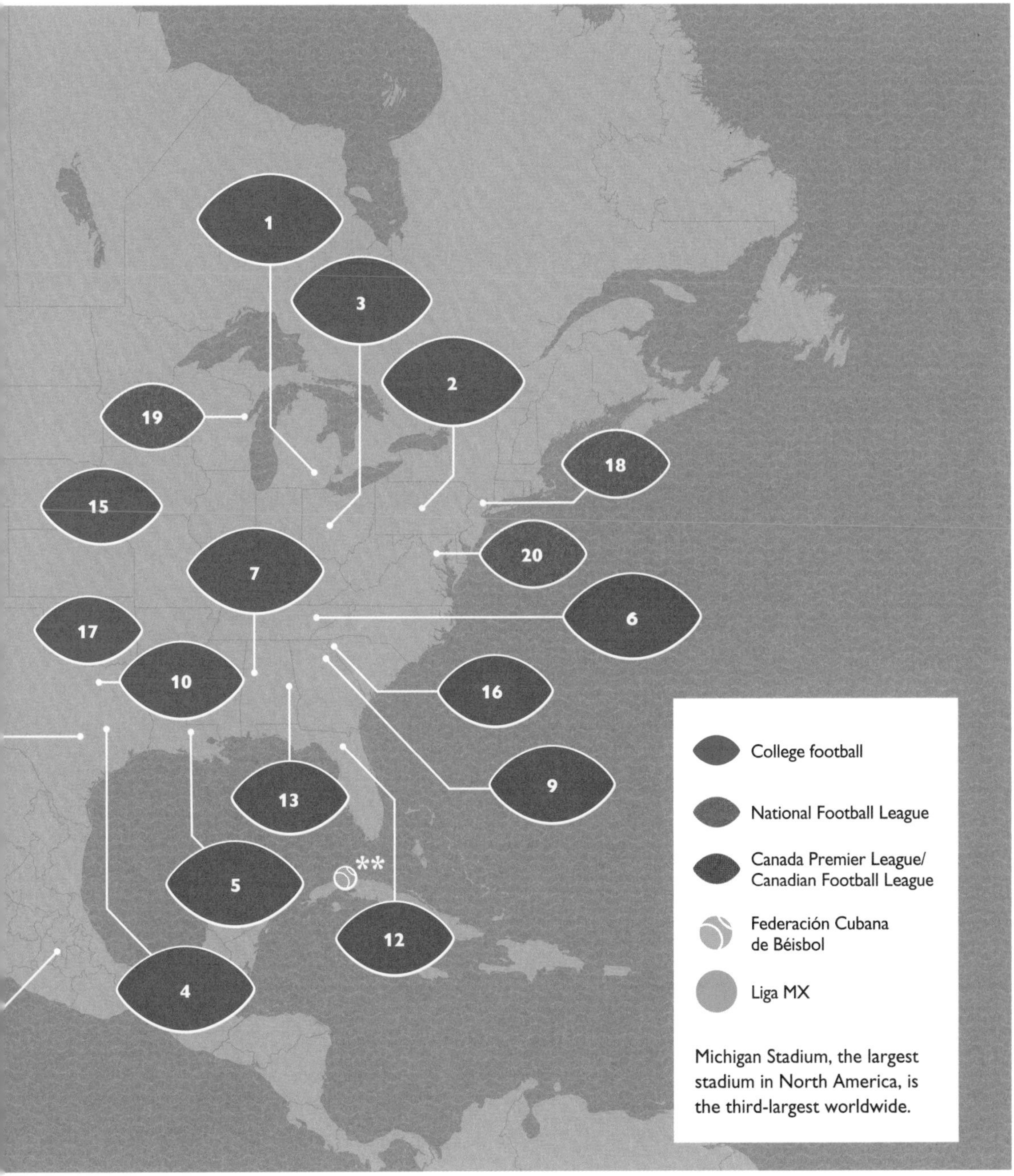
1
3
2
19
18
15
20
7
6
17
10
16
13
9
**
5
12
4
College football
National Football League
Canada Premier League/
Canadian Football League
Federación Cubana
de Béisbol
Liga MX
Michigan Stadium, the largest
stadium in North America, is
the third-largest worldwide.

50 **Topping the charts:** The place names that appear in *Billboard* No. 1 song titles

1. **Harlem Shake** — **March 2, 2013**
   Baauer
2. **California Gurls** — **June 19, 2010**
   Katy Perry featuring Snoop Dogg
3. **California Love** — **June 22, 1996**
   2Pac featuring Dr. Dre and Roger Troutman
4. **Kokomo*** — **November 5, 1988**
   The Beach Boys
5. **Miami Vice Theme** — **November 9, 1985**
   Jan Hammer
6. **Hotel California** — **May 7, 1977**
   Eagles
7. **Philadelphia Freedom** — **April 12, 1975**
   Elton John
8. **The Night Chicago Died** — **August 17, 1974**
   Paper Lace
9. **Midnight Train to Georgia** — **October 20, 1973**
   Gladys Knight & the Pips
10. **The Night the Lights Went Out in Georgia** — **April 7, 1973**
    Vicki Lawrence
11. **Georgia on My Mind** — **November 14, 1960**
    Ray Charles
12. **El Paso** — **January 4, 1960**
    Marty Robbins
13. **The Battle of New Orleans** — **June 1, 1959**
    Johnny Horton
14. **Kansas City** — **May 18, 1959**
    Wilbert Harrison
15. **The Yellow Rose of Texas** — **September 3, 1955**
    Mitch Miller
16. **The Tennessee Waltz** — **December 30, 1950**
    Patti Page
17. **Chattanoogie Shoe Shine Boy** — **February 18, 1950**
    Red Foley

* A mythical place, as the song puts it, "off the Florida Keys"

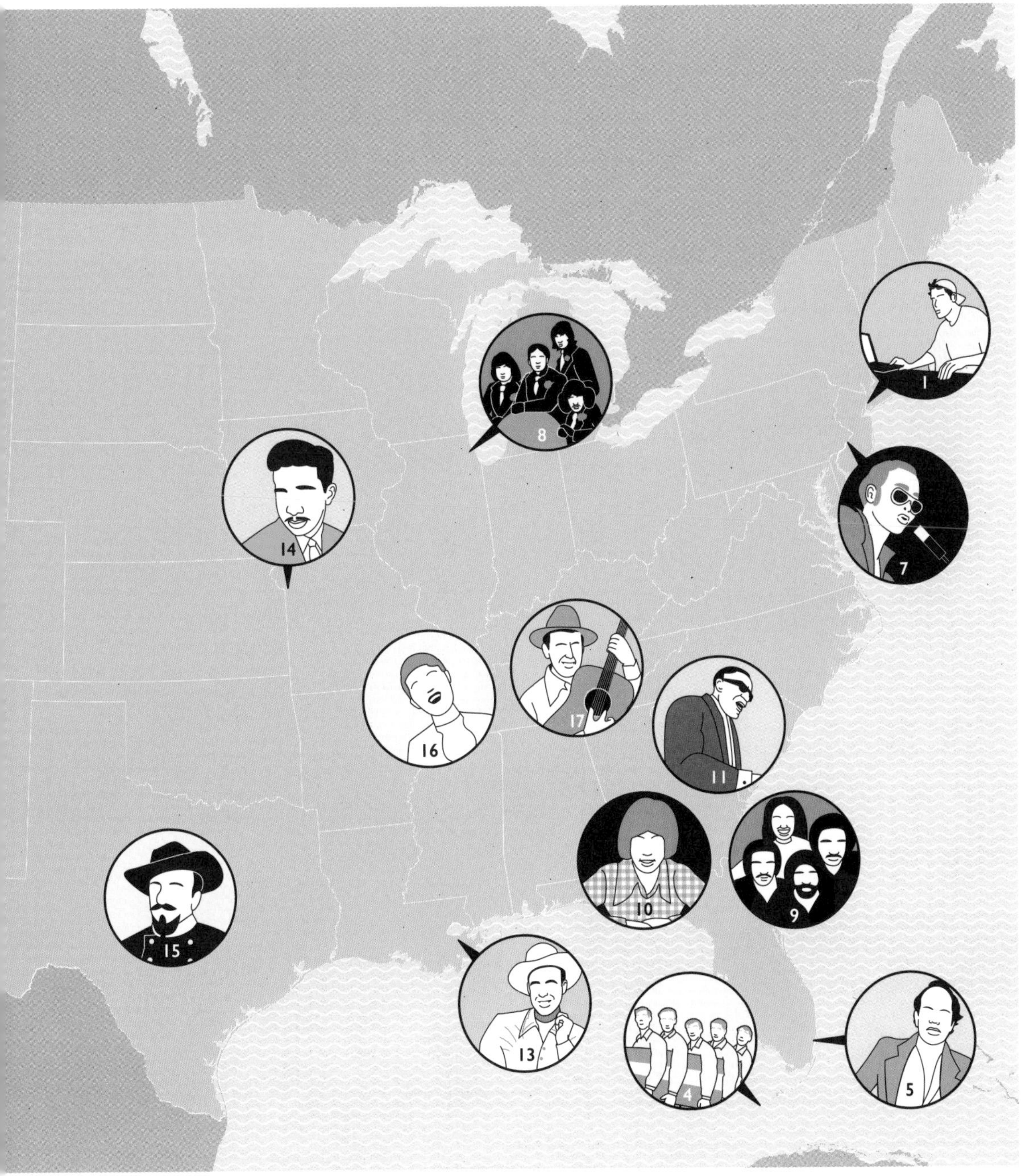
1
8
14
7
17
16
11
10
9
15
13
4
5

51 Black American firsts:
The places where they made their names

First . . .	Name	Year
1. Ivy League university president	Ruth Simmons	2001
2. Chess Grandmaster	Maurice Ashley	1999
3. Woman senator	Carol Moseley Braun	1993
4. Woman astronaut to go to space	Mae Jemison	1992
5. Woman director of a major studio movie	Julie Dash	1991
6. Woman mayor of a major US city	Sharon Pratt Dixon Kelly	1991
7. Elected governor	Lawrence Douglas Wilder	1990
8. Owner of a major metropolitan newspaper	Robert Maynard	1983
9. Astronaut to go to space	Guy Bluford	1983
10. Anchor in a network newscast	Max Robinson	1978
11. Woman commercial airline pilot	Jill Elaine Brown	1978
12. Hip-hop DJ/creator of hip-hop	DJ Kool Herc	1973
13. Director of a movie for a major studio	Gordon Parks	1969
14. Mayor of a major US city	Carl Stokes	1968
15. Elected senator	Edward Brooke III	1967
16. Graduate of the Air Force Academy	Charles Vernon Bush	1963
17. Tony Award winner	Juanita Hall	1950
18. Owner of a radio station	Jesse Blayton Sr.	1949
19. US Marine	Alfred Masters	1942
20. Academy Award winner	Hattie McDaniel	1940
21. Woman judge	Jane Bolin	1939
22. Woman legislator	Crystal Bird Fauset	1939
23. Conductor of a major US orchestra	William Grant Still	1936
24. Musician at the Grand Ole Opry	DeFord Bailey	1926
25. Radio broadcaster	Jack Leroy Cooper	1925
26. Woman self-made millionaire	Madam C. J. Walker	1919
27. Film director to make a feature film	Oscar Micheaux	1919
28. Municipal architect	Clarence Wigington	1915
29. Woman mail carrier	Mary Fields ("Stagecoach Mary")	1895
30. Woman dentist	Ida Gray Nelson Rollins	1890
31. Recognized photographer	James Conway Farley	1884
32. US congressperson (appointed)	Hiram Revels	1870
33. Elected county sheriff	Walter Moses Burton	1870
34. Town mayor	Pierre Caliste Landry	1868
35. Woman novelist	Harriet Wilson	1859
36. Published playwright	William Wells Brown	1858
37. College president	Daniel Alexander Payne	1856
38. Woman college graduate	Lucy Stanton Day Sessions	1850
39. Professor	Charles Lewis Reason	1849
40. US medical school graduate	David Peck	1847
41. Dance performer/creator of tap dance	William Henry Lane ("Master Juba")	1840s
42. Published woman author of an autobiography	Jarena Lee	1836
43. Patent holder	Thomas Jennings	1821
44. Published composer	Francis Johnson	1818
45. College graduate	John Chavis	1799
46. Recognized artist (professional painter)	Joshua Johnson	1790
47. Published author of an autobiography	Briton Hammon	1760
48. Poet	Lucy Terry	1746
49. Landowners	Anthony and Mary Johnson	1640
50. Conquistador/first known African American	Juan Garrido	1513

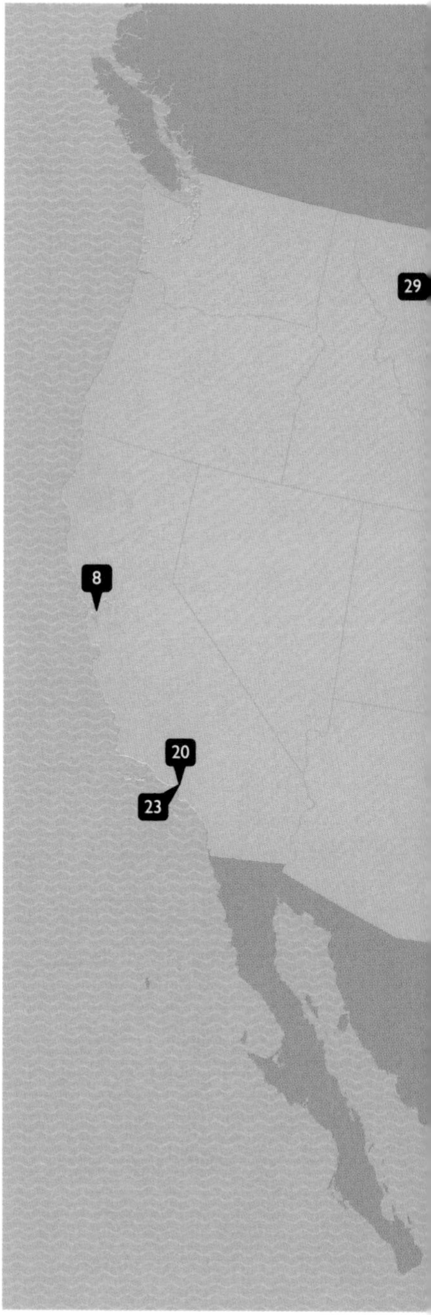

In the annals of Black American firsts, you most likely know Barack Obama (president), Oprah Winfrey (woman to host a nationally syndicated talk show), Thurgood Marshall (Supreme Court justice), and Jelly Roll Morton (inventor of jazz)—but do you know the first elected governor? Anchor in a network newscast? Woman judge? Published composer? We've opted here to highlight some names of American originals that may not as readily come to mind—and where in the US they became known for their accomplishments.

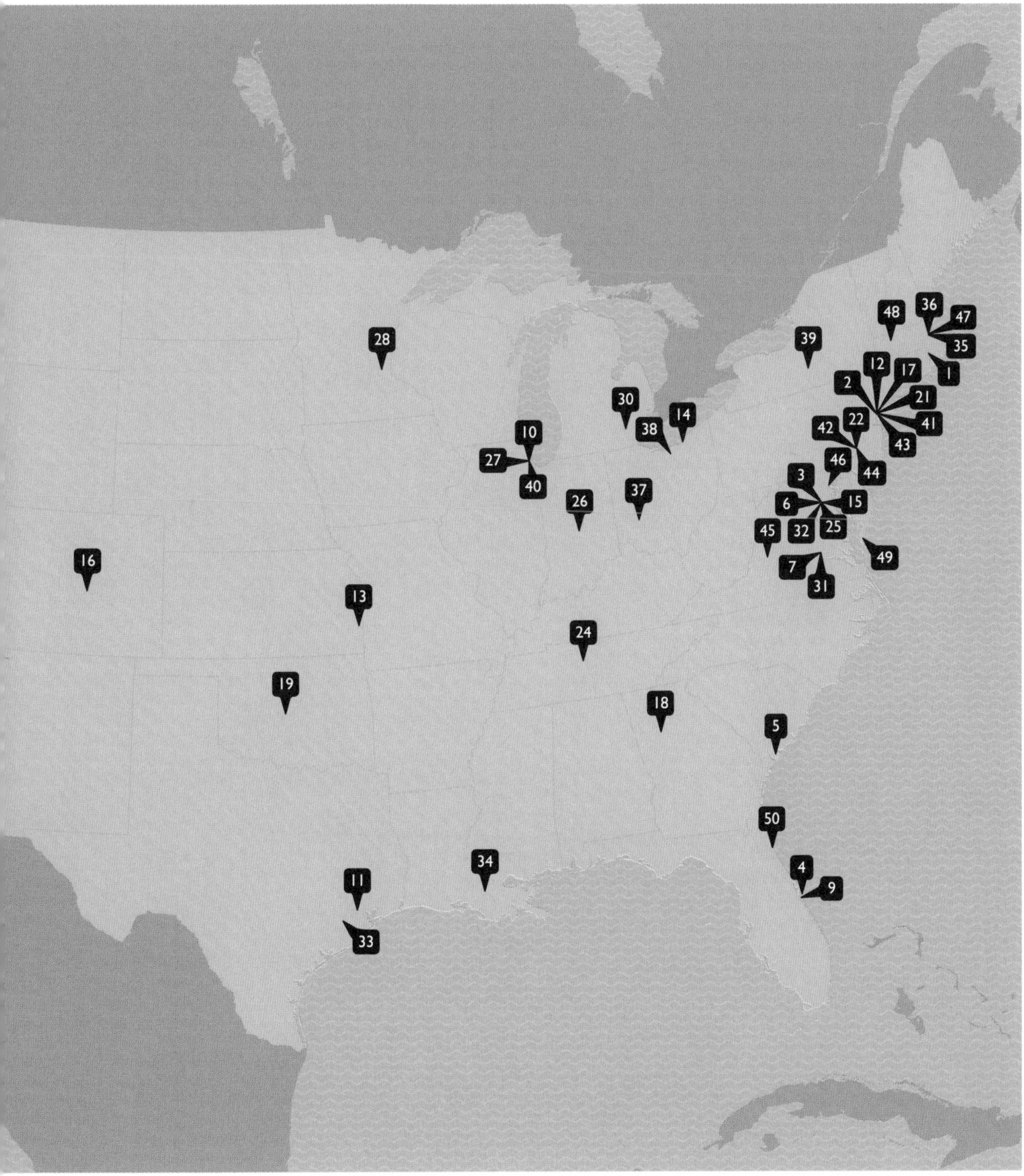

52 **Don't drop the ball:** The less-famous things that drop (or rise) on New Year's Eve

Items are generally oversize, illuminated replicas.

1. Ski gondola
2. Grapes
3. Cherry
4. Potato
5. Lettuce
6. Cowboy boot
7. Pine cone
8. Playing card
9. Chile pepper
10. Apple
11. Olive
12. Carp
13. Fleur-de-lis
14. Mirror ball
15. MoonPie
16. Cheese
17. Marlin
18. Watermelon
19. Cherry
20. Musical note
21. Apple
22. Piano
23. Meteor
24. Beach ball
25. Cherry
26. Peanut
27. Possum
28. Peach
29. Chicken
30. Buzzard
31. Popcorn ball
32. Walleye
33. Pineapple
34. Gold nugget
35. Conch shell; pirate; Pan Am airplane; drag queen; key lime; tuna
36. Possum
37. Popcorn ball
38. Crown
39. Golf ball
40. Orange
41. Anchor
42. Recycled ball
43. Star
44. Flip-flops
45. Flea
46. Acorn
47. Pickle
48. Apple
49. Doughnut
50. Key
51. Beaver
52. Pickle
53. Sled
54. Wrench
55. PAC-MAN
56. Strawberry
57. White rose
58. Hershey kiss
59. Pig
60. Coal
61. Bologna
62. Red rose
63. Beer bottle
64. Shoe
65. Duck
66. Crab
67. Mushroom
68. Muskrat
69. PEEPS chick
70. Horseshoe
71. Piñata
72. Blueberry
73. Sardine, maple leaf
74. Onion

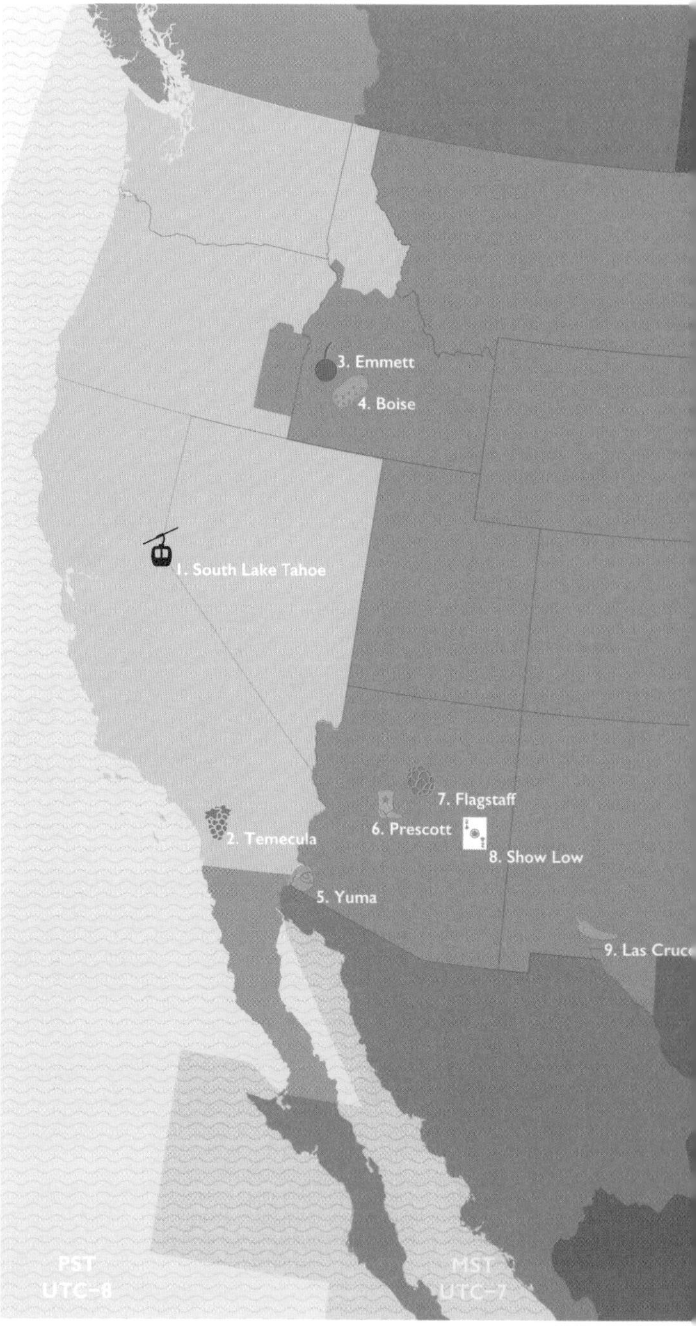

In a tradition dating back to 1907, the Times Square New Year's Eve Ball—a 12-ft, 11,875-lb (3.7 m, 5,386 kg) sphere covered in 2,688 crystal triangles illuminated by 32,256 LEDs—begins its famous descent at 11:59 PM EST. It's estimated that more than a billion people around the world tune in to watch. But if you're one of the billion, then you're missing out on some of the, shall we say, less-celebrated items dropped, raised, or lit up around the country. Maybe this is the year to ring in January 1 with a giant shoe, muskrat, or lump of coal—or one of the many other more-offbeat local traditions around the US.

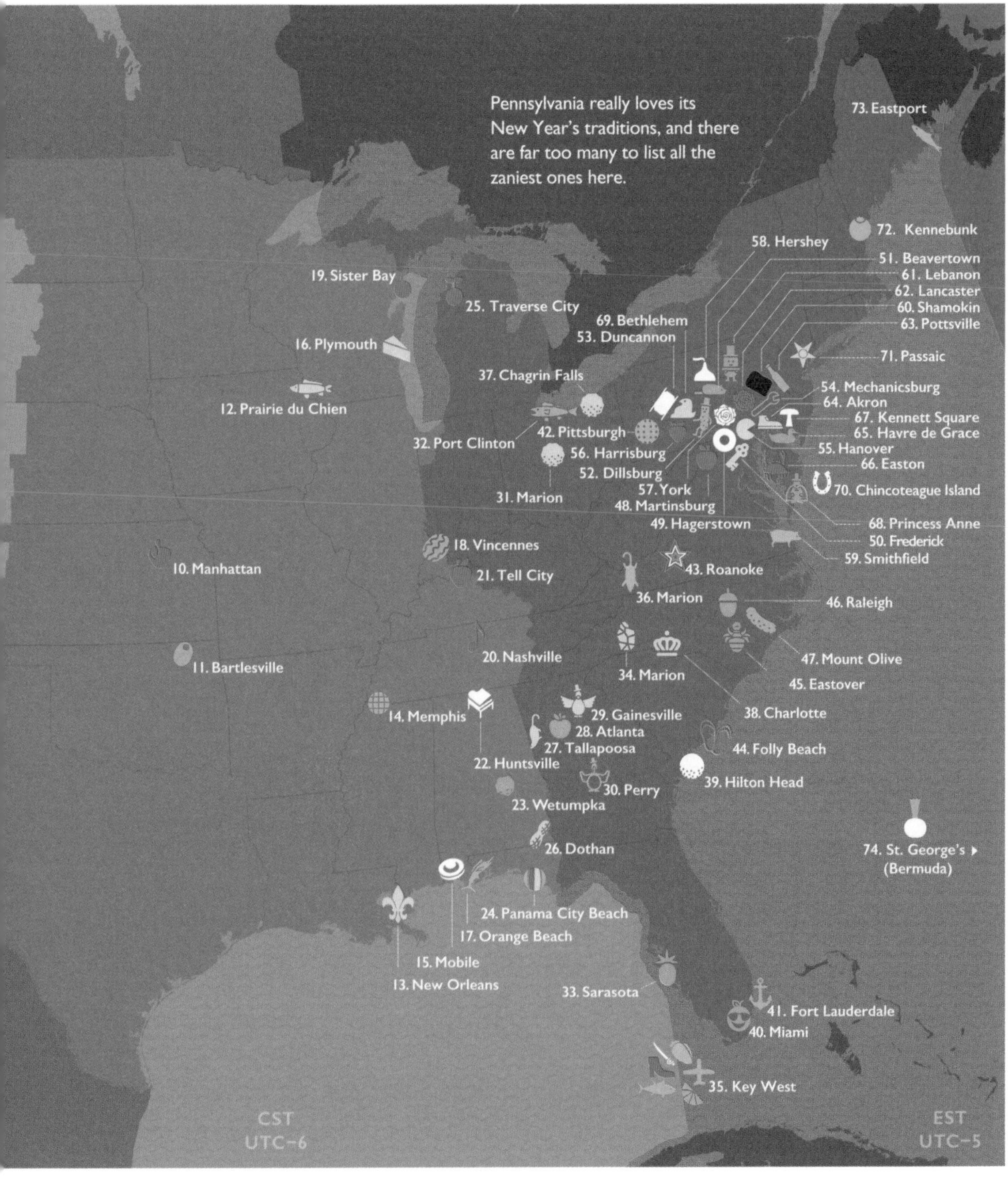

53 **How the stacks stack up:** The 20 largest physical public library collections in the US

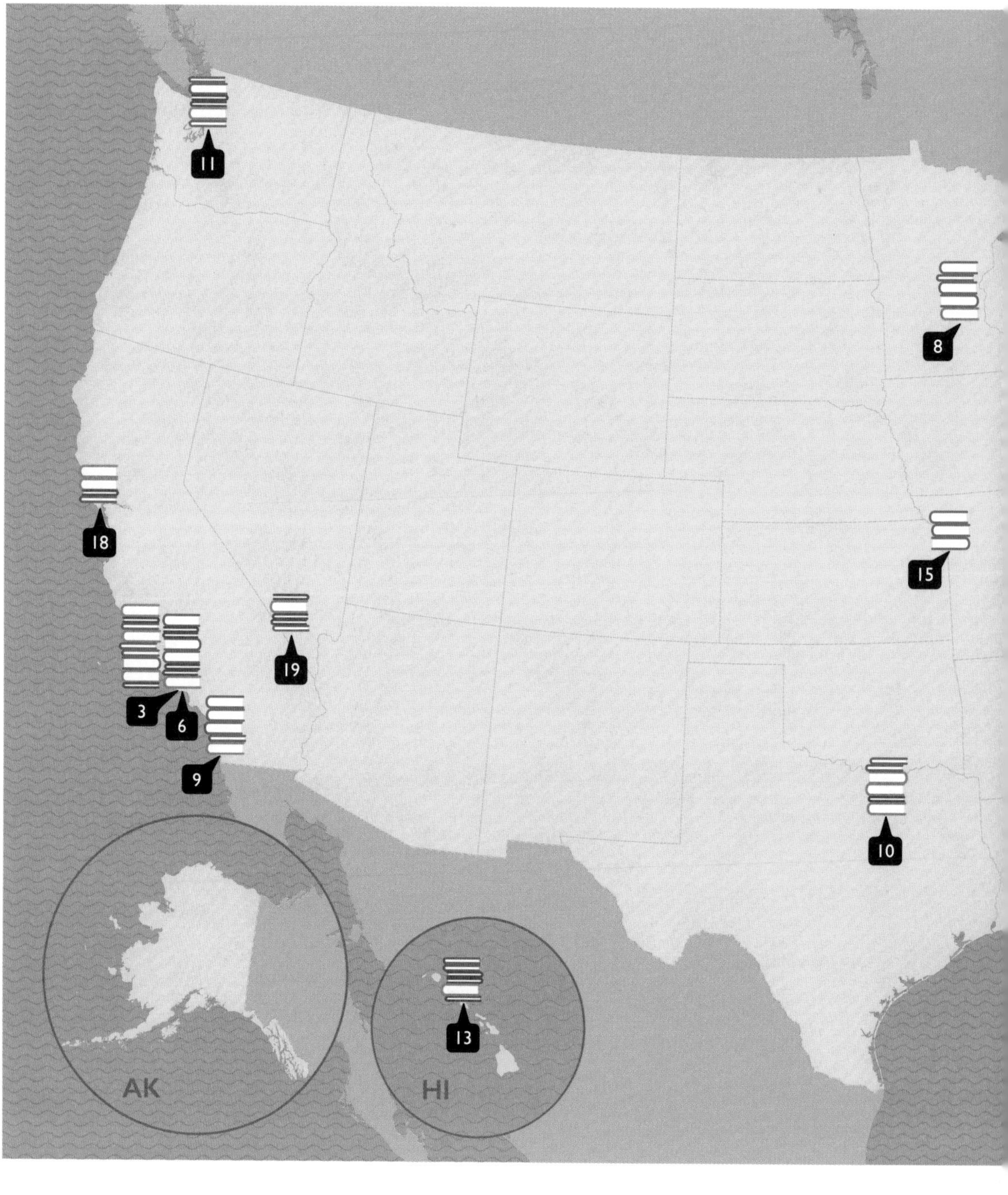

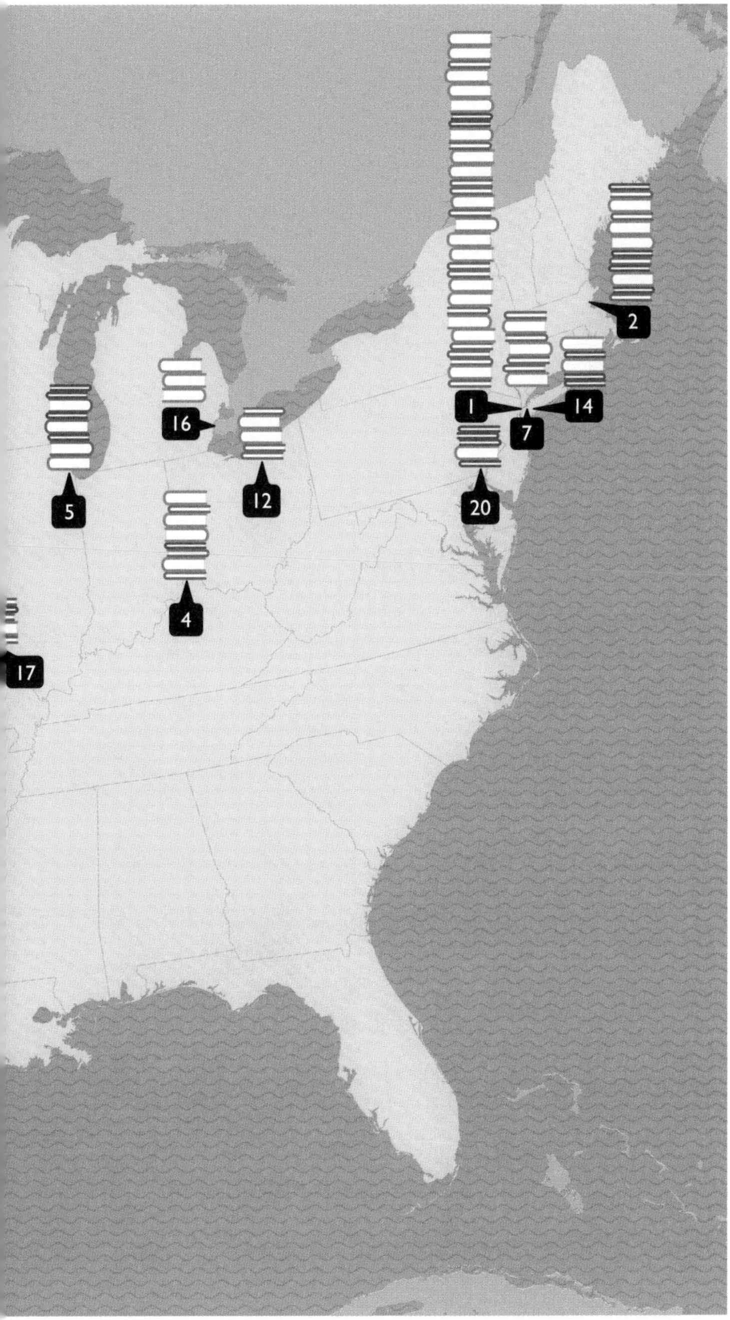

1. New York Public Library
2. Boston Public Library
3. Los Angeles Public Library
4. Cincinnati & Hamilton County Public Library
5. Chicago Public Library
6. LA County Library
7. Queens Public Library
8. Hennepin County Library (Minneapolis area)
9. San Diego Public Library
10. Dallas Public Library
11. King County Library System (Seattle area)
12. Cleveland Public Library
13. Hawaii State Public Library System
14. Brooklyn Public Library
15. Mid-Continent Public Library (Kansas City area)
16. Detroit Public Library
17. St. Louis Public Library
18. San Francisco Public Library
19. Las Vegas–Clark County Library District
20. Free Library of Philadelphia

Total physical collection, in millions

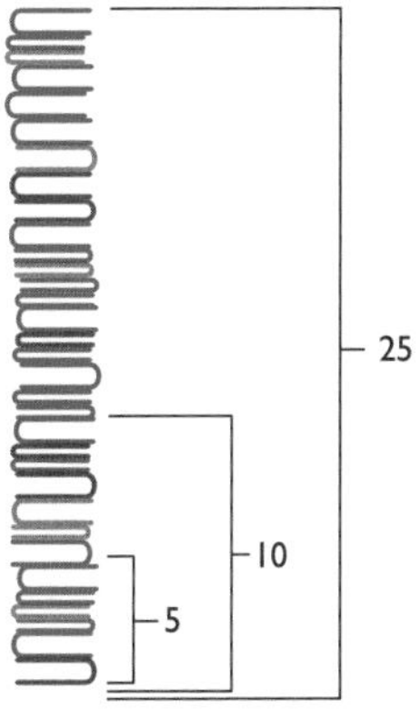

54 **Magnificent museums:** The 20 most popular museums in North America by attendance

Museum name — Annual visitors, in millions

1. The Metropolitan Museum of Art — 6.8
2. American Museum of Natural History — 5.0
3. National Museum of Natural History — 4.2
4. National Gallery of Art — 4.1
5. National Air and Space Museum — 3.2
6. National Museum of American History — 2.8
7. California Science Center — 2.2
8. National Museum of African American History and Culture — 2.0
9. Smithsonian American Art Museum — 2.0
10. The Museum of Modern Art — 2.0
11. Houston Museum of Natural Science — 2.0
12. National Portrait Gallery — 1.7
13. The Art Institute of Chicago — 1.7
14. United States Holocaust Memorial Museum — 1.6
15. Field Museum of Natural History — 1.5
16. Denver Museum of Nature and Science — 1.5
17. Museum of Science — 1.5
18. J. Paul Getty Museum — 1.4
19. Museum of Science and Industry — 1.4
20. California Academy of Sciences — 1.3

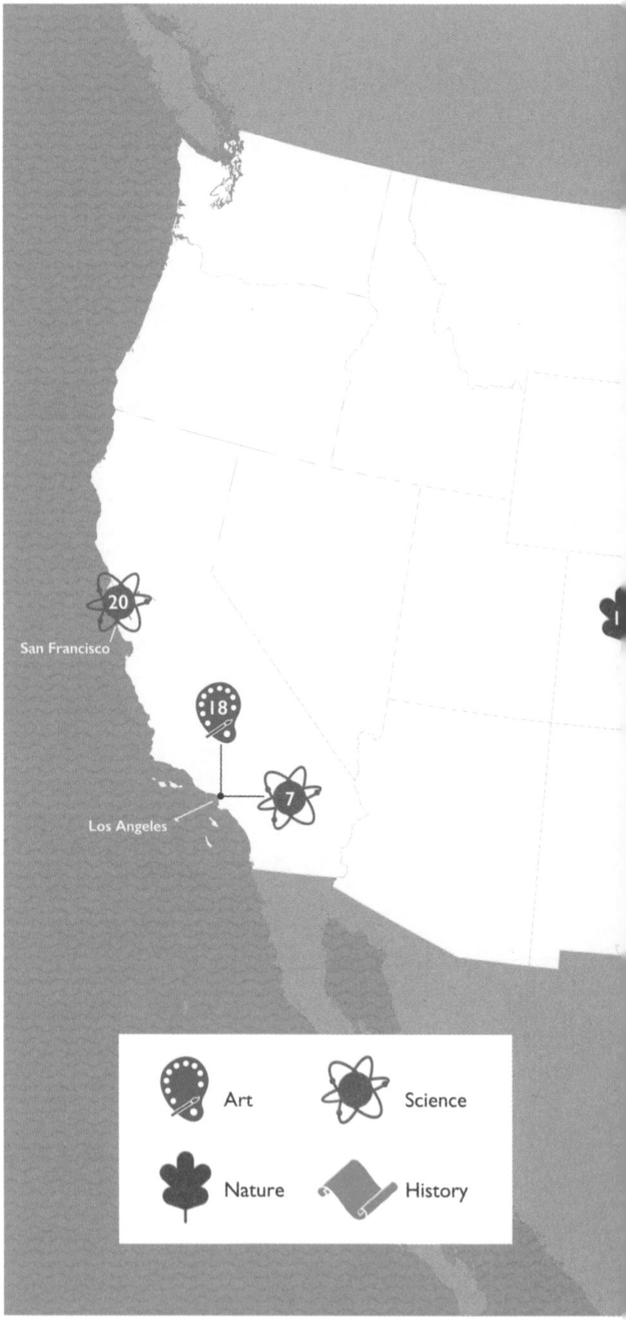

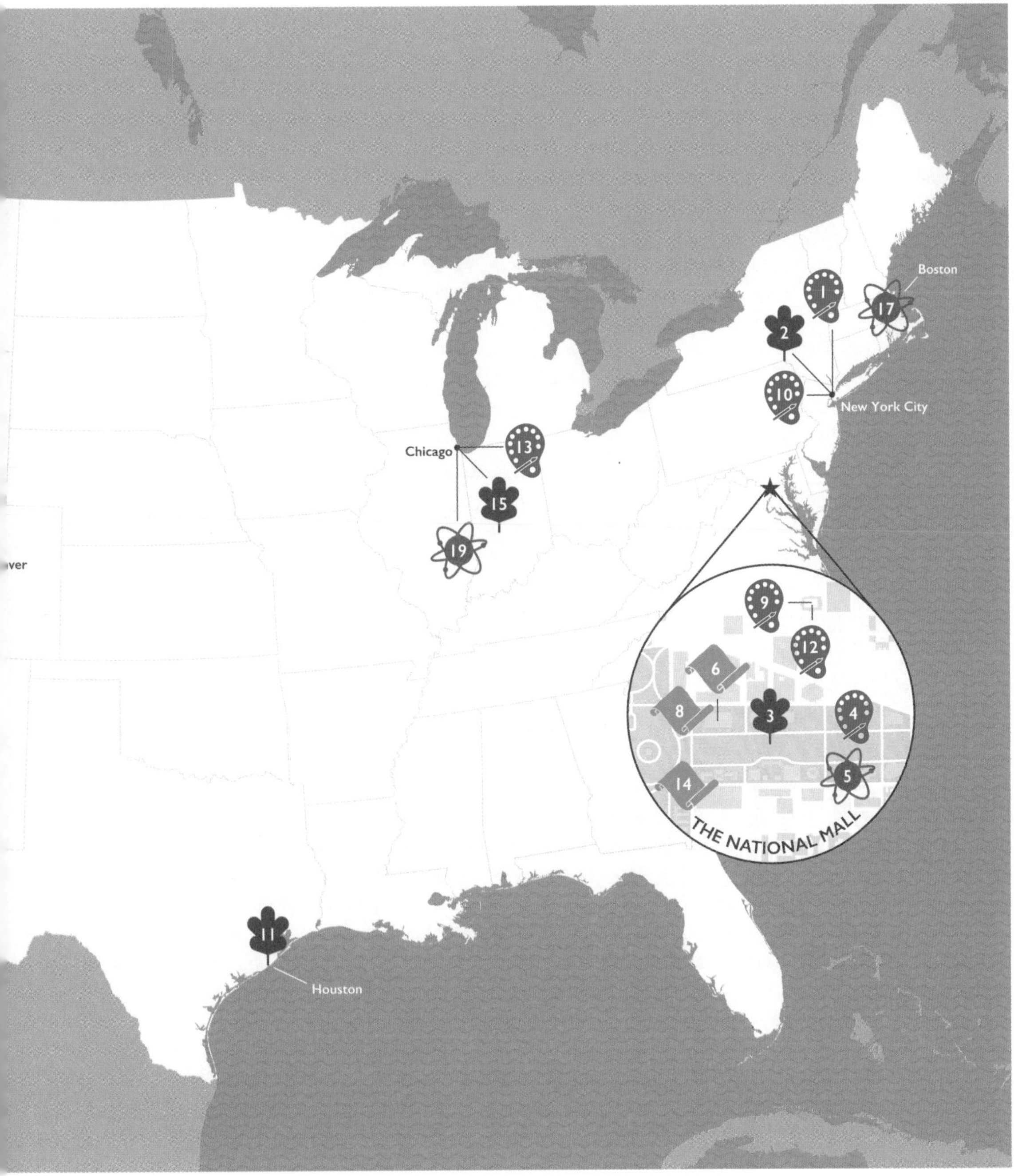

ver
Chicago
13
15
19
1
2
17
10
Boston
New York City
9
12
6
8
3
4
5
14
THE NATIONAL MALL
11
Houston

55 **Hometown heroes:** Hall-of-famer birthplaces* in the 4 major sports leagues

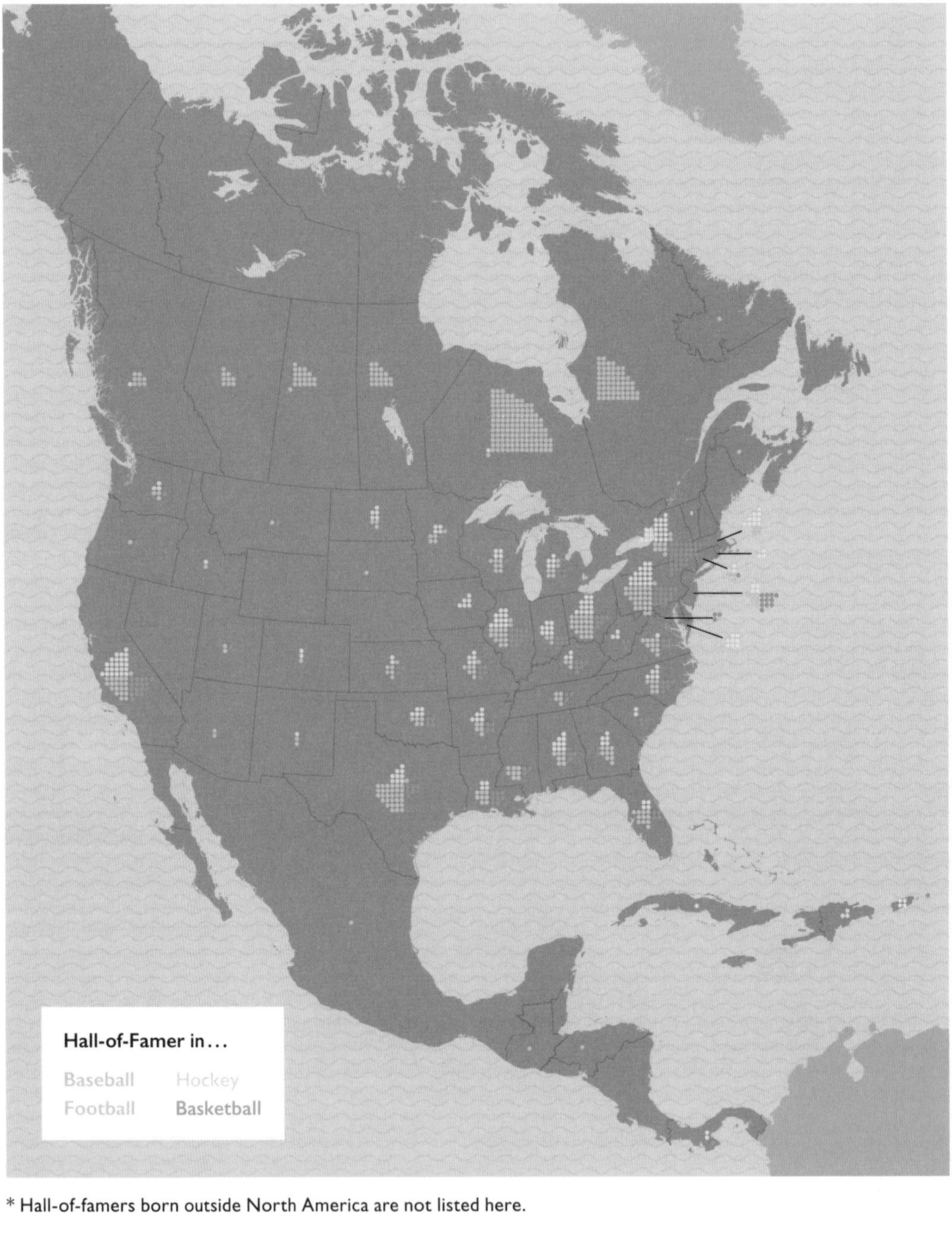

* Hall-of-famers born outside North America are not listed here.

56 **The Midas brush:**

The most expensive paintings of North America

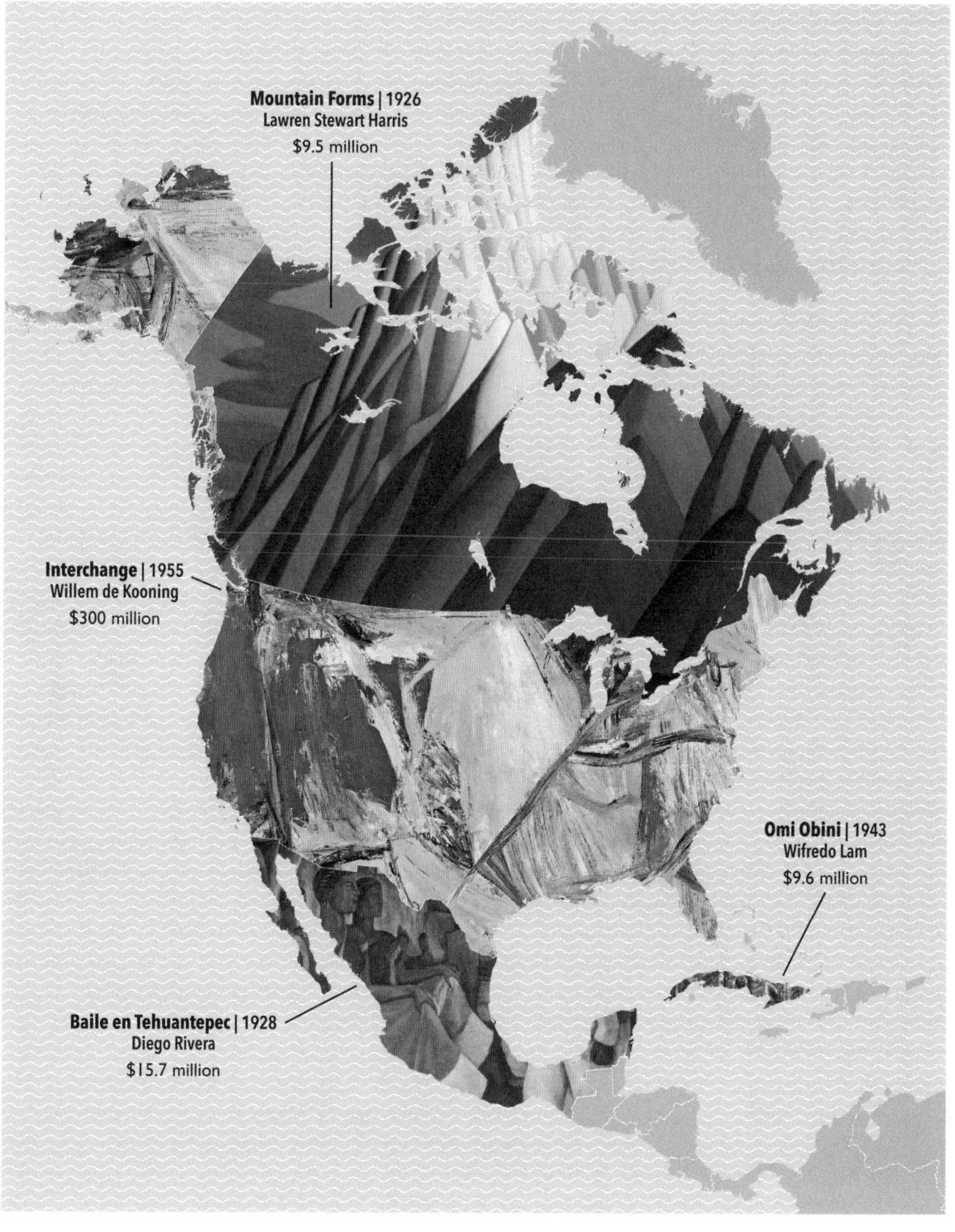

57 Go see a Frank Lloyd Wright building: Every one of his publicly accessible works

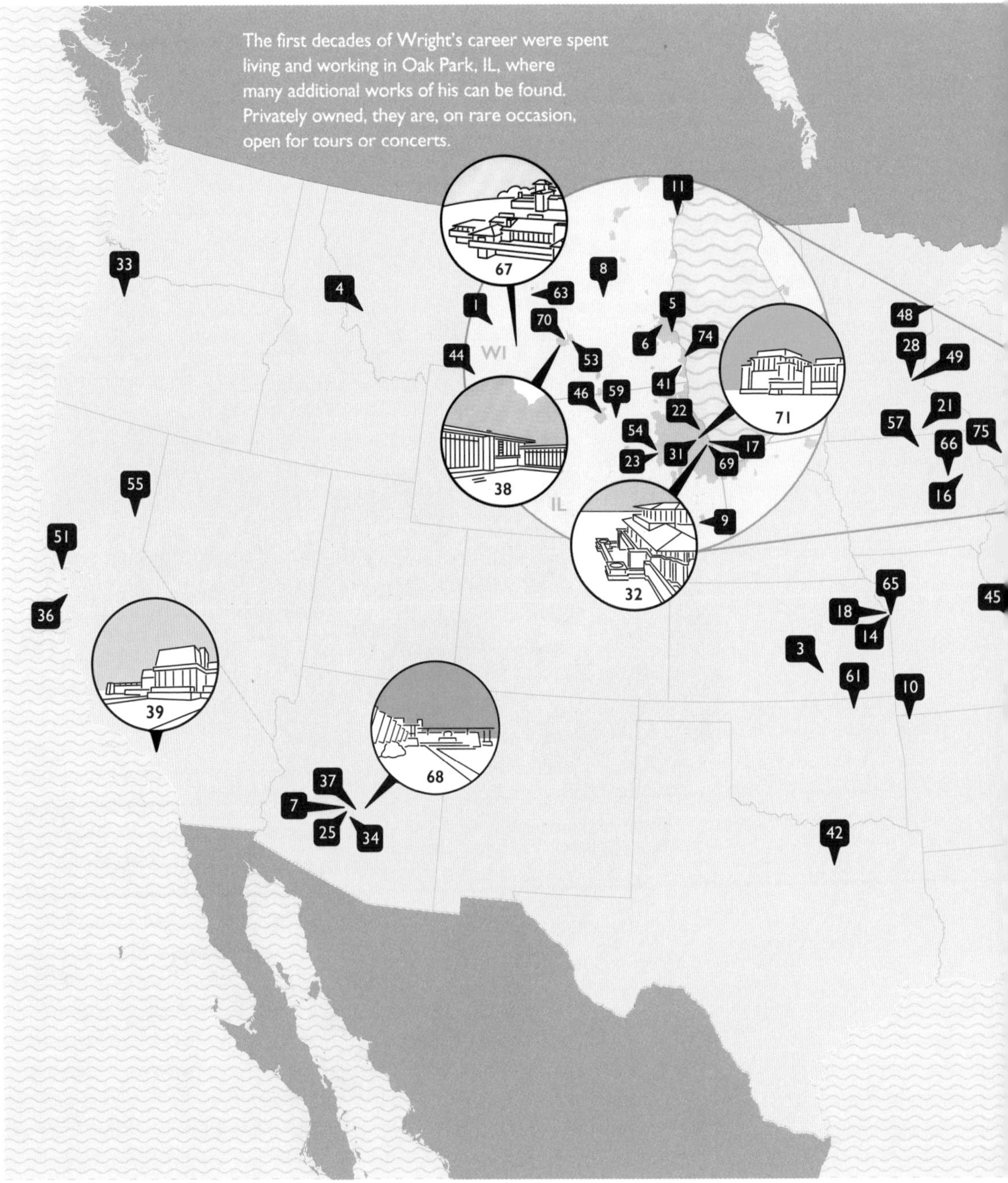

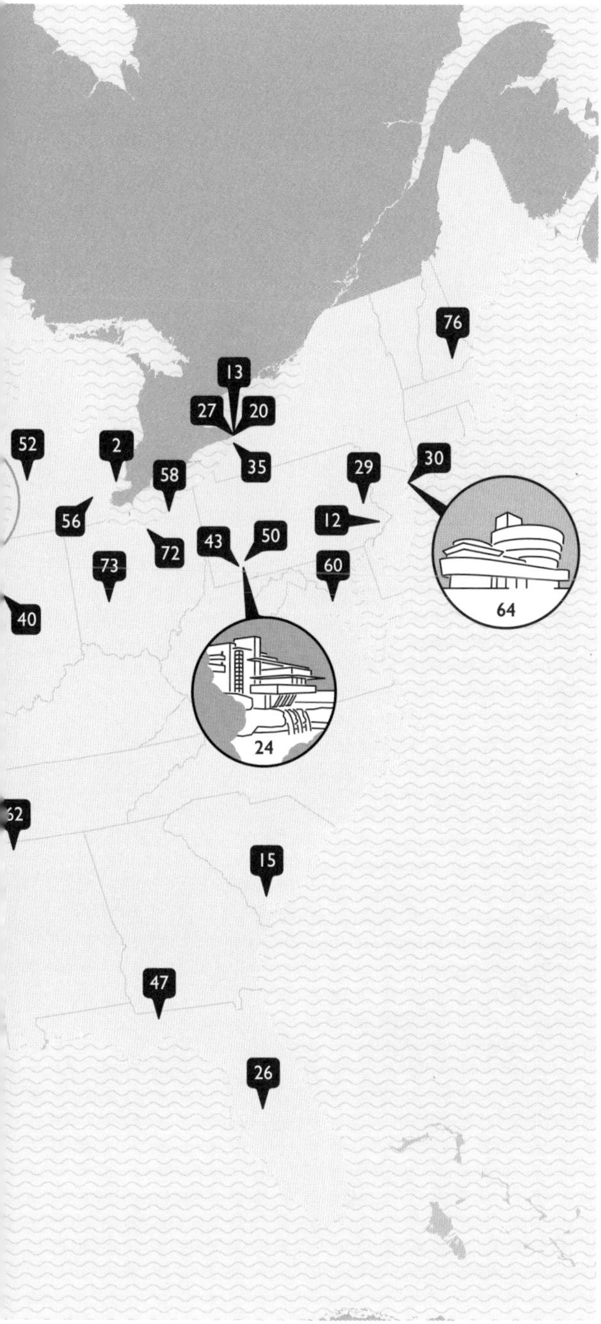

1. A. D. German Warehouse
2. Affleck House
3. Allen-Lambe House
4. Alpine Meadows Ranch
5. American System-Built Homes, Model B1
6. Annunciation Greek Orthodox Church
7. Arizona Biltmore Hotel
8. Arnold Jackson House*
9. B. Harley Bradley House
10. Bachman-Wilson House
11. Bernard Schwartz House
12. Beth Sholom Synagogue
13. Blue Sky Mausoleum
14. Bott House*
15. C. Leigh Stevens House, "Auldbrass Plantation"*
16. Cedar Rock
17. Charnley-Perskey House
18. Community Christian Church
19. Dana-Thomas House
20. Darwin Martin House
21. Elam House*
22. Emil Bach House
23. Fabyan Villa
24. Fallingwater
25. First Christian Church
26. Florida Southern College
27. Fontana Boathouse
28. Francis Little House (hallway)
29. Francis Little House II (library)
30. Francis Little House (living room)
31. Frank Lloyd Wright Home and Studio
32. Frederick C. Robie House
33. Gordon House
34. Grady Gammage Memorial Auditorium
35. Graycliff
36. Hanna House
37. Harold Price, Sr. House*
38. Herbert and Katherine Jacobs House*
39. Hollyhock House
40. John and Catherine Christian House (SAMARA)
41. Johnson Administration and Research Tower
42. Kalita Humphreys Theater
43. Kentuck Knob*
44. Kinney House*
45. Kraus House/Frank Lloyd Wright House in Ebsworth Park
46. Laurent House
47. Lewis Spring House
48. Lindholm Oil Company Service Station
49. Malcolm and Nancy Willey House*
50. Mäntylä/R. W. Lindholm House & Duncan House
51. Marin County Civic Center
52. Meyer May House
53. Monona Terrace
54. Muirhead Farmhouse
55. Nakoma Golf Resort
56. Palmer House*
57. Park Inn Hotel
58. Penfield House*
59. Pettit Chapel
60. Pope-Leighey House
61. Price Tower
62. Rosenbaum House
63. Seth Peterson Cottage
64. Solomon R. Guggenheim Museum
65. Sondern-Adler House*
66. Stockman House
67. Taliesin & Hillside
68. Taliesin West
69. The Rookery
70. Unitarian Meeting House
71. Unity Temple
72. Weltzheimer-Johnson House
73. Westcott House
74. Wingspread
75. Wyoming Valley School Cultural Arts Center
76. Zimmerman House

* Limited access, either by renting out or occasional tours

58 **Bees expertise:** Winning Scripps National Spelling Bee words by winner's home state

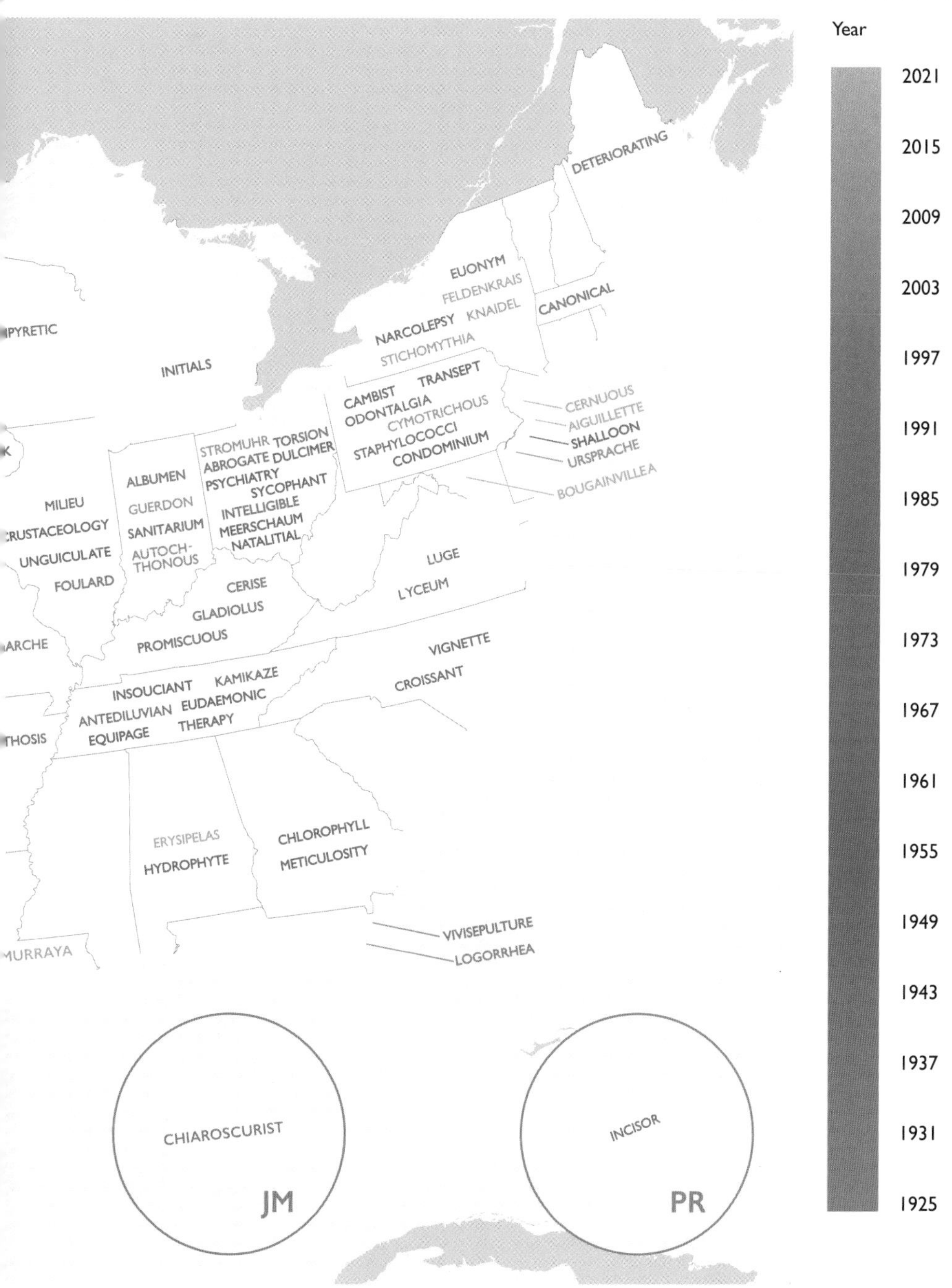
Year
2021
2015
2009
2003
1997
1991
1985
1979
1973
1967
1961
1955
1949
1943
1937
1931
1925
DETERIORATING
EUONYM
FELDENKRAIS
KNAIDEL
CANONICAL
NARCOLEPSY
STICHOMYTHIA
INITIALS
CAMBIST
TRANSEPT
ODONTALGIA
CYMOTRICHOUS
STAPHYLOCOCCI
CONDOMINIUM
CERNUOUS
AIGUILLETTE
SHALLOON
URSPRACHE
BOUGAINVILLEA
STROMUHR
TORSION
ABROGATE
DULCIMER
PSYCHIATRY
SYCOPHANT
INTELLIGIBLE
MEERSCHAUM
NATALITIAL
ALBUMEN
GUERDON
SANITARIUM
AUTOCH-
THONOUS
MILIEU
UNGUICULATE
FOULARD
CERISE
GLADIOLUS
PROMISCUOUS
LUGE
LYCEUM
VIGNETTE
CROISSANT
INSOUCIANT
KAMIKAZE
ANTEDILUVIAN
EUDAEMONIC
EQUIPAGE
THERAPY
ERYSIPELAS
HYDROPHYTE
CHLOROPHYLL
METICULOSITY
VIVISEPULTURE
LOGORRHEA
CHIAROSCURIST
JM
INCISOR
PR

59 **All together now:** The all-time largest gatherings these cities have ever seen

1. **New York City, NY** — **April 20, 1951**
   Fifth Avenue ticker tape parade to honor General Douglas MacArthur after Truman relieves him of Far East command

2. **Chicago, IL** — **November 4, 2016**
   Cubs World Series parade and rally; first championship since 1908

3. **Boston, MA** — **October 30, 2004**
   Red Sox World Series parade; first championship since 1918, ending the Curse of the Bambino

4. **Mexico City, Mexico** — **October 27, 2019**
   Largest annual Day of the Dead festival ever

5. **Philadelphia, PA** (Tie) — **May 21, 1974**
   NHL Stanley Cup champion Philadelphia Flyers parade

6. **Philadelphia, PA** (Tie) — **October 3, 1979**
   Pope John Paul II's outdoor mass

7. **Washington, DC** — **January 20, 2009**
   First inauguration of Barack Obama

8. **Detroit, MI** — **June 6, 2008**
   NHL Stanley Cup champion Detroit Red Wings parade

9. **Cleveland, OH** — **June 22, 2016**
   NBA champion Cleveland Cavaliers parade; the victory ended a 52-year drought in major professional sport championships for the city

10. **Ecatepec, Mexico** — **February 14, 2016**
    Welcoming Pope Francis to Mexico

11. **Los Angeles, CA** — **March 26, 2006**
    March for immigrant rights in a "national day of action"

12. **Houston, TX** — **April 5, 1986**
    Outdoor concert for Houston's 150th anniversary, NASA's 25th, and as tribute to the Challenger astronauts

13. **San Francisco, CA** — **June 30, 2013**
    43rd annual San Francisco Pride; 4 days after Supreme Court ruled against Proposition 8, a state ballot initiative to ban same-sex marriage

14. **Toronto, ON, Canada** — **June 17, 2019**
    NBA champion Toronto Raptors parade

15. **Denver, CO** — **February 9, 2016**
    NFL champion Denver Broncos parade

16. **San Juan, Puerto Rico** — **July 17 | July 22, 2019**
    Peak days of months-long protest over Telegramgate, a political scandal over then Gov. Ricardo Rosselló's leaked racist and homophobic Telegram messages

17. **Oakland, CA** — **June 19, 2015 | June 15, 2017 | June 12, 2018**
    NBA champion Golden State Warriors parades

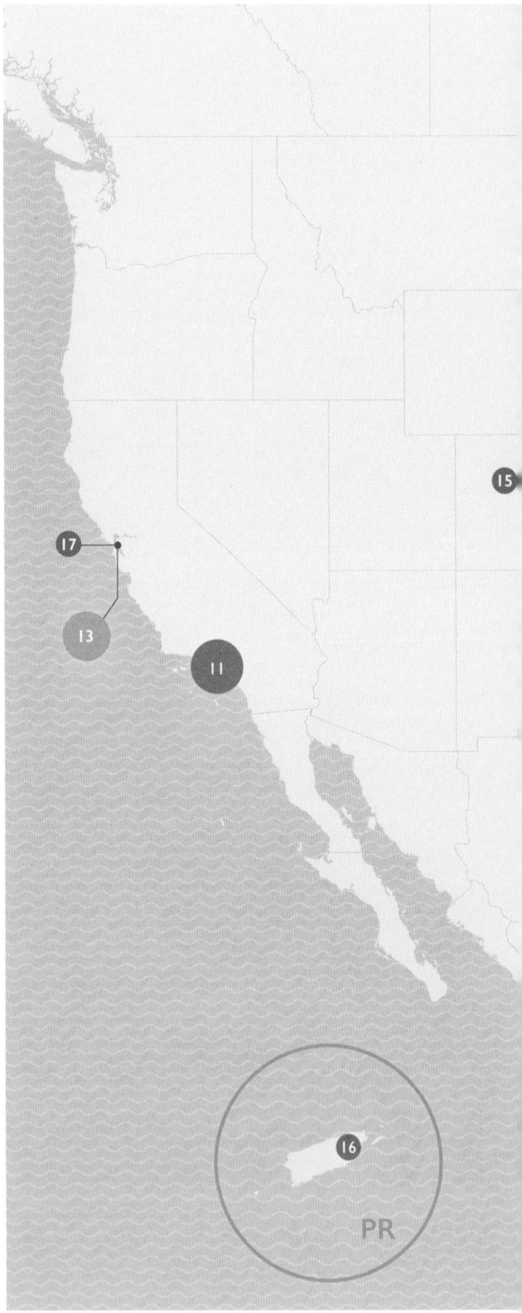

Only 16 cities around North America have had gatherings of a million people or more. The largest-ever mass assemblies in each of those cities represents a wide range of reasons we get together—from celebrating our sports championships to fighting for equal rights.

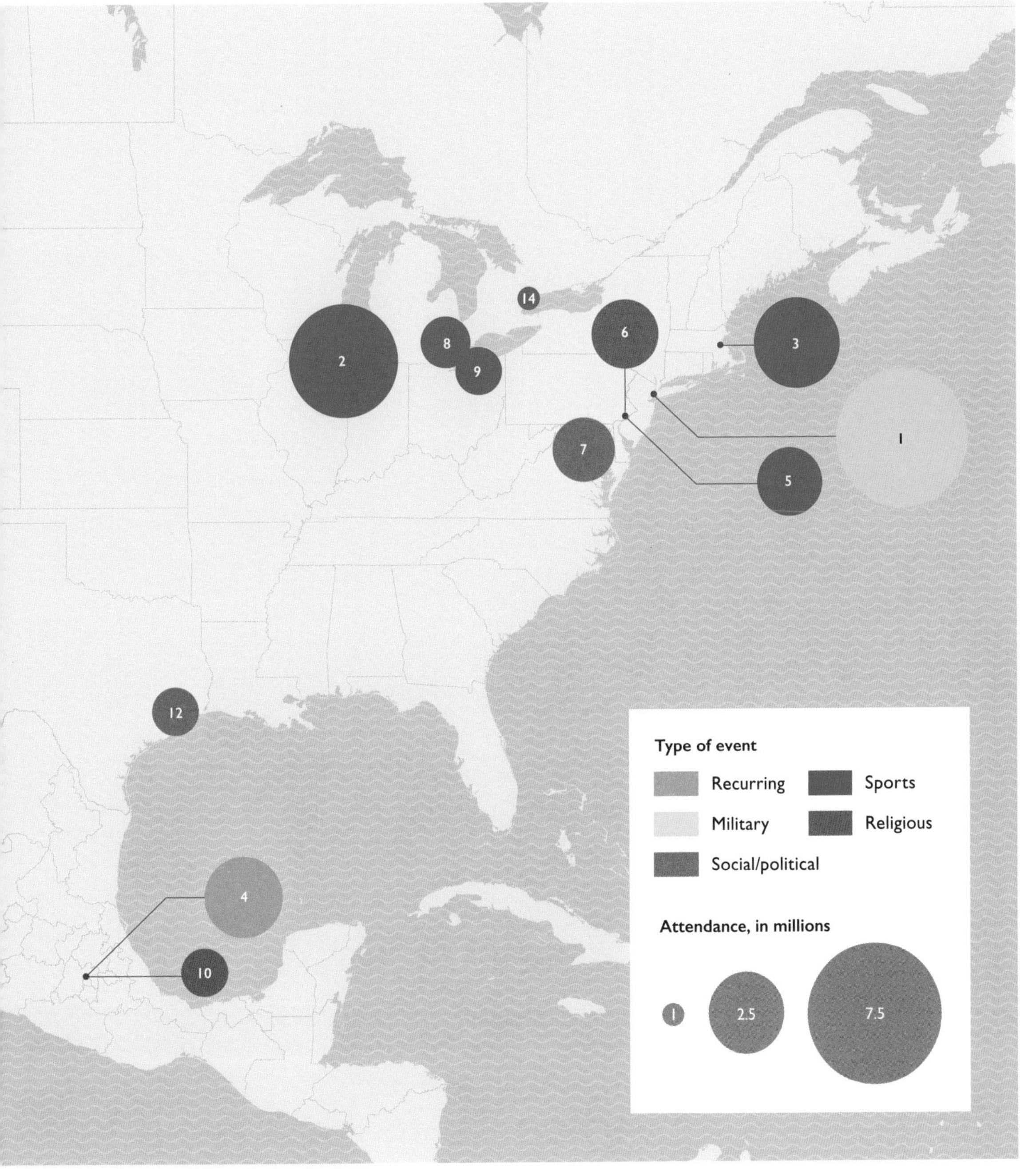

60 **Where the stars are born:** Where every Best Actor and Actress Oscar winner* hails from

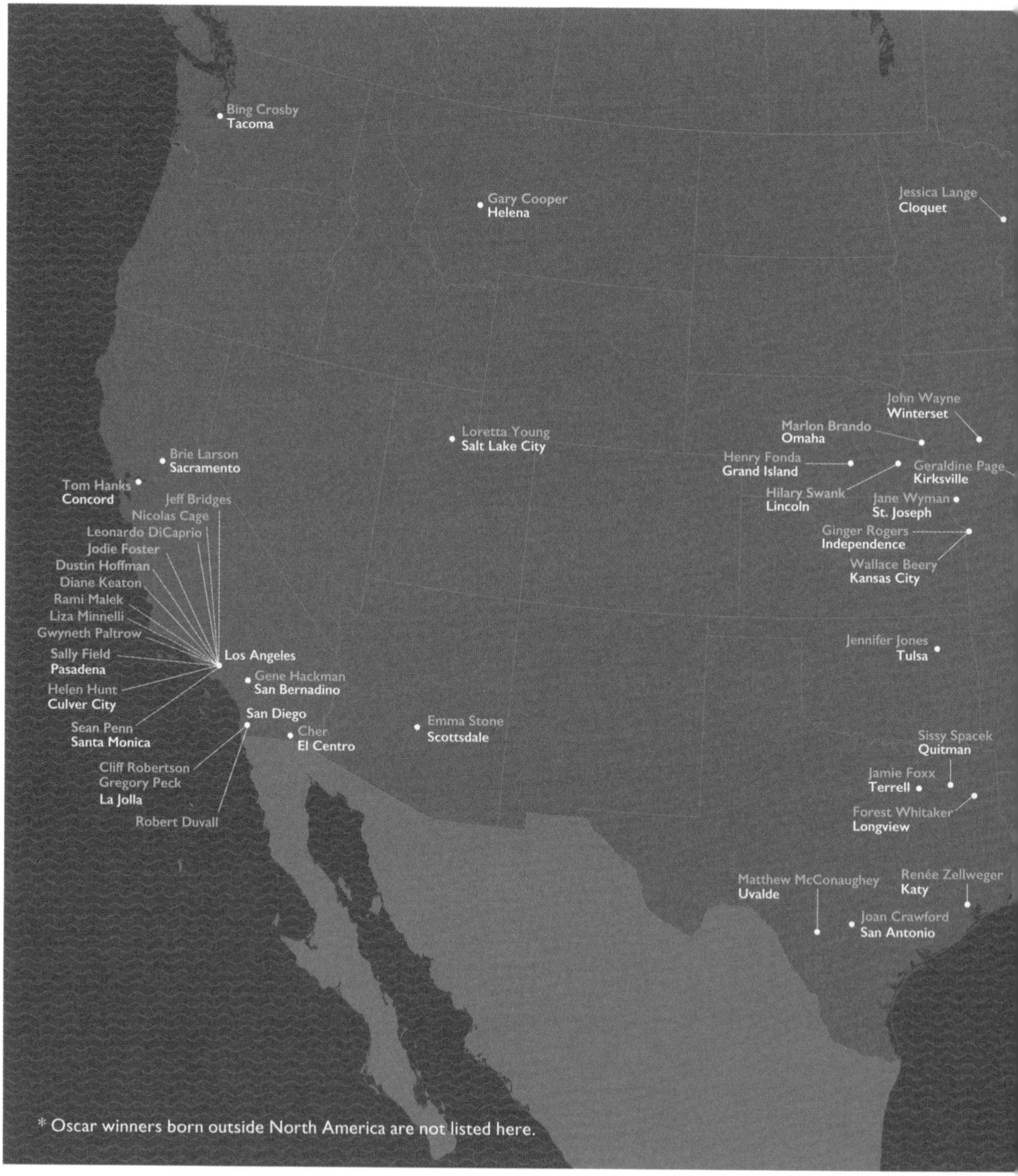

* Oscar winners born outside North America are not listed here.

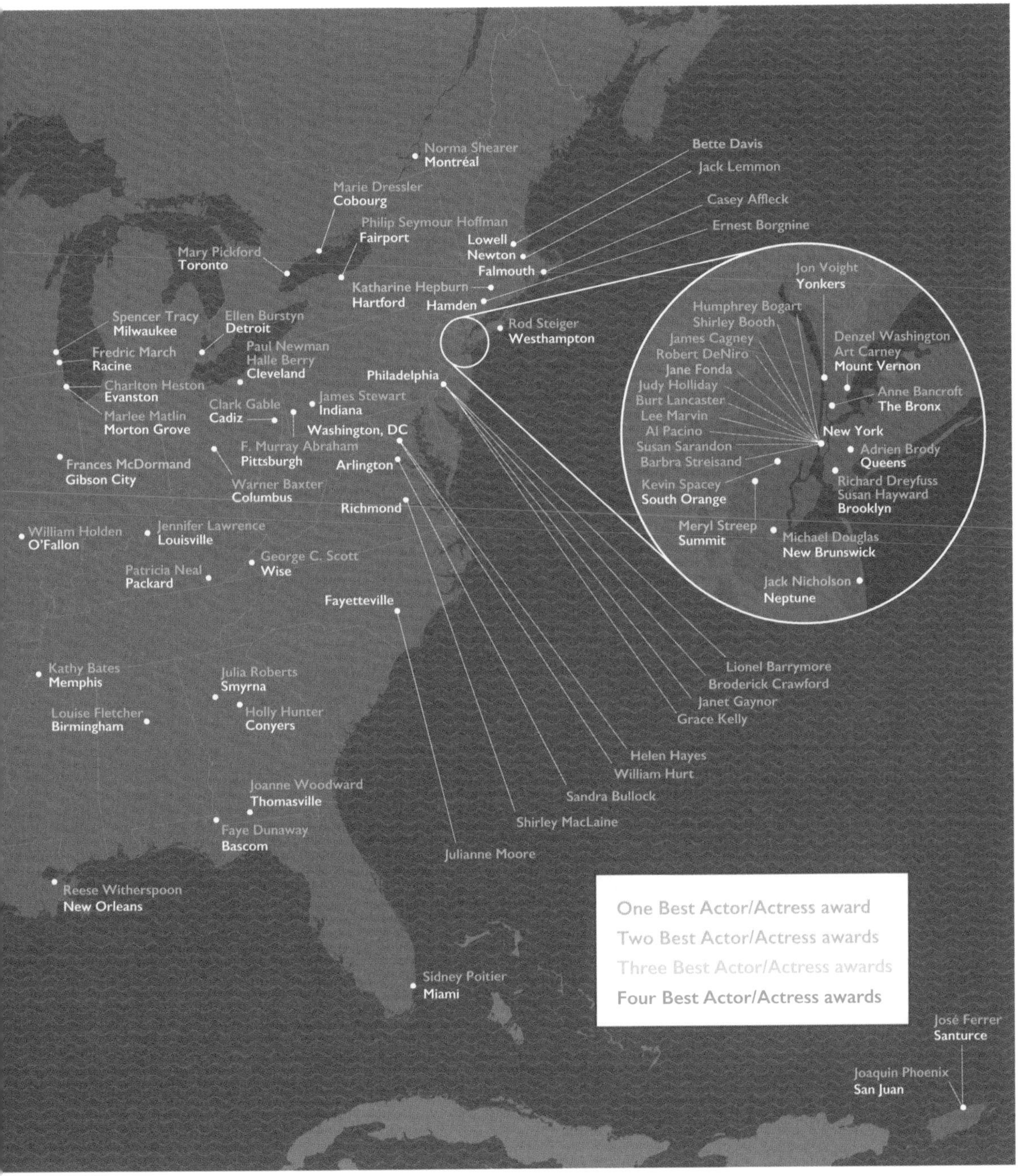
Norma Shearer
Montréal
Marie Dressler
Cobourg
Philip Seymour Hoffman
Fairport
Mary Pickford
Toronto
Bette Davis
Jack Lemmon
Casey Affleck
Ernest Borgnine
Lowell
Newton
Falmouth
Katharine Hepburn
Hartford
Hamden
Rod Steiger
Westhampton
Spencer Tracy
Milwaukee
Ellen Burstyn
Detroit
Fredric March
Racine
Paul Newman
Halle Berry
Cleveland
Charlton Heston
Evanston
Marlee Matlin
Morton Grove
Clark Gable
Cadiz
James Stewart
Indiana
Philadelphia
Washington, DC
F. Murray Abraham
Pittsburgh
Arlington
Frances McDormand
Gibson City
Warner Baxter
Columbus
Richmond
William Holden
O'Fallon
Jennifer Lawrence
Louisville
George C. Scott
Wise
Patricia Neal
Packard
Fayetteville
Jon Voight
Yonkers
Humphrey Bogart
Shirley Booth
James Cagney
Robert DeNiro
Jane Fonda
Judy Holliday
Burt Lancaster
Lee Marvin
Al Pacino
Susan Sarandon
Barbra Streisand
Denzel Washington
Art Carney
Mount Vernon
Anne Bancroft
The Bronx
New York
Adrien Brody
Queens
Kevin Spacey
South Orange
Richard Dreyfuss
Susan Hayward
Brooklyn
Meryl Streep
Summit
Michael Douglas
New Brunswick
Jack Nicholson
Neptune
Kathy Bates
Memphis
Julia Roberts
Smyrna
Louise Fletcher
Birmingham
Holly Hunter
Conyers
Lionel Barrymore
Broderick Crawford
Janet Gaynor
Grace Kelly
Helen Hayes
William Hurt
Sandra Bullock
Shirley MacLaine
Joanne Woodward
Thomasville
Faye Dunaway
Bascom
Julianne Moore
Reese Witherspoon
New Orleans
Sidney Poitier
Miami
One Best Actor/Actress award
Two Best Actor/Actress awards
Three Best Actor/Actress awards
Four Best Actor/Actress awards
José Ferrer
Santurce
Joaquin Phoenix
San Juan

61 What are you looking at?
The highest-rated programs on US TV

Program	Telecast date
1. *M*A*S*H* (final episode)	2/28/83
2. *Dallas* ("Who Done It?" episode)	11/21/80
3. *Roots*, "Part VIII"	1/30/77
4. *Super Bowl XVI*	1/24/82
5. *Super Bowl XVII*	1/30/83
6. 17th Winter Olympics (in Norway)	2/23/94
7. *Super Bowl XX*	1/26/86
8. *Super Bowl XLIX*	2/1/15
9. *Gone With the Wind Pt. 1*	11/7/76
10. *Gone With the Wind Pt. 2*	11/8/76
11. *Super Bowl XII*	1/15/78
12. *Super Bowl XLVIII*	2/2/14
13. *Super Bowl XIII*	1/21/79
14. *Super Bowl 50*	2/7/16
15. *Super Bowl XLVI*	2/5/12
16. *Super Bowl XLVII*	2/3/13
17. *The Bob Hope Christmas Special: Around the World with the USO*	1/15/70
18. *Super Bowl XVIII*	1/22/84
19. *Super Bowl XIX*	1/20/85
20. *Super Bowl XIV*	1/20/80
21. *Super Bowl XLV*	2/6/11
22. *Super Bowl XXX*	1/28/96
23. ABC Sunday Night Movie (*The Day After*)	11/20/83
24. *Roots*, "Part VI"	1/28/77
25. *The Fugitive* (final episode)	8/29/67
26. *Super Bowl LI*	2/5/17
27. *Super Bowl XXI*	1/25/87
28. *Roots*, "Part V"	1/27/77
29. *Super Bowl XXVIII*	1/30/94
30. *Cheers* (final episode)	5/20/93

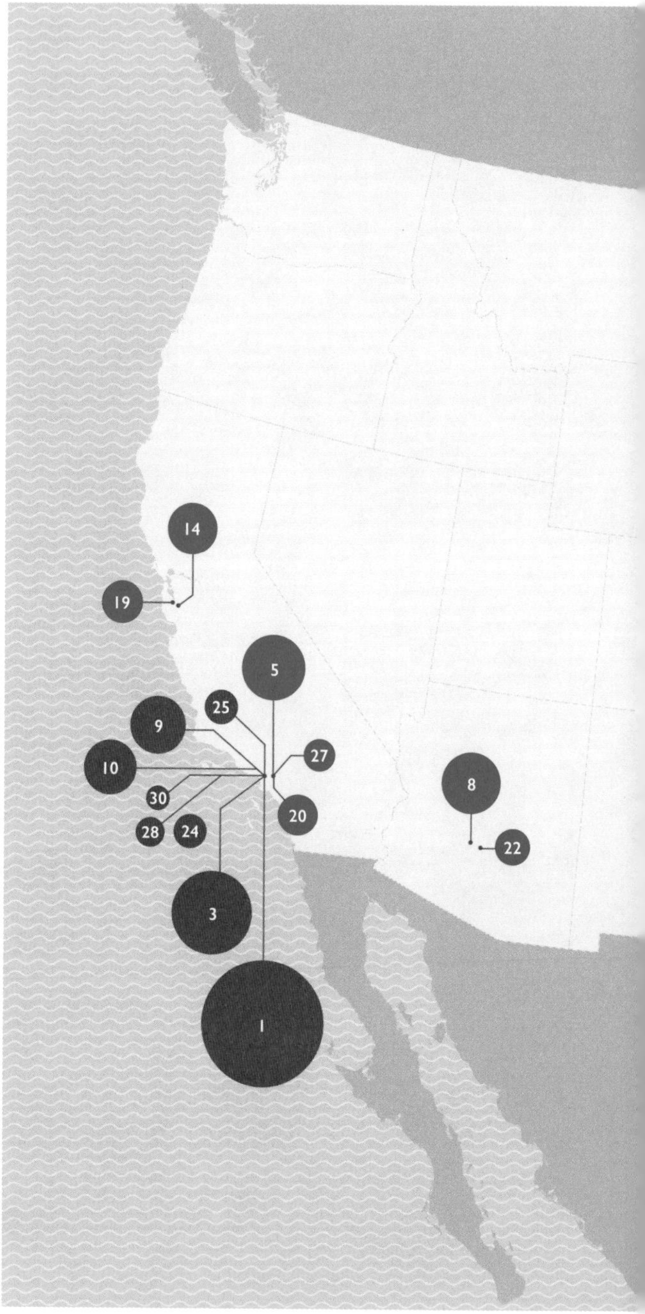

Ratings

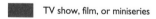

Sports event

TV show, film, or miniseries

The Bob Hope Christmas Special

The most-watched television event in history is the Apollo 11 Moon landing, with a 90+ percent rating (the percentage of US, TV-owning households that tuned in—which is the metric we've used in this map). The rest of the all-time list (since 1961, when ratings began being recorded) consists of earthbound programs, all but one of which was broadcast or filmed within North America. Note that ratings aren't as reliably established for unsponsored or joint network telecasts, as the Moon landing was, so we've excluded such events, along with programs shorter than 30 minutes.

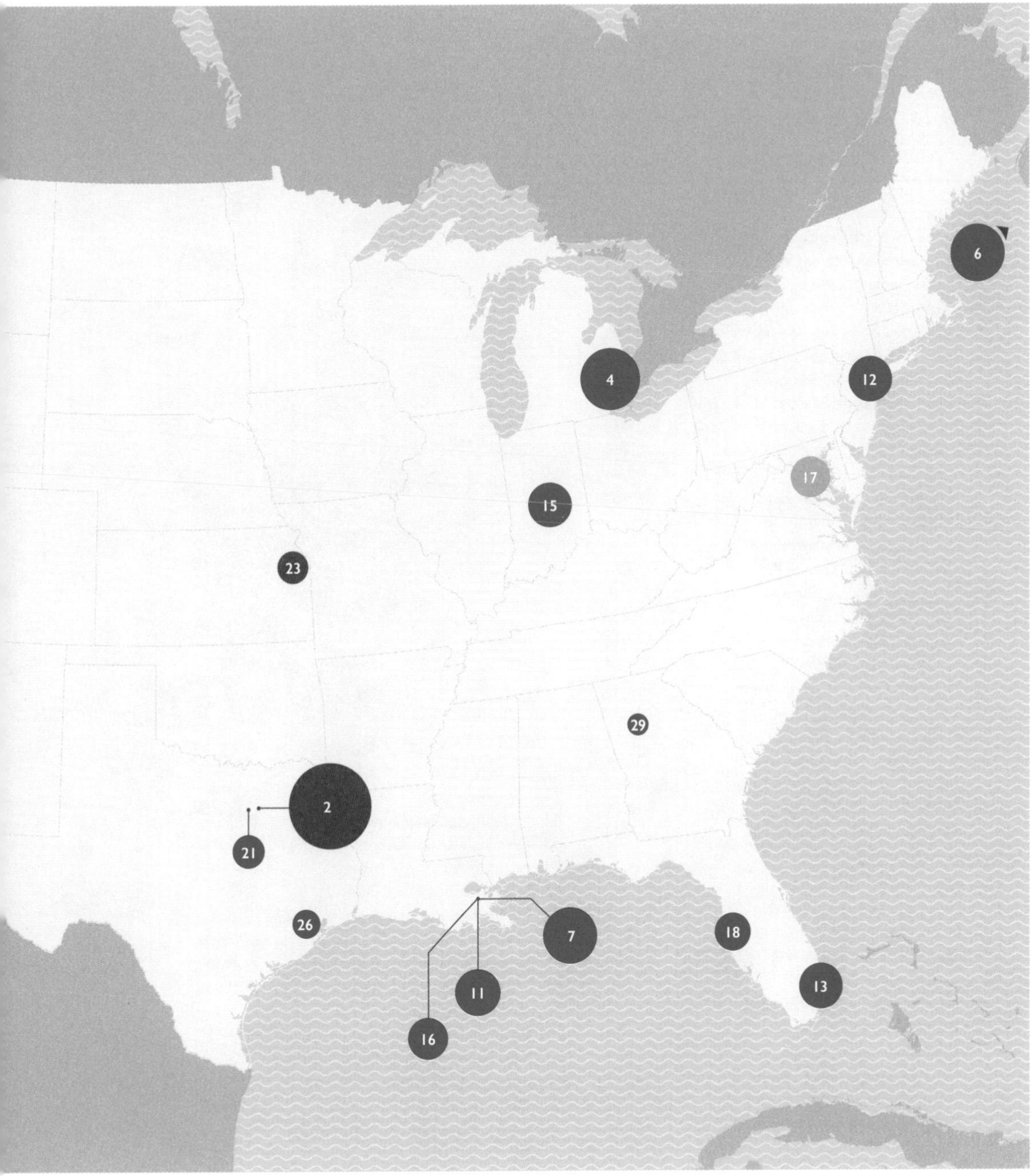

62 Ladies and gentlemen, the Beatles!: Every show they played in North America

1st Trip to US

1.	Studio 50, New York*	2/9/64
2.	Washington Coliseum, DC	2/11/64
3.	Carnegie Hall, New York	2/12/64
4.	Deauville Hotel, Miami†	2/16/64

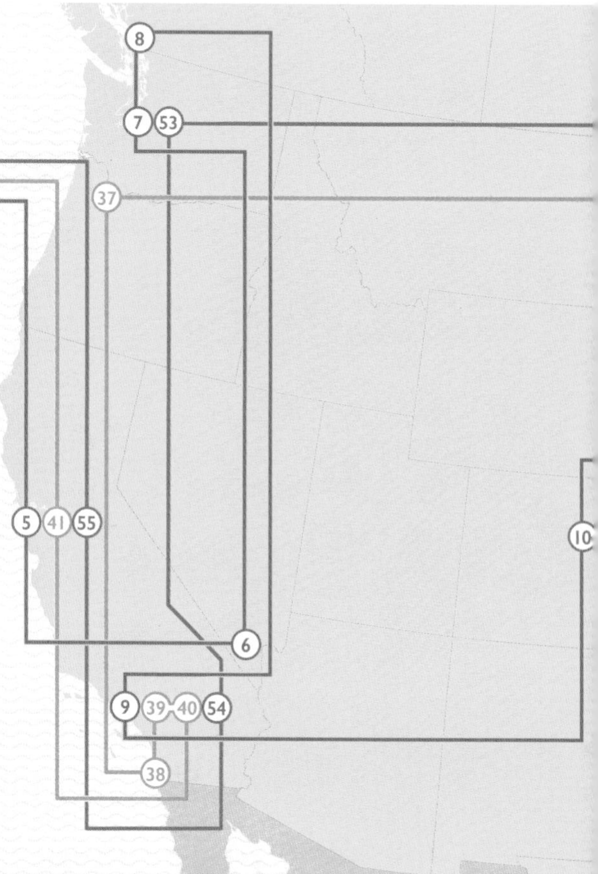

1st Concert Tour

5.	Cow Palace, San Francisco	8/19/64
6.	Convention Center, Las Vegas	8/20/64
7.	Seattle Center Coliseum, Seattle	8/21/64
8.	Empire Stadium, Vancouver	8/22/64
9.	Hollywood Bowl, Los Angeles	8/23/64
10.	Red Rocks Amphitheatre, Denver	8/26/64
11.	Cincinnati Gardens, Cincinnati	8/27/64
12.	Forest Hills Tennis Stadium, New York‡	8/28/64
13.	Forest Hills Tennis Stadium, New York	8/29/64
14.	Atlantic City Convention Hall, Atlantic City	8/30/64
15.	Philadelphia Convention Hall, Philadelphia	9/2/64
16.	Indiana State Fair, Indianapolis	9/3/64
17.	Milwaukee Arena, Milwaukee	9/4/64
18.	International Amphitheatre, Chicago	9/5/64
19.	Olympia Stadium, Detroit	9/6/64
20.	Maple Leaf Gardens, Toronto	9/7/64
21.	Montreal Forum, Montreal	9/8/64
22.	Gator Bowl, Jacksonville**	9/11/64
23.	Boston Garden, Boston	9/12/64
24.	Baltimore Civic Center, Baltimore	9/13/64
25.	Civic Arena, Pittsburgh	9/14/64
26.	Public Auditorium, Cleveland	9/15/64
27.	City Park Stadium, New Orleans	9/16/64
28.	Municipal Stadium, Kansas City	9/17/64
29.	Memorial Auditorium, Dallas	9/18/64
30.	Paramount Theatre, New York	9/20/64

2nd Concert Tour

31.	Shea Stadium, New York††	8/15/65
32.	Maple Leaf Gardens, Toronto	8/17/65
33.	Fulton County Stadium, Atlanta	8/18/65
34.	Sam Houston Coliseum, Houston	8/19/65
35.	Comiskey Park, Chicago	8/20/65
36.	Metropolitan Stadium, Minneapolis	8/21/65
37.	Memorial Coliseum, Portland	8/22/65
38.	Balboa Stadium, San Diego‡‡	8/28/65
39.	Hollywood Bowl, Los Angeles	8/29/65
40.	Hollywood Bowl, Los Angeles	8/30/65
41.	Cow Palace, San Francisco	8/31/65

3rd Concert Tour

42.	International Amphitheatre, Chicago	8/12/66
43.	Olympia Stadium, Detroit	8/13/66
44.	Municipal Stadium, Cleveland	8/14/66
45.	Washington Stadium, DC	8/15/66
46.	John F. Kennedy Stadium, Philadelphia	8/16/66
47.	Maple Leaf Gardens, Toronto	8/17/66
48.	Suffolk Downs Racecourse, Boston	8/18/66
49.	Memphis Coliseum, Memphis	8/19/66
50.	Crosley Field, Cincinnati	8/21/66
51.	Busch Stadium, St. Louis	8/21/66
52.	Shea Stadium, New York	8/23/66
53.	Seattle Center Coliseum, Seattle	8/25/66
54.	Dodger Stadium, Los Angeles	8/28/66
55.	Candlestick Park, San Francisco	8/29/66

The Beatles opened their famous first appearance on *The Ed Sullivan Show* by performing "All My Loving." With a TV rating of 45.3, that broadcast would have been number 31 on the all-time most-watched list (see page 112). However, "Twist and Shout" was actually the first song they performed in North America, in a set taped hours beforehand and broadcast on February 23, 1964. They closed their last North American tour at San Francisco's Candlestick Park—a performance that would go down in history as their final paid concert—with Little Richard's "Long Tall Sally."

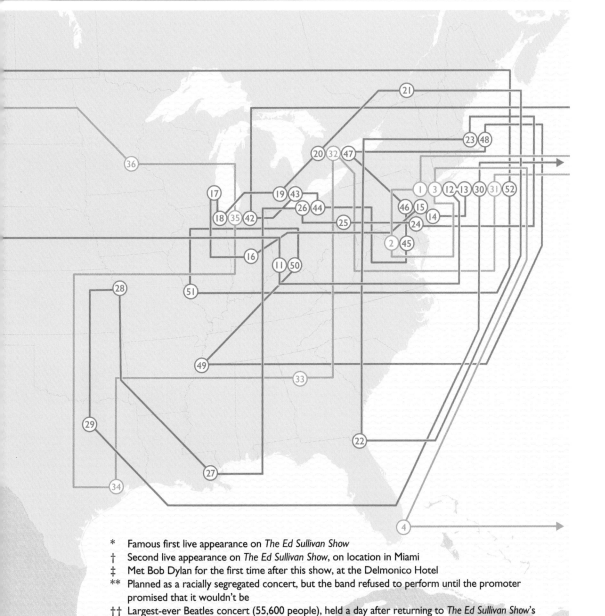

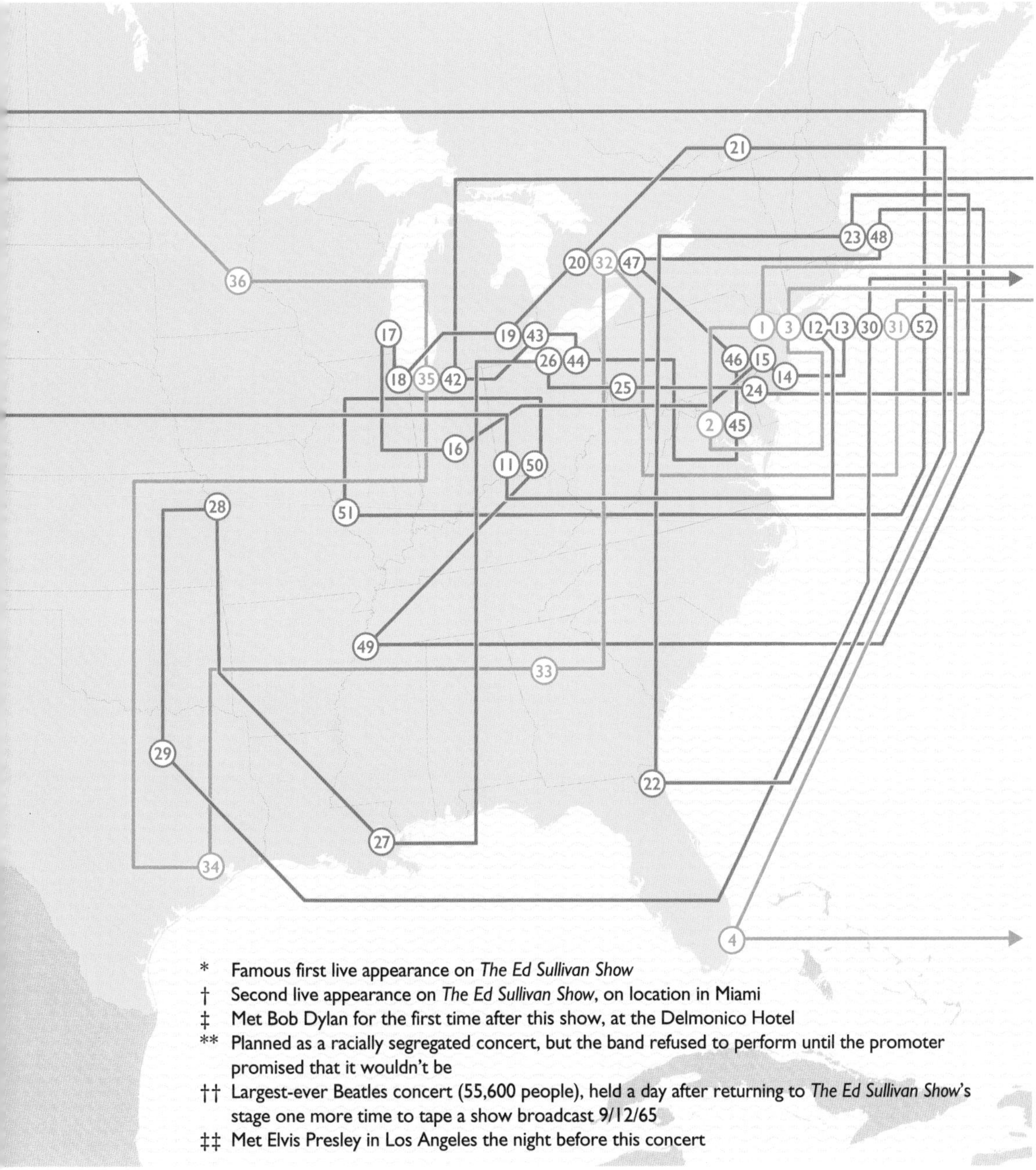

* Famous first live appearance on *The Ed Sullivan Show*

† Second live appearance on *The Ed Sullivan Show*, on location in Miami

‡ Met Bob Dylan for the first time after this show, at the Delmonico Hotel

** Planned as a racially segregated concert, but the band refused to perform until the promoter promised that it wouldn't be

†† Largest-ever Beatles concert (55,600 people), held a day after returning to *The Ed Sullivan Show*'s stage one more time to tape a show broadcast 9/12/65

‡‡ Met Elvis Presley in Los Angeles the night before this concert

5

PEOPLE AND POPULATIONS

63 His name is my name, too:
Most common surnames by state or province

Smith
Hernandez
Johnson
Garcia
Lopez
Martinez
Rodriguez
Brown
Friesen
Gonzalez
Jean
Leblanc
Lee
MacDonald
Olsen
Power
Rolle
Sanchez
Tremblay
Williams
Chan

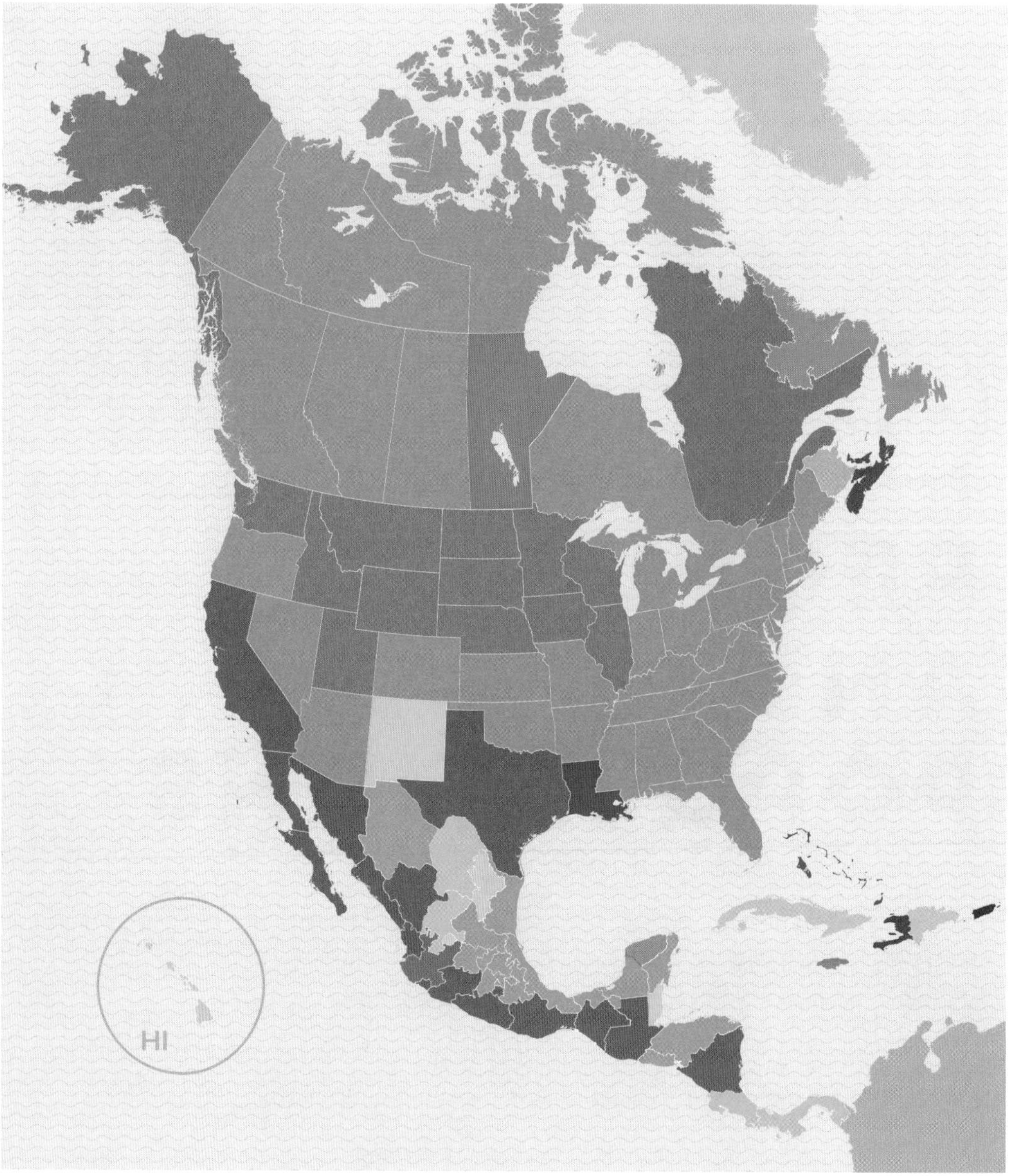

64 Native lands in native hands:
Indigenous population by state or province

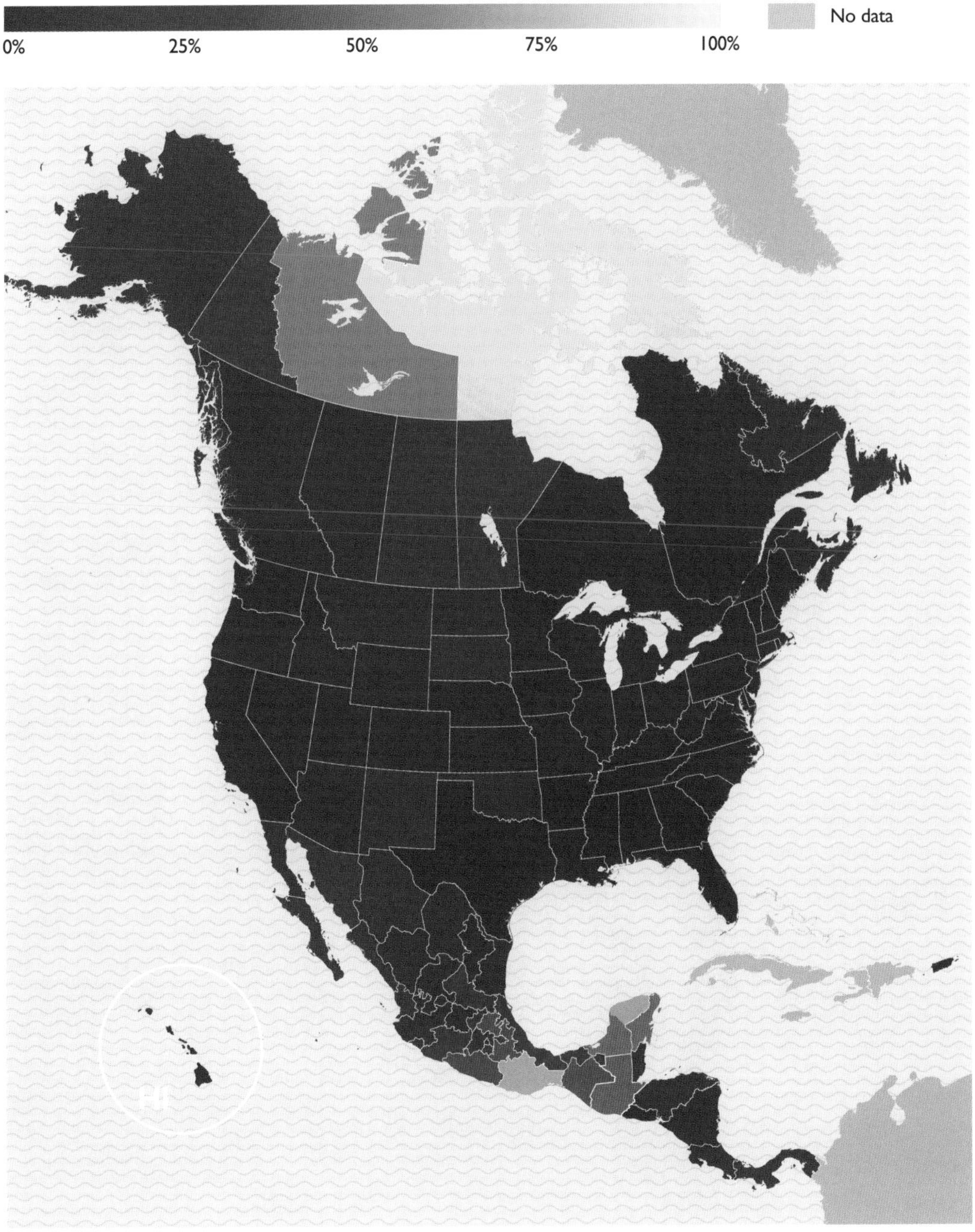

65 **Where we're coming from:** Most common countries of origin outside the US* by state

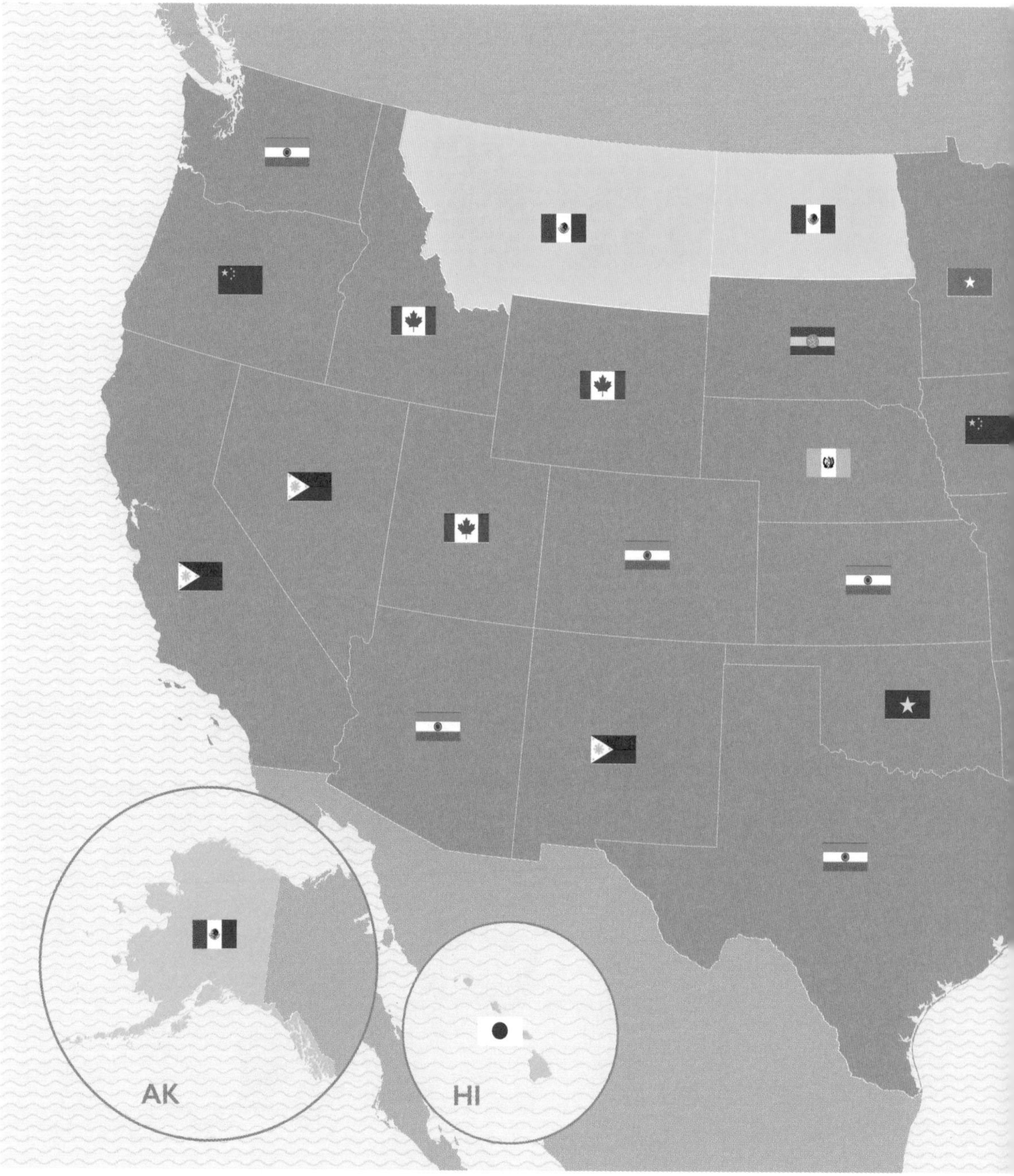

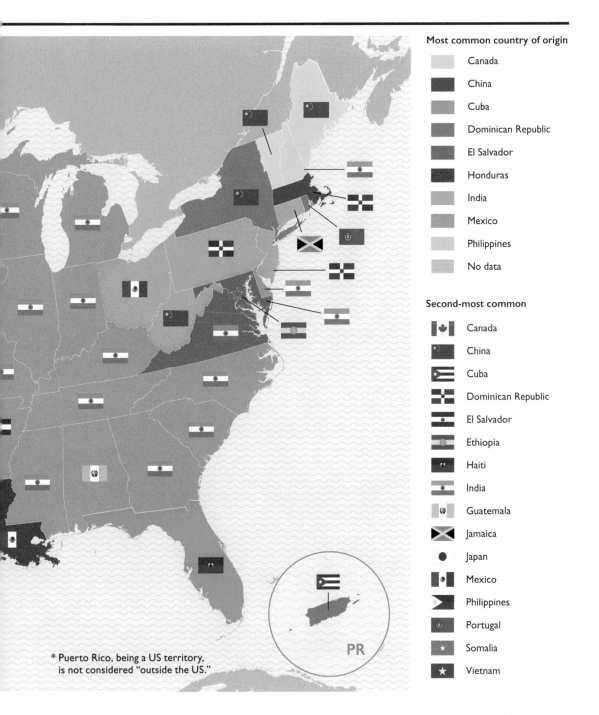

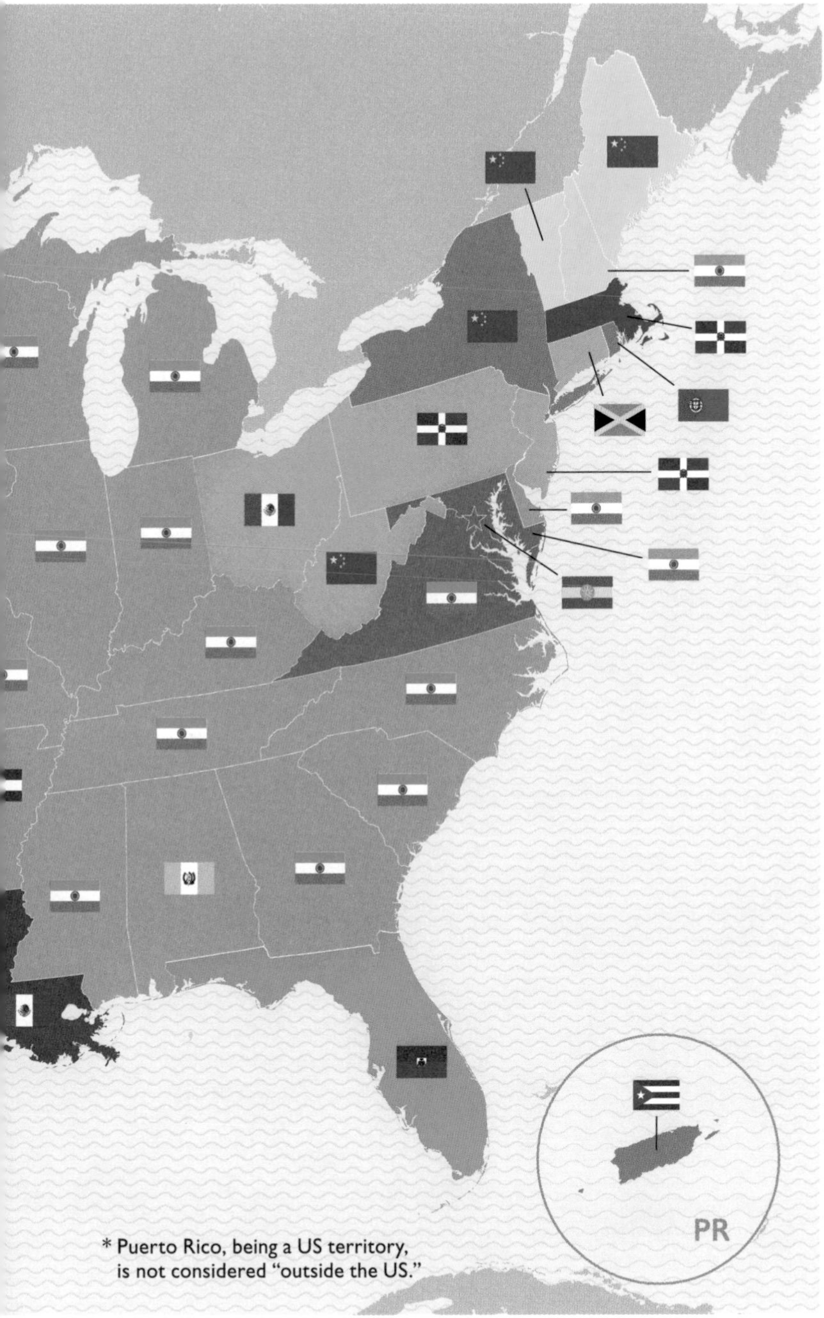

* Puerto Rico, being a US territory, is not considered "outside the US."

Most common country of origin

- Canada
- China
- Cuba
- Dominican Republic
- El Salvador
- Honduras
- India
- Mexico
- Philippines
- No data

Second-most common

- Canada
- China
- Cuba
- Dominican Republic
- El Salvador
- Ethiopia
- Haiti
- India
- Guatemala
- Jamaica
- Japan
- Mexico
- Philippines
- Portugal
- Somalia
- Vietnam

66 American pie charts:
Origin of residents by state

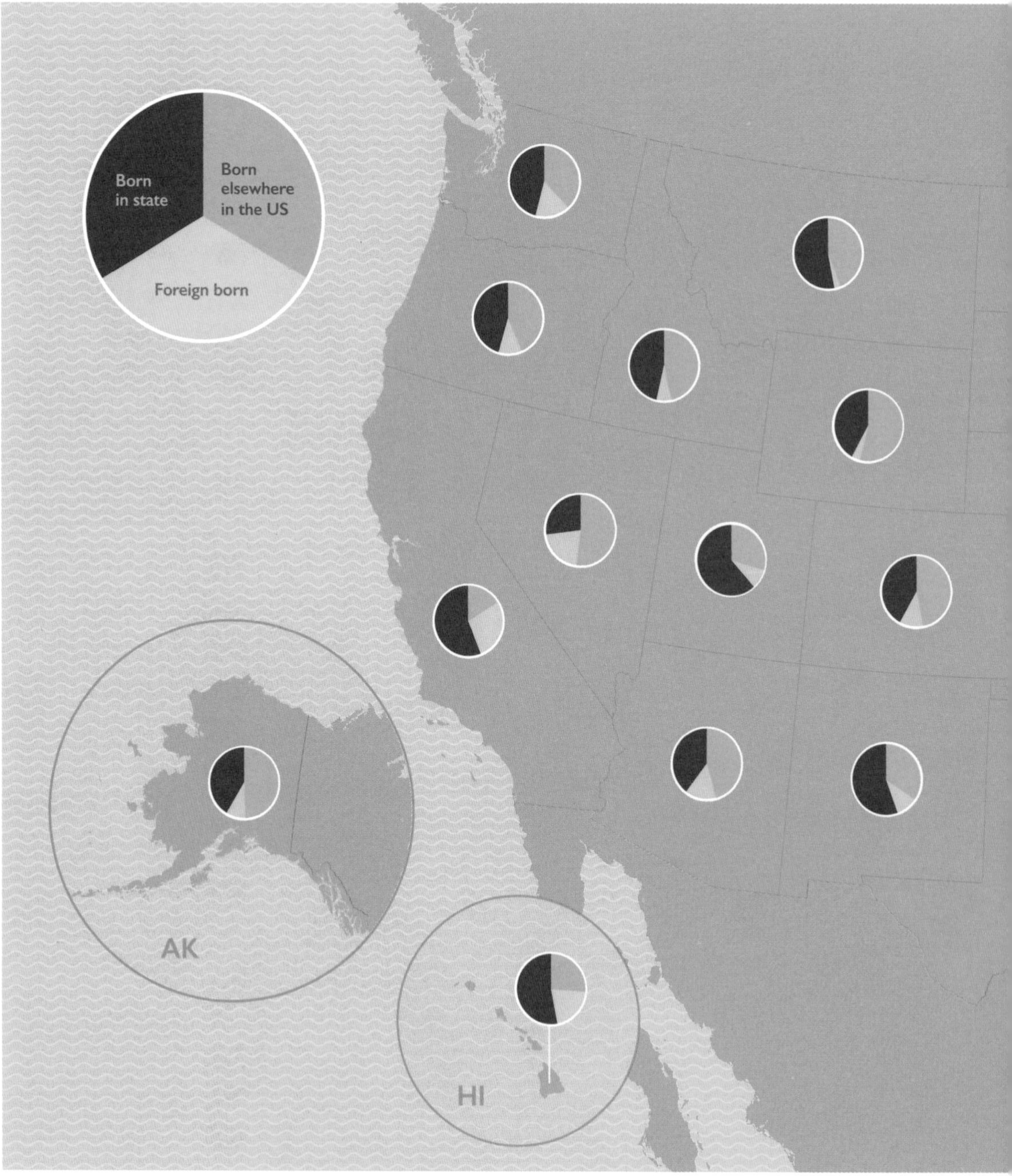

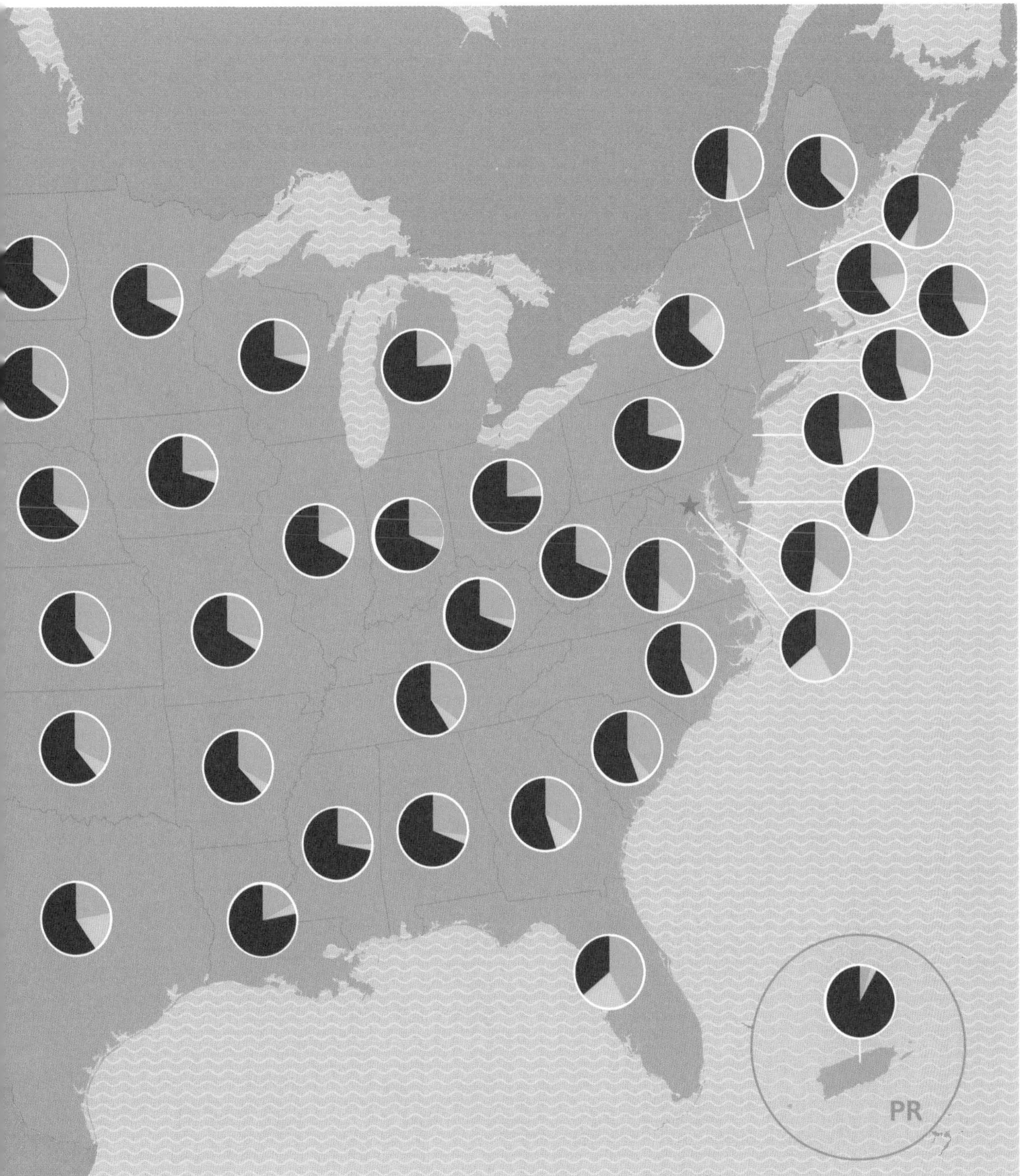
PR

67 Birds of a feather:
Most common cross-state migrations

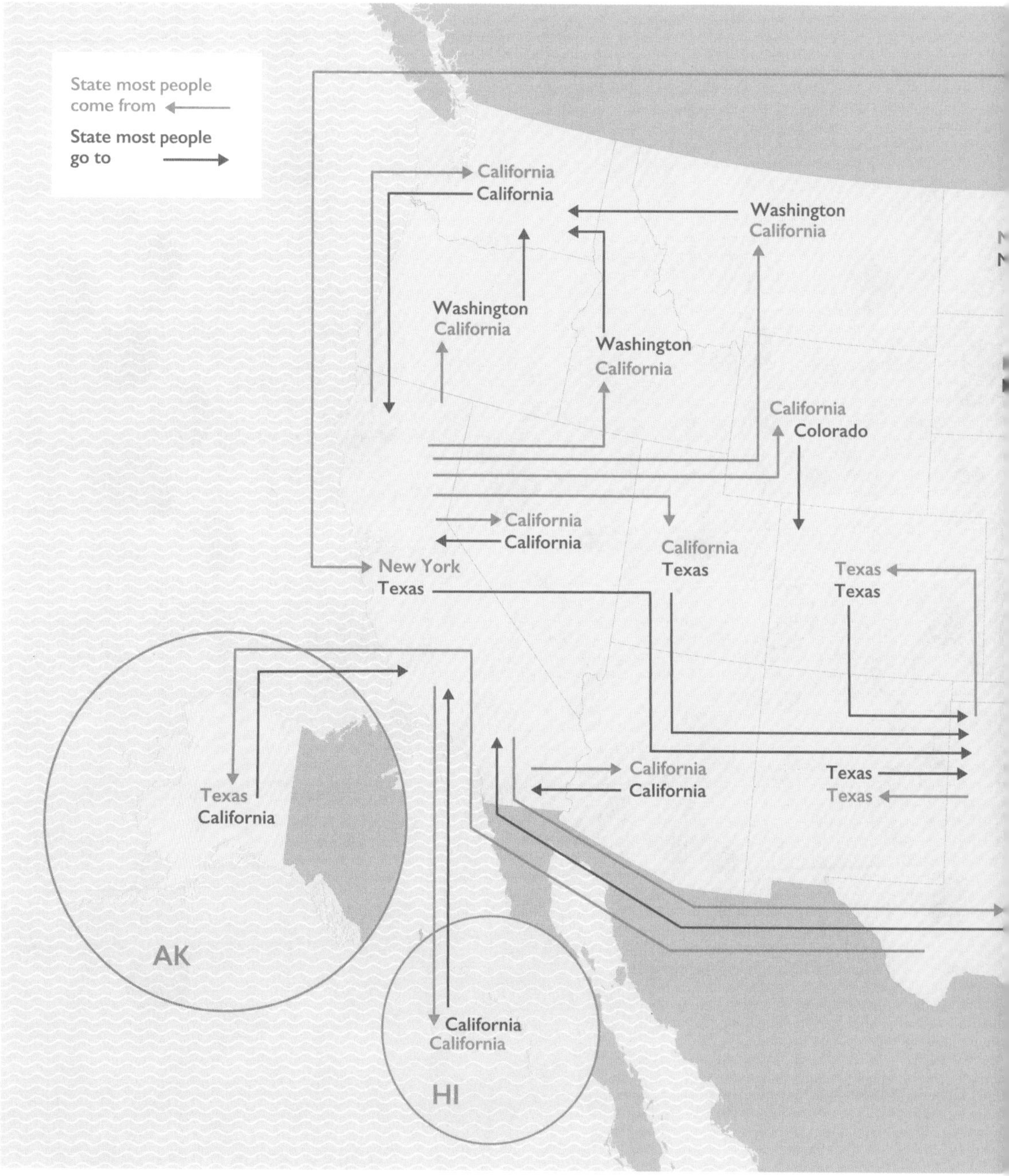

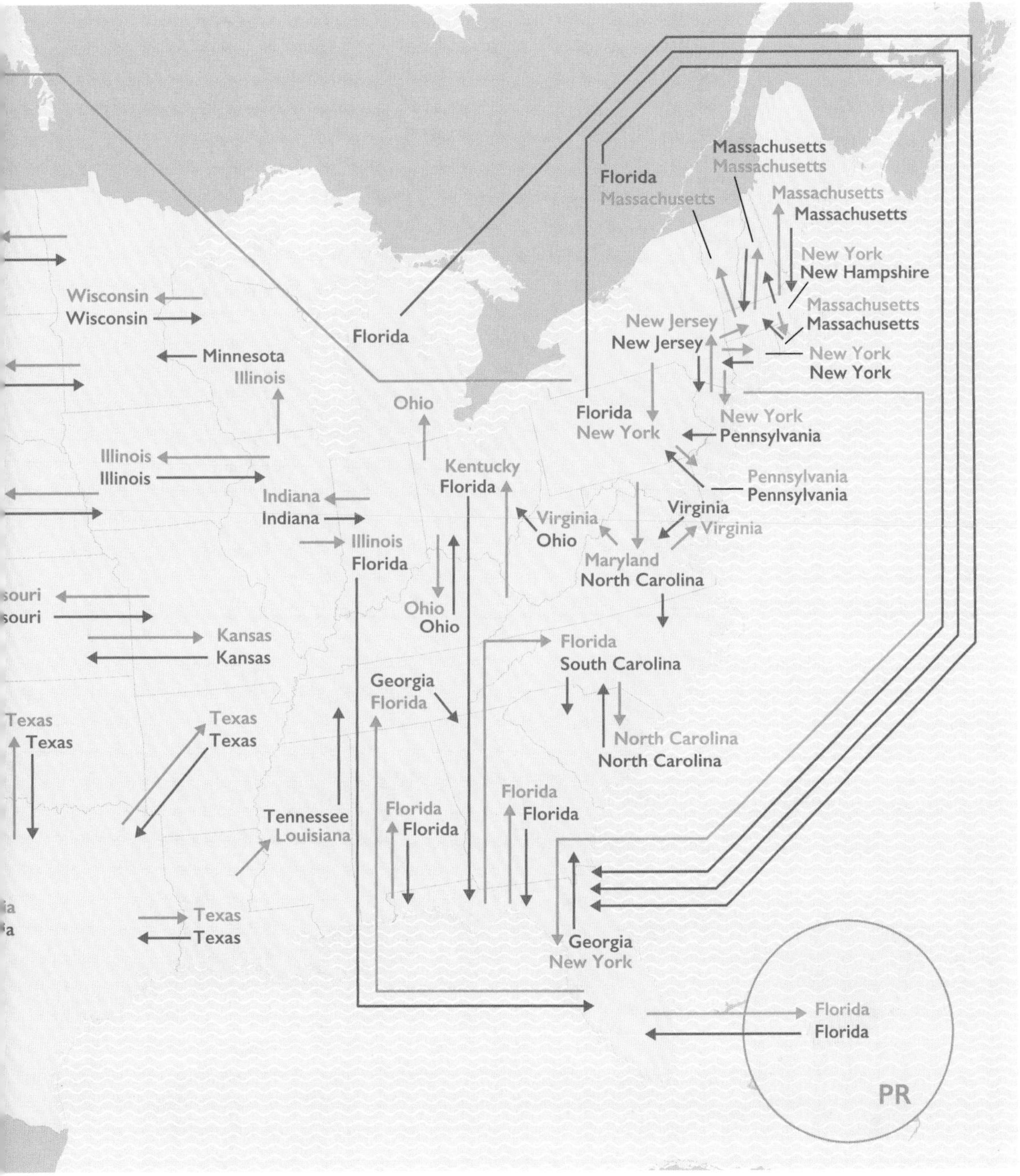

Massachusetts
Massachusetts
Florida
Massachusetts
Massachusetts
Massachusetts
New York
New Hampshire
Massachusetts
Massachusetts
Wisconsin
Wisconsin
Florida
New Jersey
New Jersey
New York
New York
Minnesota
Illinois
Ohio
Florida
New York
New York
Pennsylvania
Illinois
Illinois
Kentucky
Florida
Pennsylvania
Pennsylvania
Indiana
Indiana
Virginia
Ohio
Virginia
Virginia
Illinois
Florida
Maryland
North Carolina
Ohio
Ohio
Kansas
Kansas
Florida
South Carolina
Georgia
Florida
Texas
Texas
Texas
Texas
North Carolina
North Carolina
Tennessee
Louisiana
Florida
Florida
Florida
Florida
Texas
Texas
Georgia
New York
Florida
Florida
PR

68 **Brightest lights, biggest cities:**
The top 10 most populous US cities over time

1790

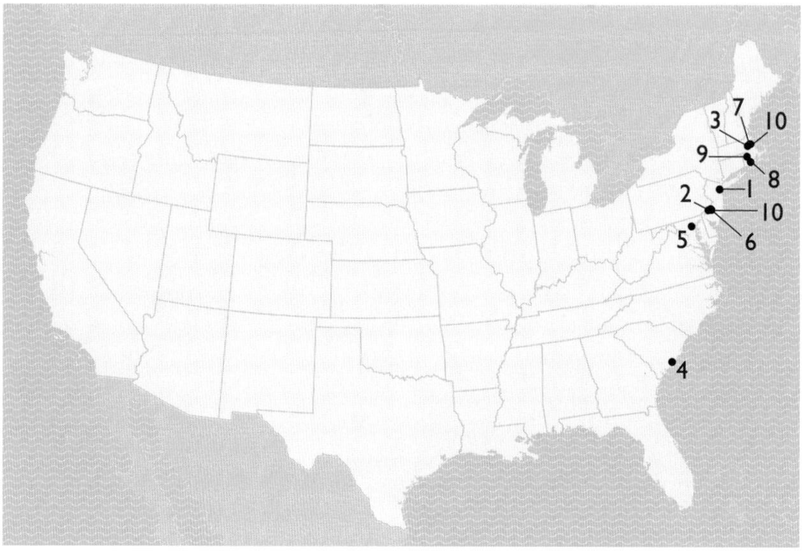

1. New York City
2. Philadelphia
3. Boston
4. Charleston
5. Baltimore
6. Northern Liberties
7. Salem
8. Newport
9. Providence
10. Southwark ⎤—TIED
10. Marblehead ⎦

1870

1. New York City
2. Philadelphia
3. Brooklyn
4. St. Louis
5. Chicago
6. Baltimore
7. Boston
8. Cincinnati
9. New Orleans
10. San Francisco

1950

1. New York City
2. Chicago
3. Philadelphia
4. Los Angeles
5. Detroit
6. Baltimore
7. Cleveland
8. St. Louis
9. Washington, DC
10. Boston

2018

1. New York City
2. Los Angeles
3. Chicago
4. Houston
5. Phoenix
6. Philadelphia
7. San Antonio
8. San Diego
9. Dallas
10. San Jose

69 Balancing act:
America's population centers of gravity

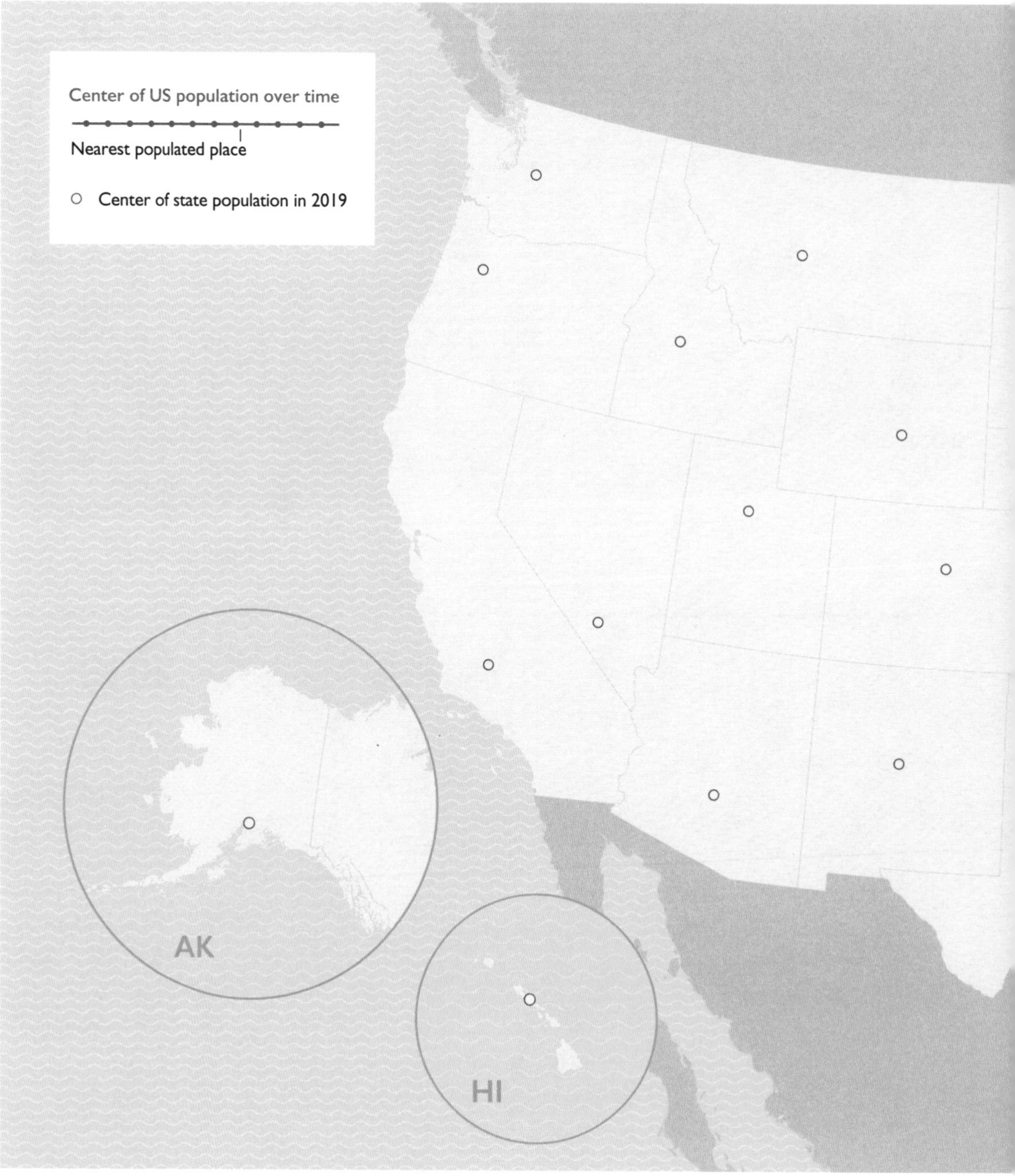

The US Census Bureau considers the center (or mean) of population to be the point at which there would be a perfect center of balance if the country were an imaginary, flat surface, and every inhabitant a weight placed on that surface. Here we see the country's center of gravity migrate westward over time, a phenomenon that's still ongoing today but slowing down. Also shown is the present-day center of population for each state.

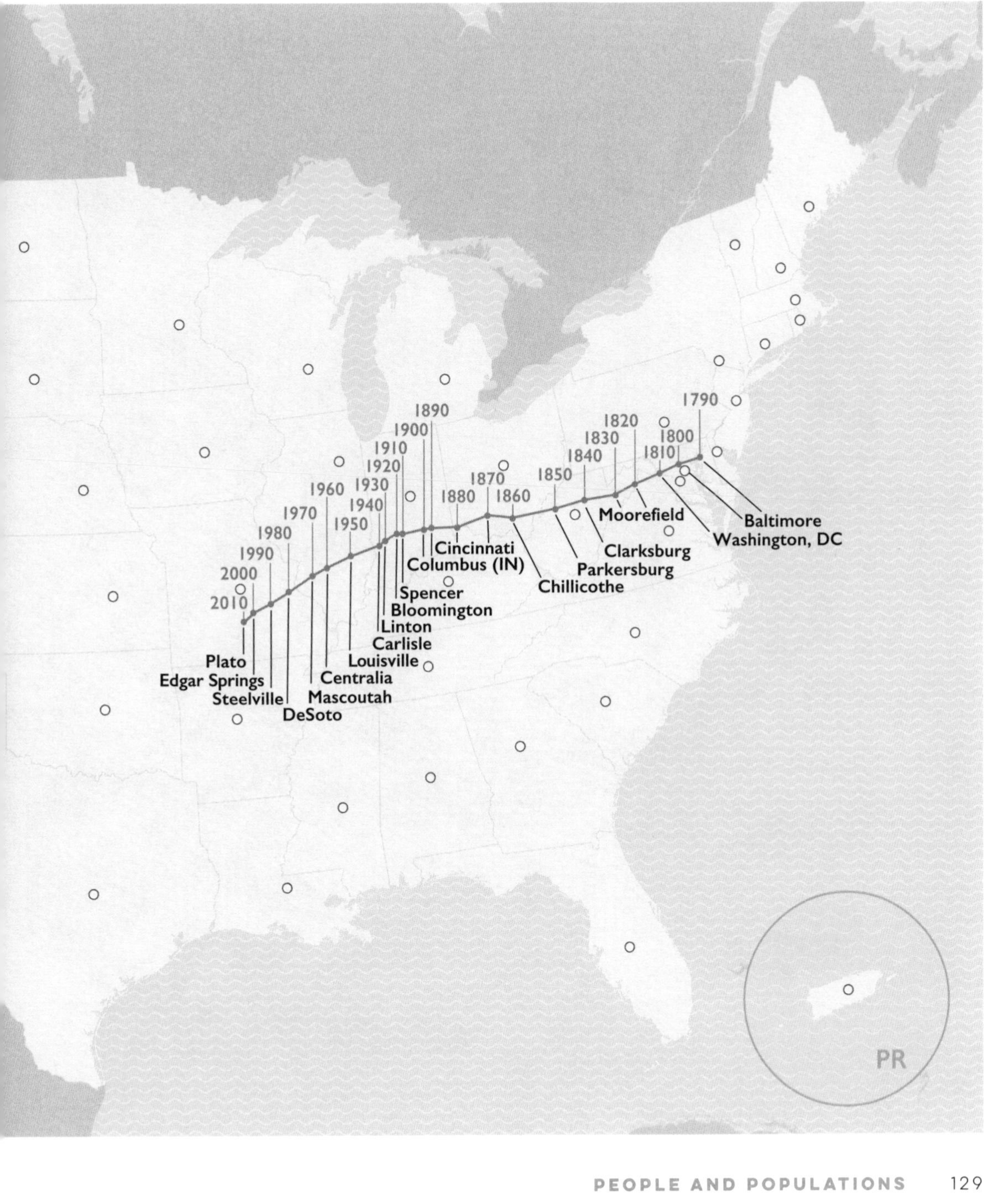

70 Homes away from homes:
Where people have second homes

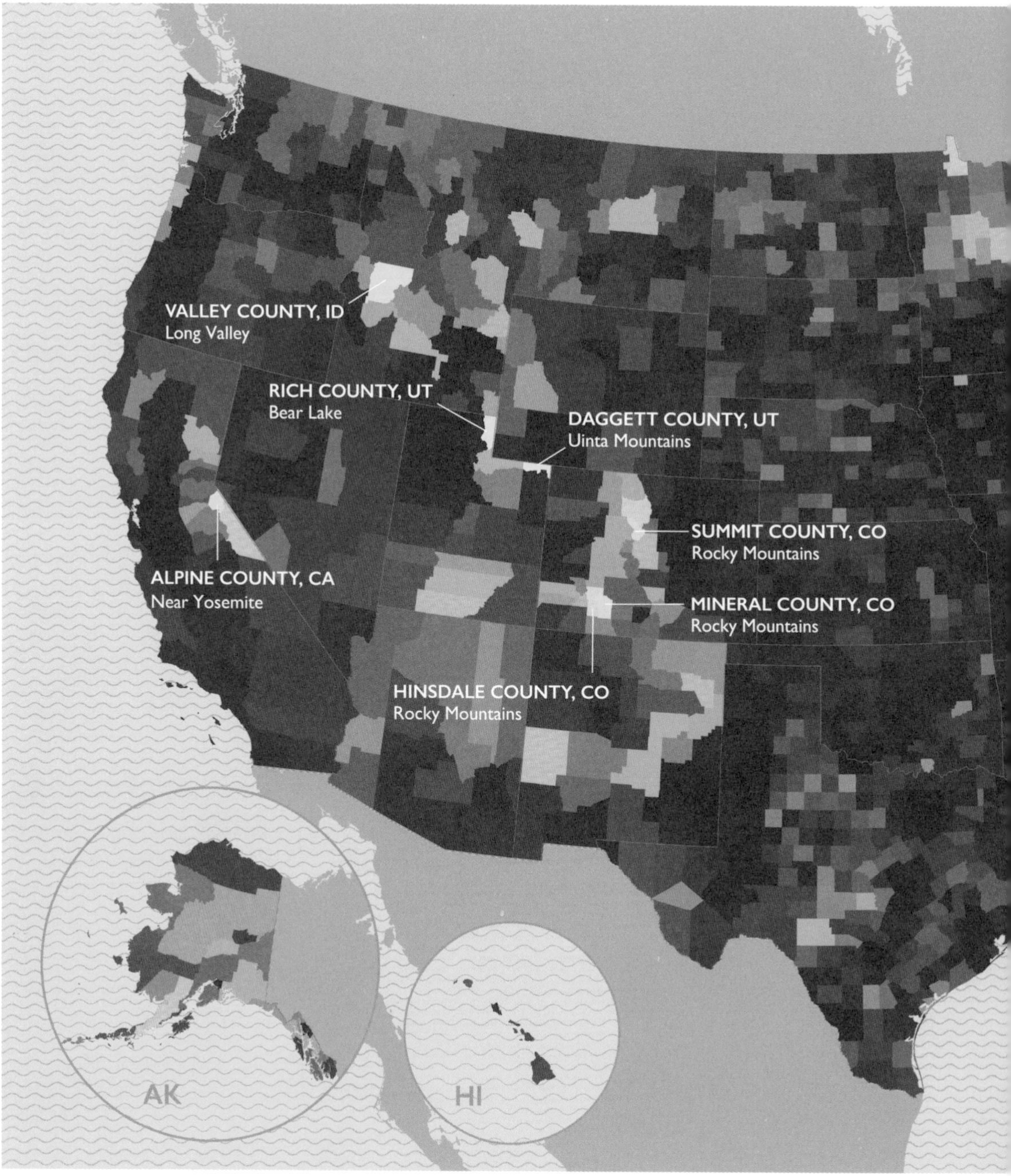

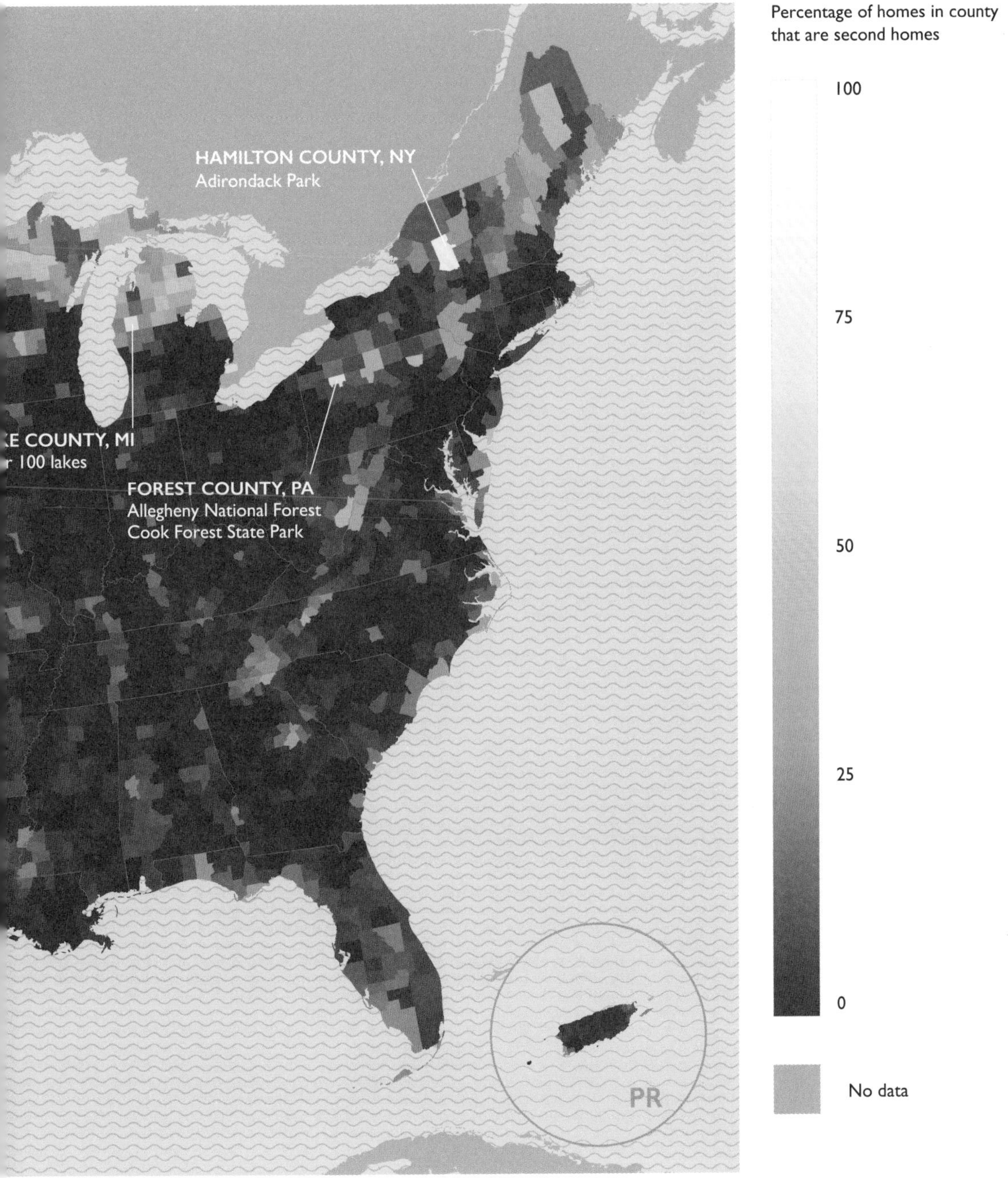
HAMILTON COUNTY, NY
Adirondack Park
E COUNTY, MI
r 100 lakes
FOREST COUNTY, PA
Allegheny National Forest
Cook Forest State Park
PR
Percentage of homes in county that are second homes
100
75
50
25
0
No data

71 **Homes all around, but not a place to stay:** Number of second homes per unhoused person

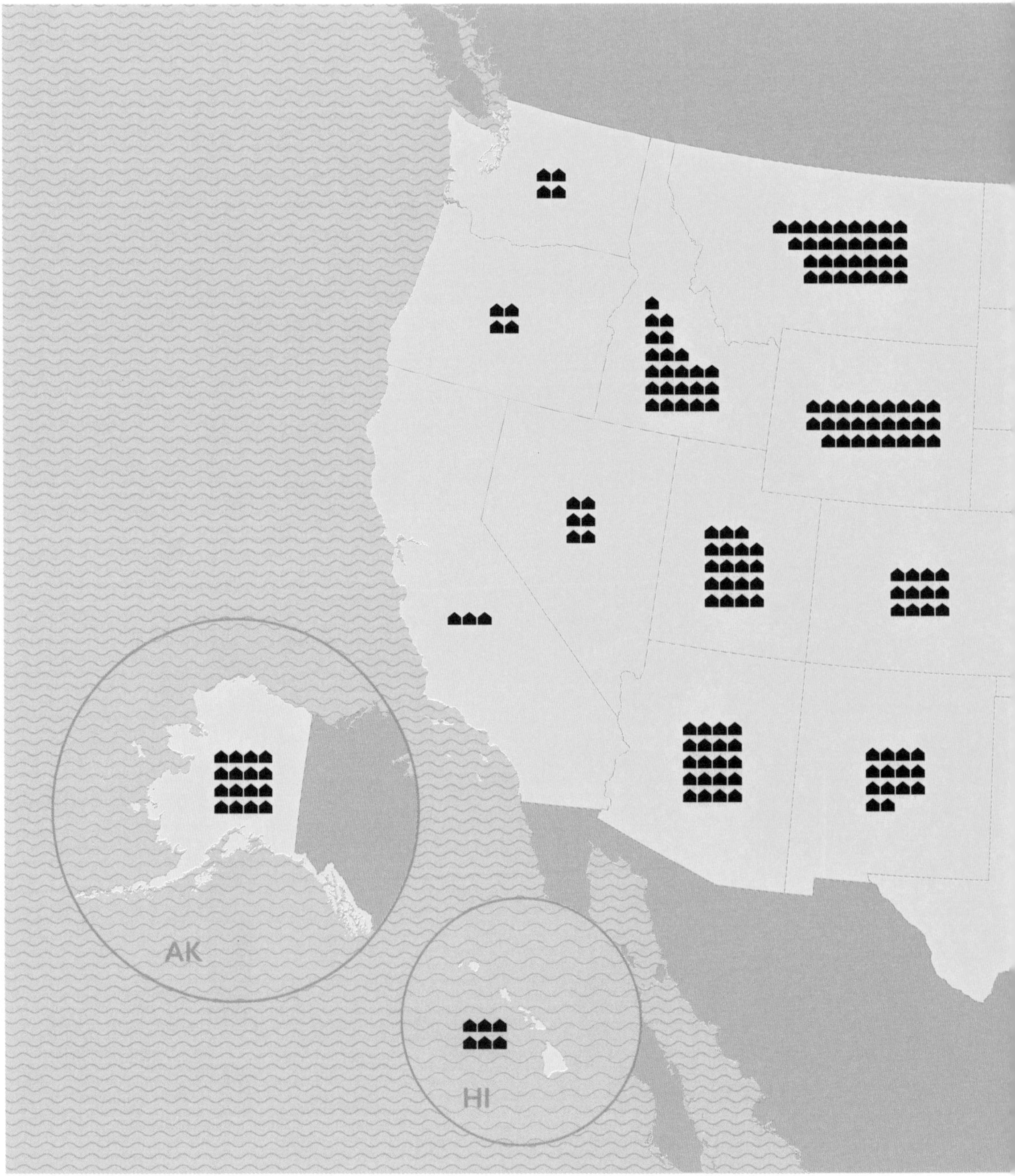

There's a wealth of second homes in the US (see page 130); California alone has around 400,000 vacation homes. The number of these homes exceeds California's population of people experiencing homelessness—at roughly 150,000 people, the highest in the nation—with nearly three largely vacant homes for every unhoused Californian. Lack of available housing is one factor in a range of complex factors that may lead to homelessness, but that there are more vacation homes than people with no homes in every US state is, to us, eye-opening.

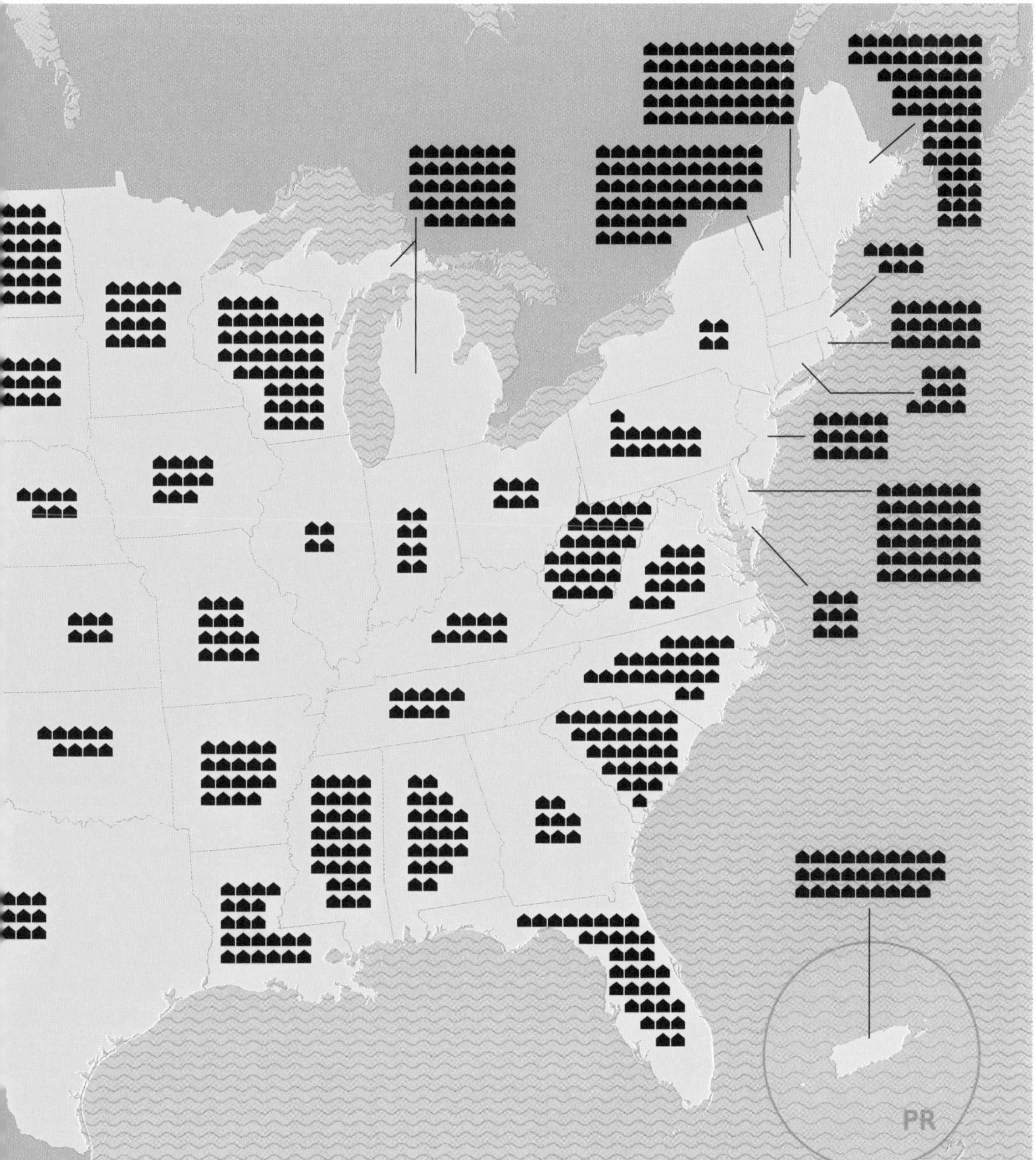

72 **The extremely Big Apple:** 38 states have smaller populations than NYC

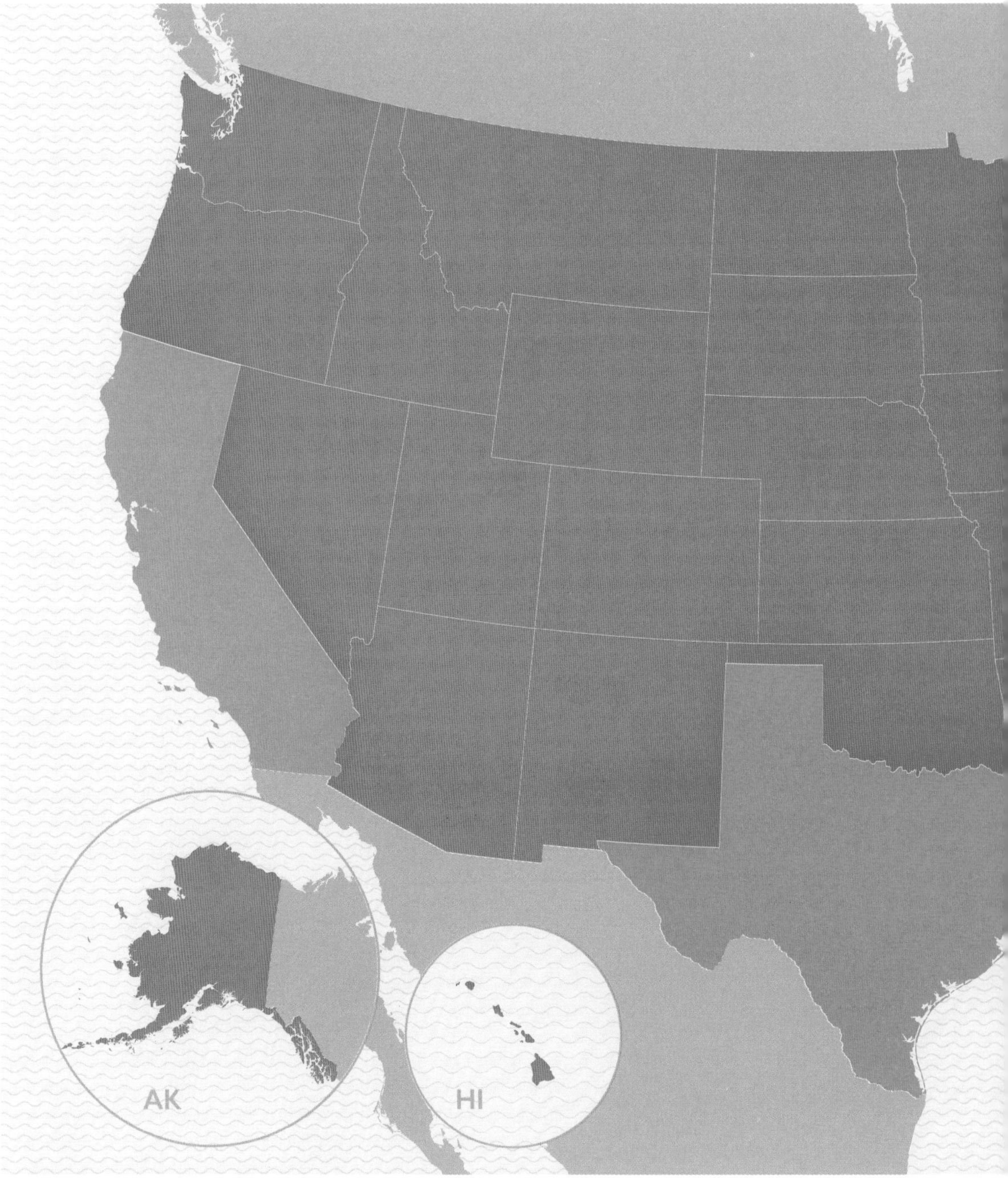

NYC's population is 8,336,817 people, per Census Bureau estimate of July 1, 2019.

Larger population than NYC

Smaller population than NYC

73 **The language barrier:** Percentage of households speaking only English by county

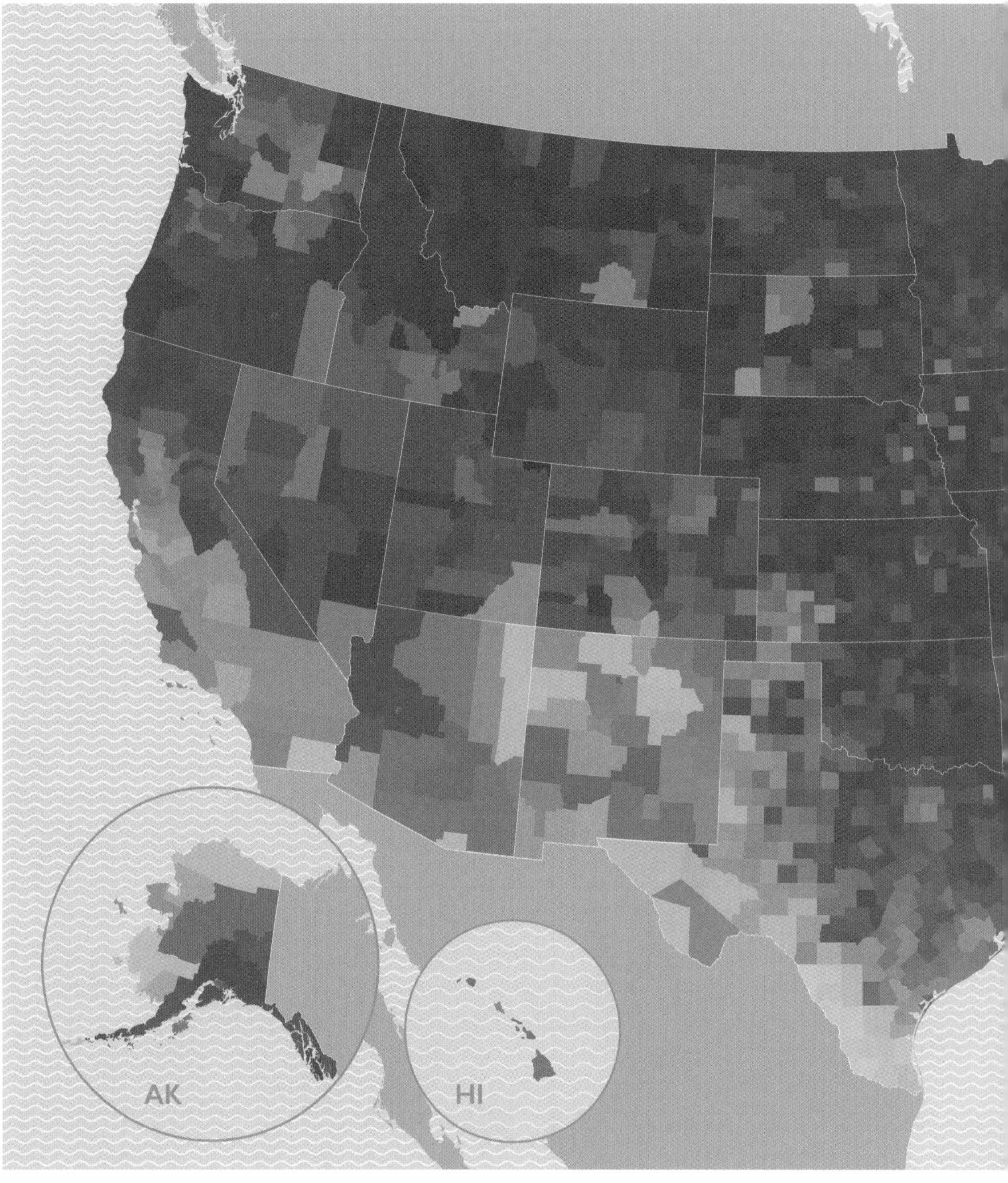

Perhaps unsurprisingly, all of Puerto Rico's 78 municipios (their version of counties) have a lower proportion of English-only households than all but 7 counties in the US, and those 7 are all in Texas: Starr, Maverick, Webb, Zapata, Zavala, Kenedy, and Hidalgo counties. At the other extreme, in 5 counties every household reported speaking only English: Faulk County, SD; Loup County, NE; Daggett County, UT; Carroll County, MS; and Issaquena County, MS.

74 Turtle Island:
Indigenous homelands in 1491

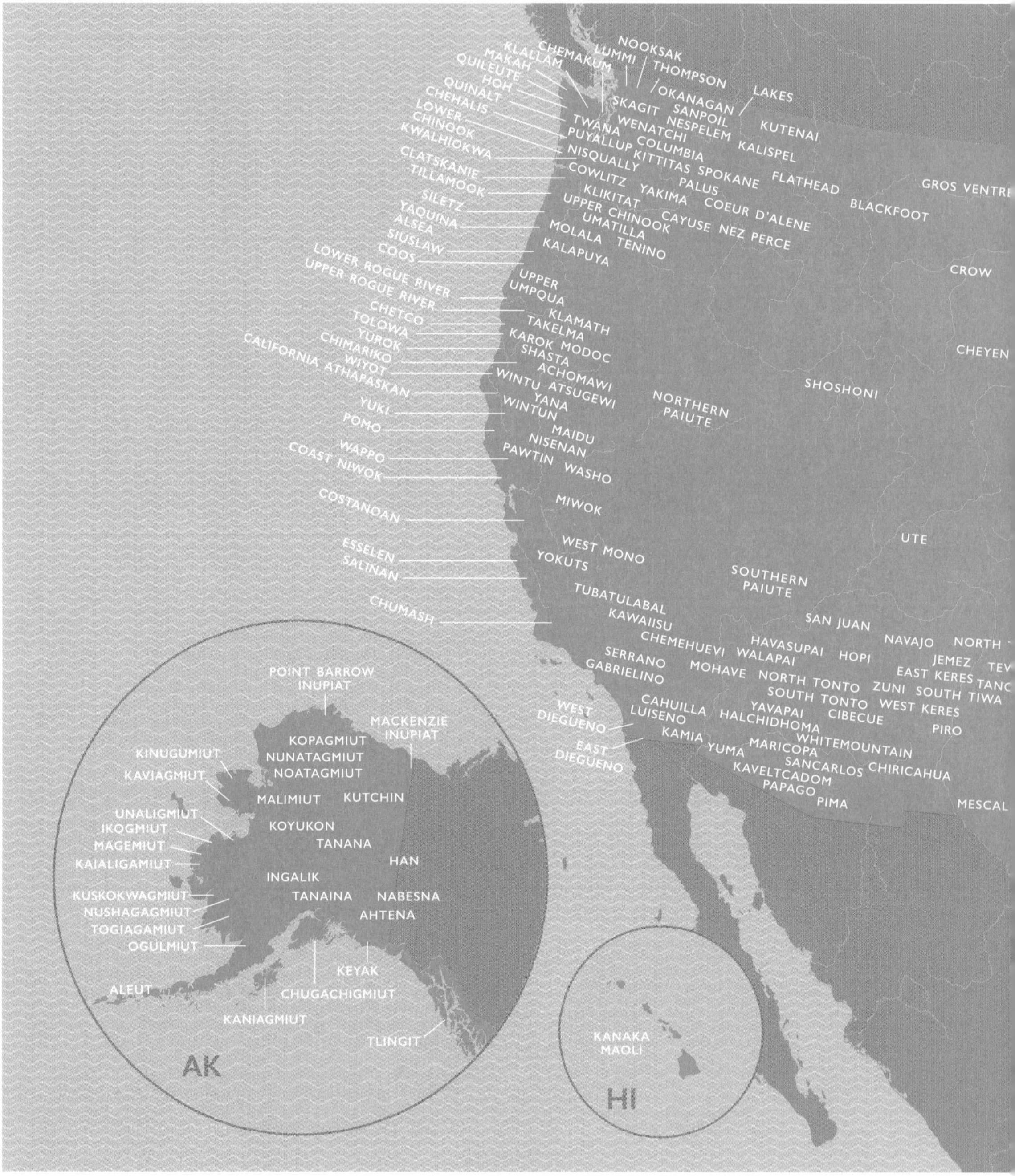
NOOKSAK
LUMMI
CHEMAKUM
KLALLAM
MAKAH
QUILEUTE
HOH
QUINALT
CHEHALIS
LOWER CHINOOK
KWALHIOKWA
THOMPSON
OKANAGAN
LAKES
SKAGIT
SANPOIL
NESPELEM
KUTENAI
TWANA
WENATCHI
PUYALLUP
COLUMBIA
KALISPEL
NISQUALLY
KITTITAS
SPOKANE
FLATHEAD
COWLITZ
PALUS
YAKIMA
COEUR D'ALENE
BLACKFOOT
CLATSKANIE
TILLAMOOK
SILETZ
YAQUINA
ALSEA
SIUSLAW
COOS
KLIKITAT
UPPER CHINOOK
UMATILLA
CAYUSE
NEZ PERCE
MOLALA
TENINO
KALAPUYA
CROW
LOWER ROGUE RIVER
UPPER ROGUE RIVER
UPPER UMPQUA
CHETCO
TOLOWA
YUROK
CHIMARIKO
WIYOT
CALIFORNIA ATHAPASKAN
KLAMATH
TAKELMA
KAROK
MODOC
SHASTA
ACHOMAWI
WINTU
ATSUGEWI
YANA
WINTUN
NORTHERN PAIUTE
SHOSHONI
YUKI
POMO
MAIDU
NISENAN
WAPPO
COAST NIWOK
PAWTIN
WASHO
MIWOK
COSTANOAN
UTE
ESSELEN
SALINAN
WEST MONO
YOKUTS
SOUTHERN PAIUTE
TUBATULABAL
KAWAIISU
CHUMASH
SAN JUAN
CHEMEHUEVI
HAVASUPAI
WALAPAI
HOPI
NAVAJO
NORTH
JEMEZ
SERRANO
GABRIELINO
MOHAVE
NORTH TONTO
ZUNI
EAST KERES
SOUTH TIWA
SOUTH TONTO
WEST KERES
WEST DIEGUENO
CAHUILLA
LUISENO
YAVAPAI
HALCHIDHOMA
CIBECUE
PIRO
EAST DIEGUENO
KAMIA
YUMA
MARICOPA
WHITEMOUNTAIN
SANCARLOS
CHIRICAHUA
KAVELTCADOM
PAPAGO
PIMA
POINT BARROW INUPIAT
MACKENZIE INUPIAT
KOPAGMIUT
NUNATAGMIUT
NOATAGMIUT
KINUGUMIUT
KAVIAGMIUT
MALIMIUT
KUTCHIN
UNALIGMIUT
IKOGMIUT
KOYUKON
MAGEMIUT
TANANA
KAIALIGAMIUT
HAN
INGALIK
KUSKOKWAGMIUT
TANAINA
NABESNA
NUSHAGAGMIUT
AHTENA
TOGIAGAMIUT
OGULMIUT
KEYAK
ALEUT
CHUGACHIGMIUT
KANIAGMIUT
TLINGIT
AK
KANAKA MAOLI
HI

In 1492, Columbus sailed the ocean blue . . . and colonists began their centuries-long campaign to wrest North America (or Turtle Island, as it's known by some Indigenous peoples) from the American Indians who had long called these lands home. A fully comprehensive map is impossible at this scale—only the more major tribes are depicted here—and no record exists that would allow total certainty about tribe locations and names as they referred to themselves. But one thing is certain: A year before Columbus set sail, a rich and uninterrupted constellation of tribes spanned from sea to sea.

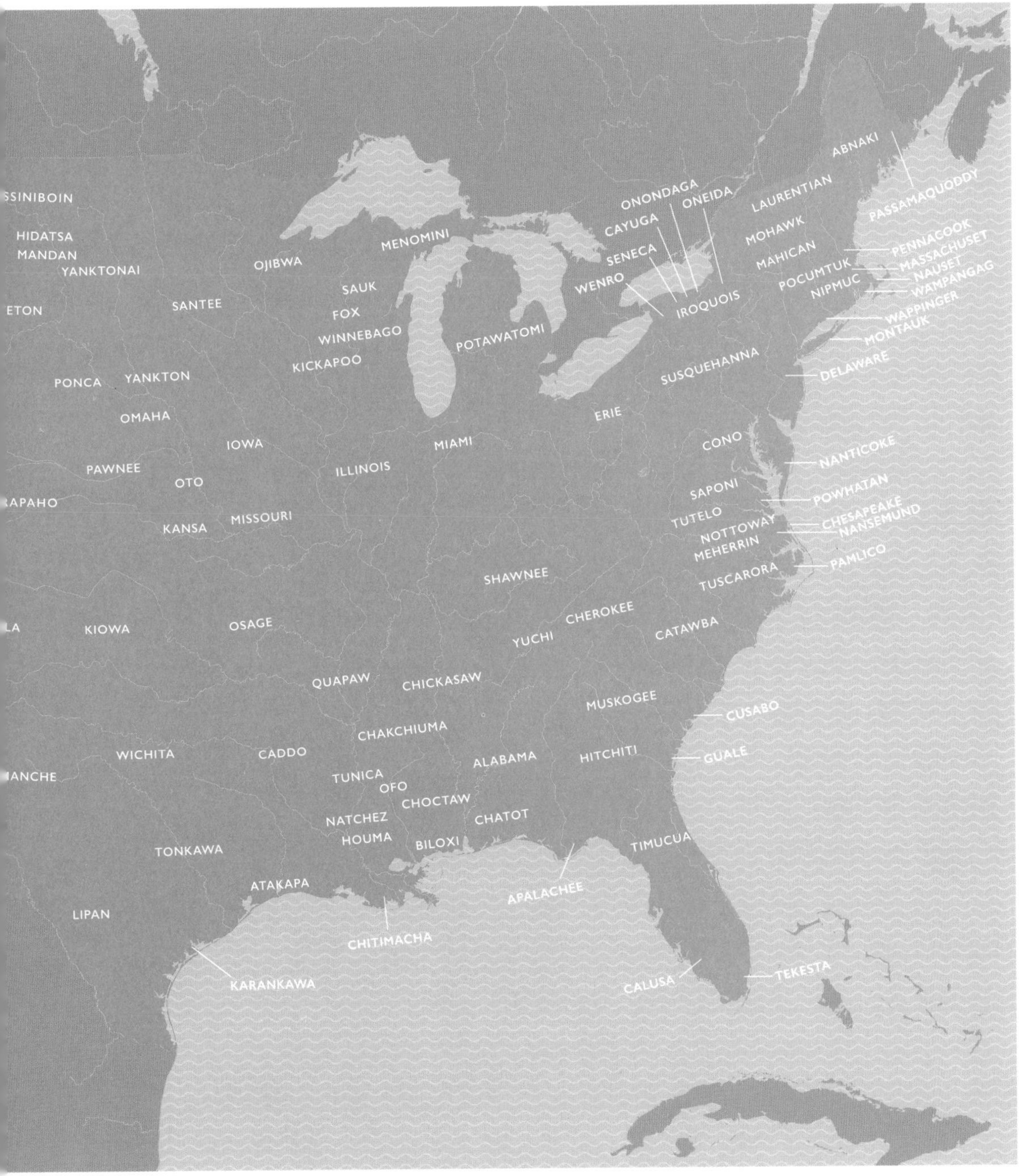

75 Strangers in their own land:
Indigenous homelands today

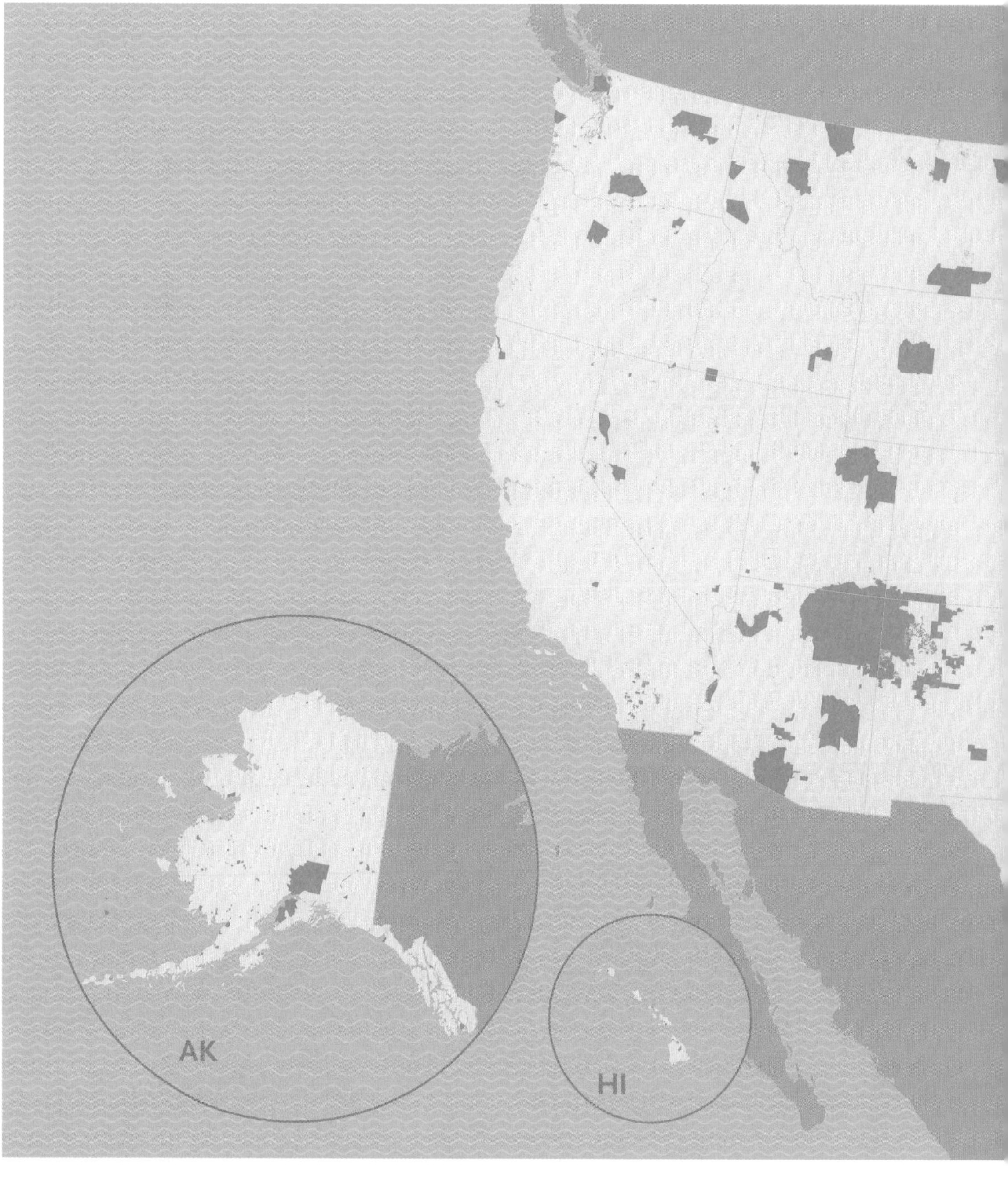

Today, the total area of the US is roughly 3.8 million square miles (9.8 million km^2), while the total area of all federally recognized American Indian reservations and off-reservation trust land areas, state-recognized American Indian reservations, and Hawaiian homelands—along with Alaska Native village and Oklahoma tribal, tribal designated, and state-designated tribal statistical areas—doesn't quite reach 200,000 square miles (about 500,000 km^2).

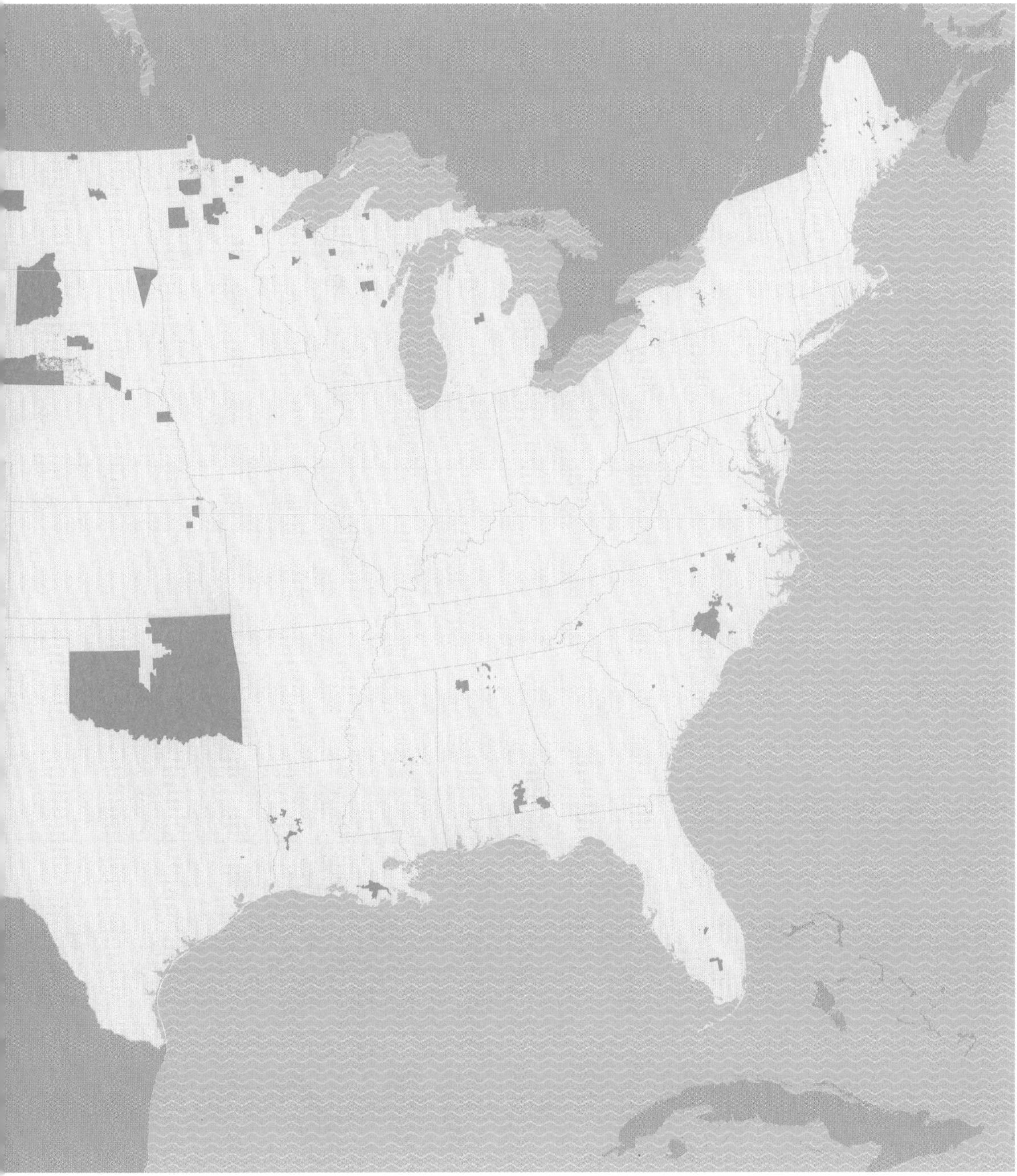

6

LIFESTYLE AND HEALTH

76 Wafflography:
Number of Waffle Houses by latitude

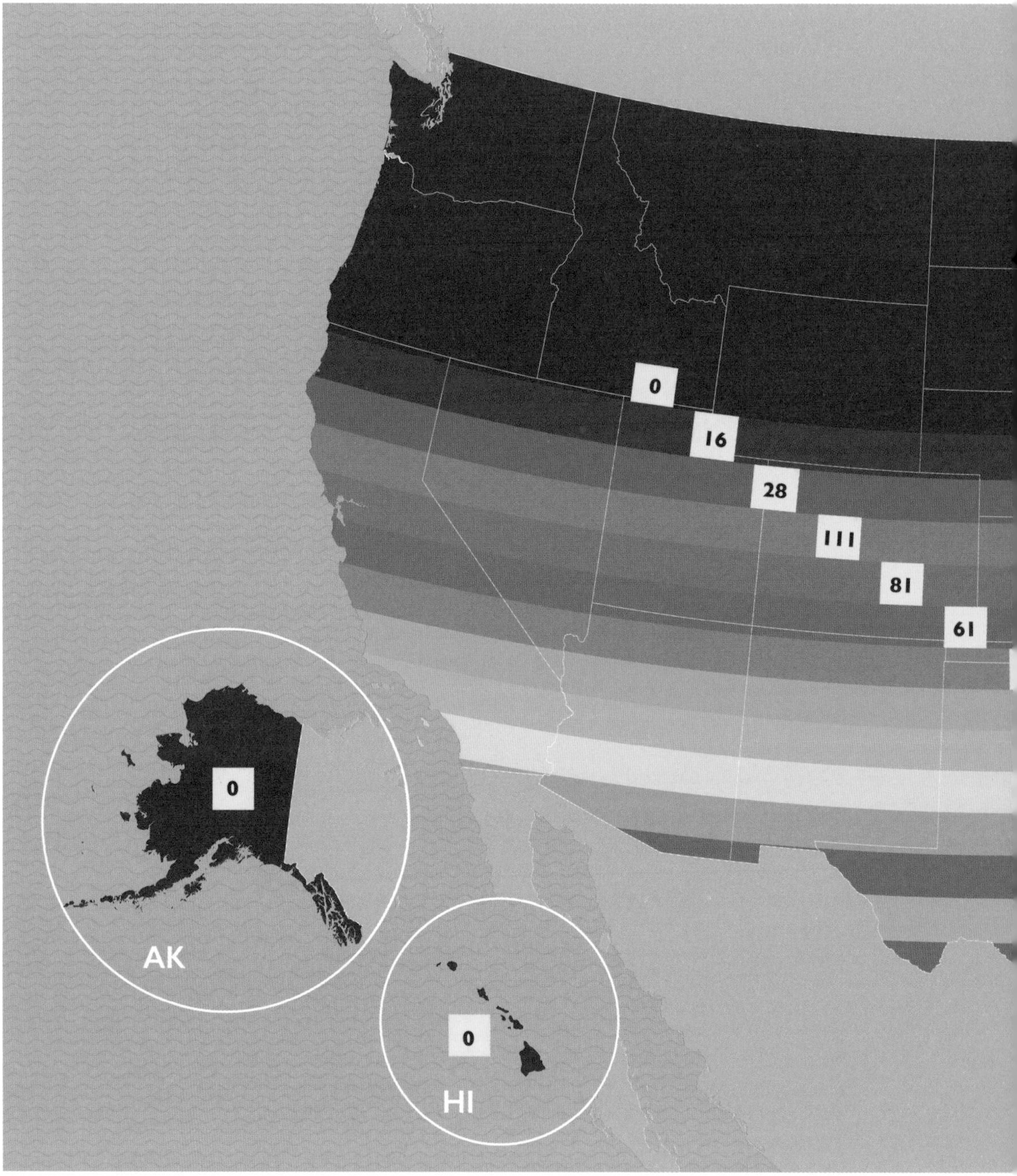

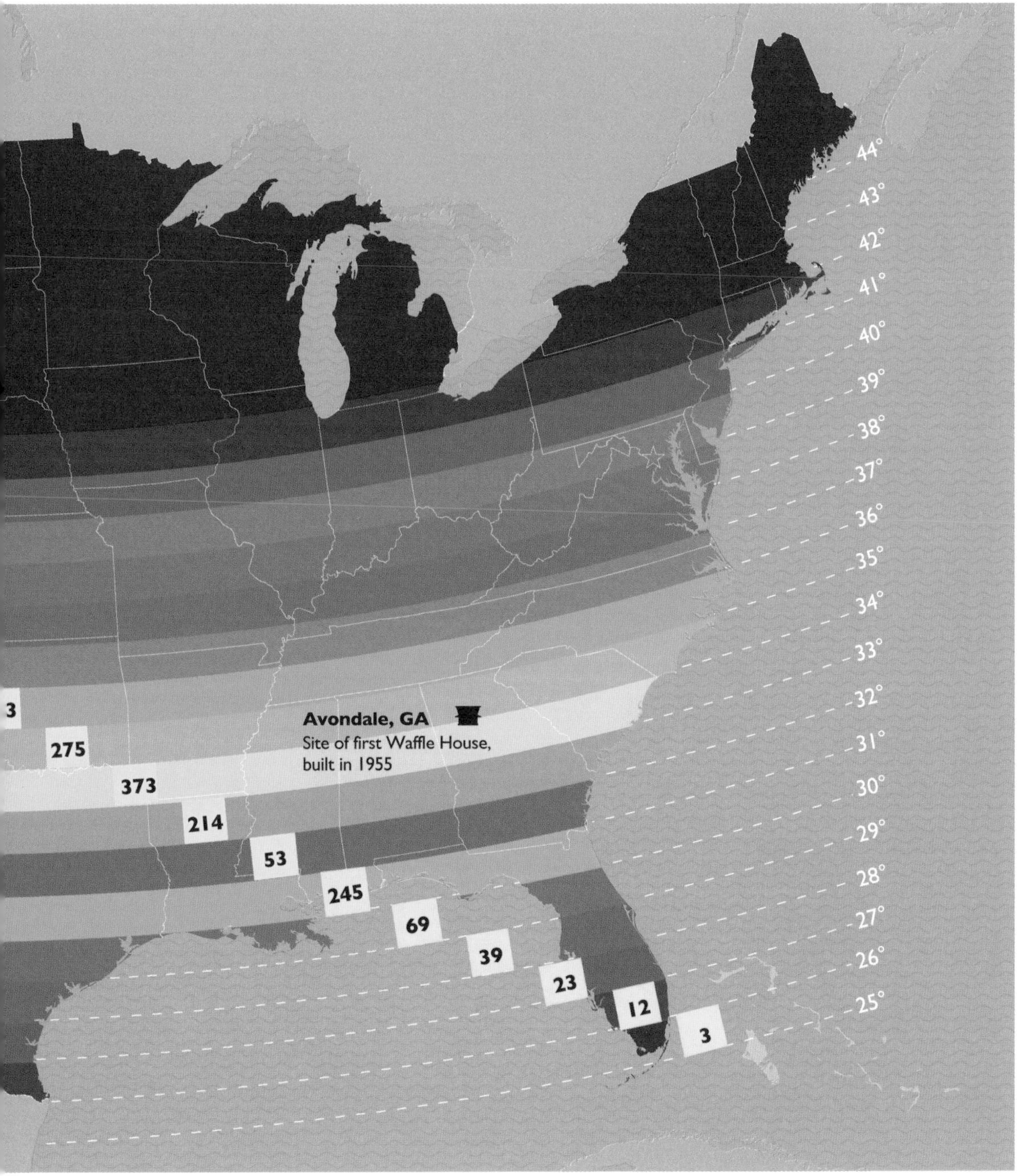
44°
43°
42°
41°
40°
39°
38°
37°
36°
35°
34°
33°
32°
31°
30°
29°
28°
27°
26°
25°
3
275
373
214
53
245
69
39
23
12
3
Avondale, GA
Site of first Waffle House, built in 1955

77 **Fast food diet:** Which states are a drive-through lover's paradise?

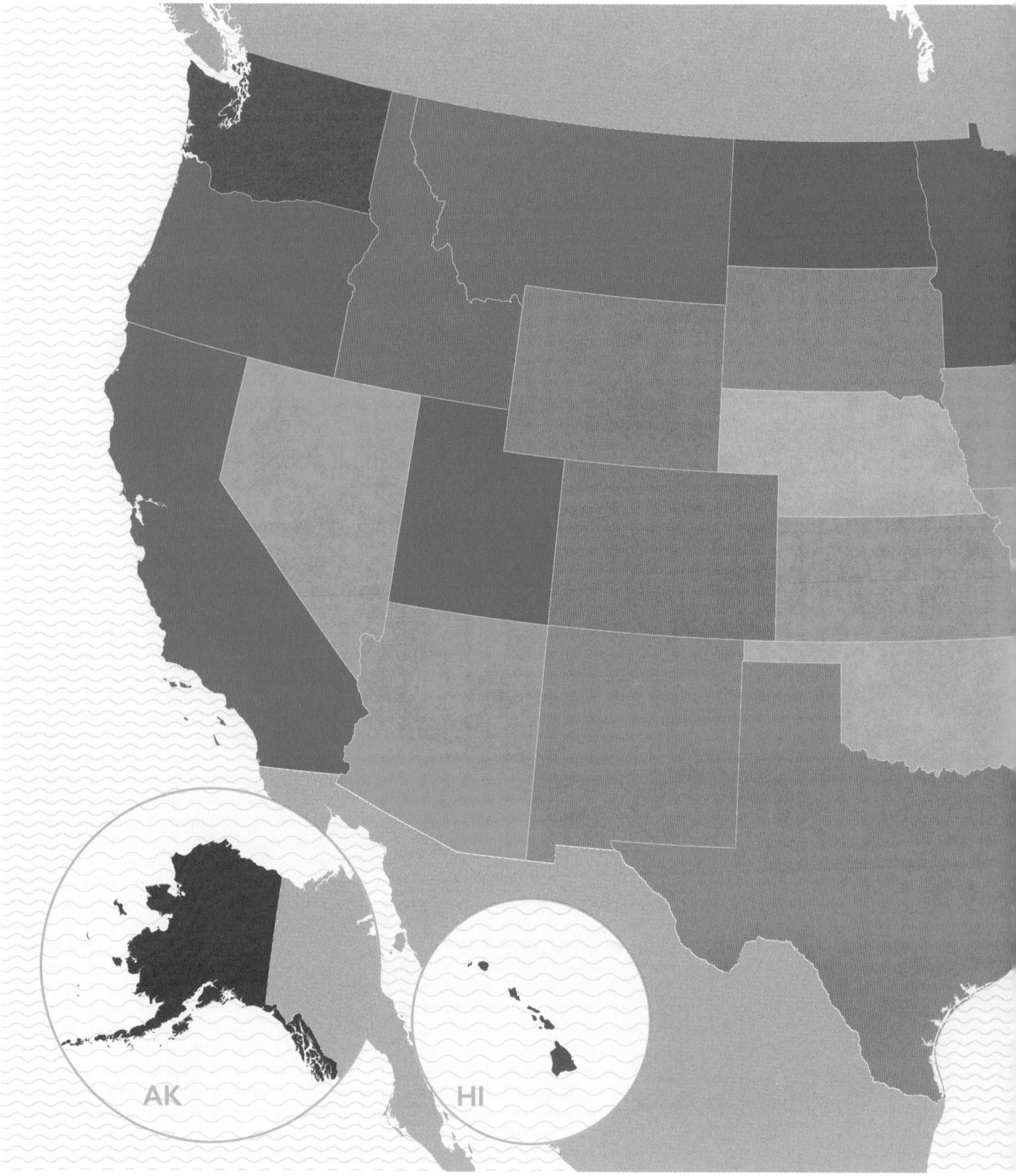

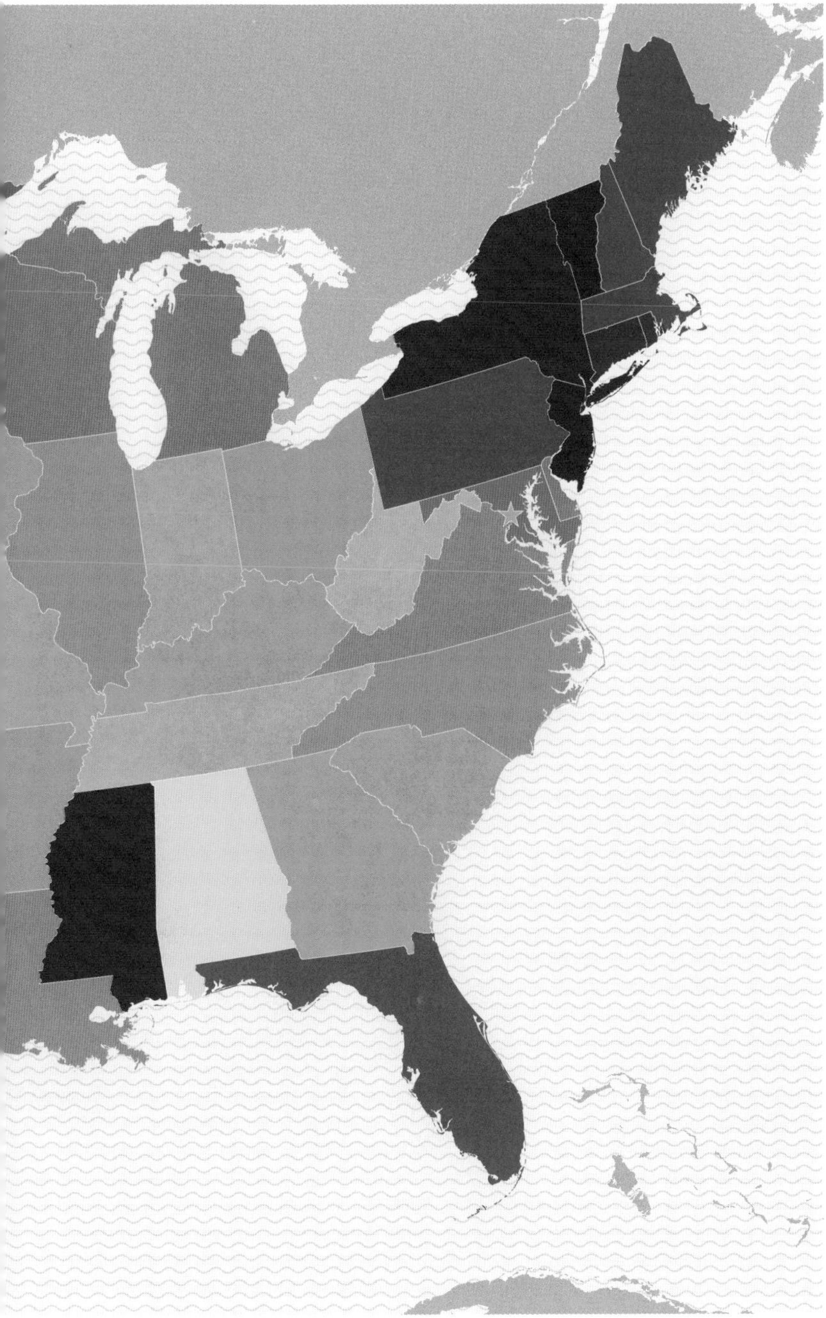

Fast food restaurants
per 10,000 people

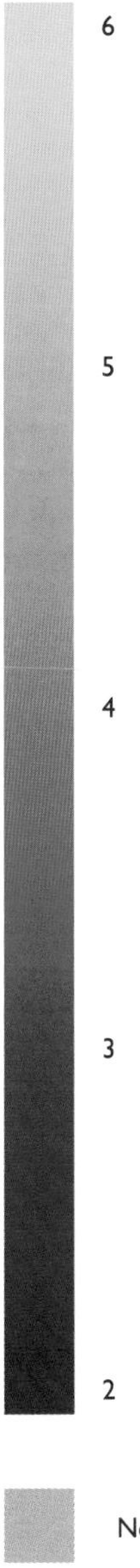

78 **Getting the runaround:** How fast recreational runners finish a marathon by state

Massachusetts, home to the famous Boston Marathon, is perhaps unsurprisingly home to the US's fastest marathon runners, who cross the finish line at an average time of 4:04:20—a full 2 hours faster than marathoners from Hawaii, the nation's slowest state. Each state's finish time reflects the times run by their residents regardless of where they run their races.

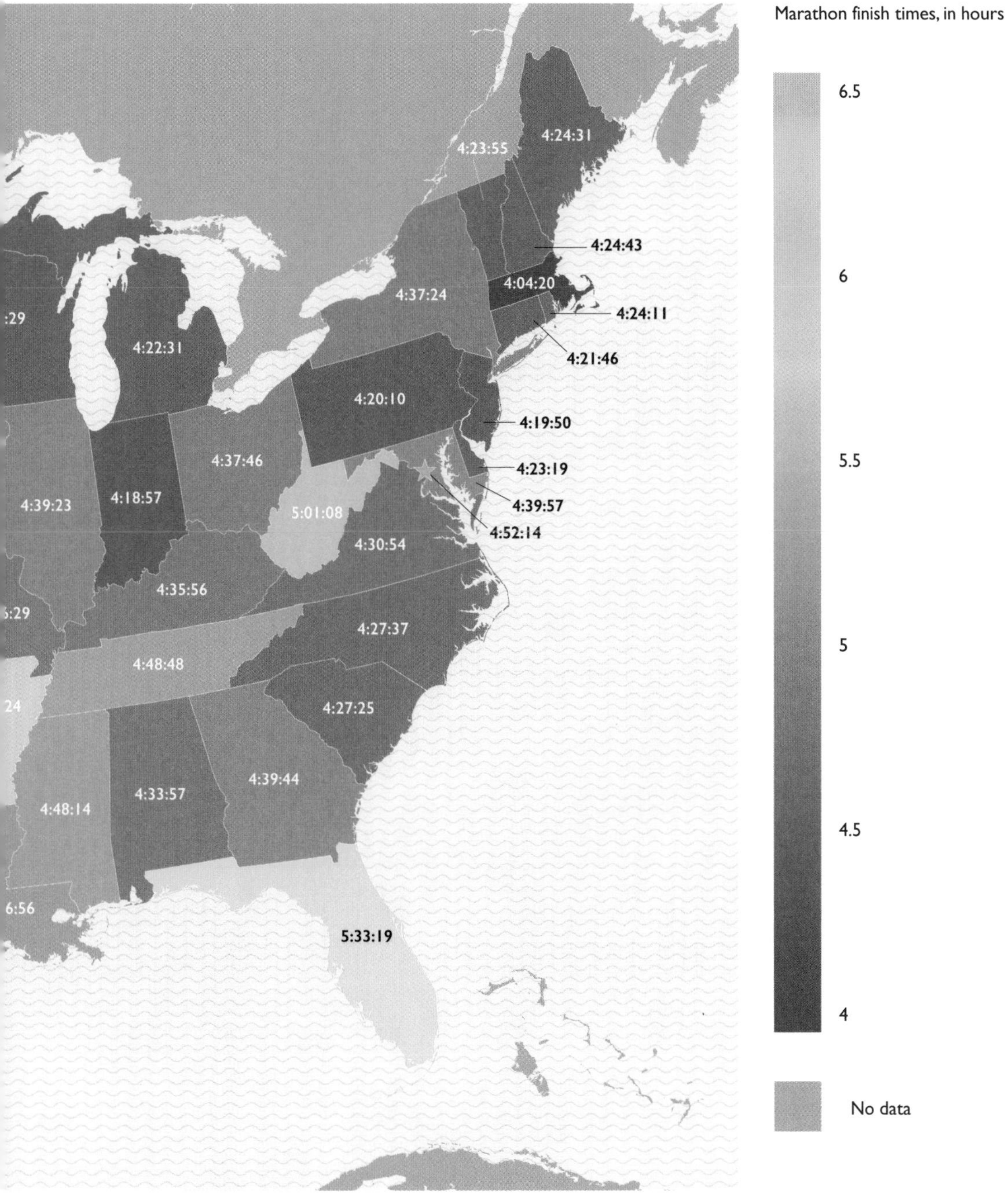

79 **One for the ages:** Median age by state

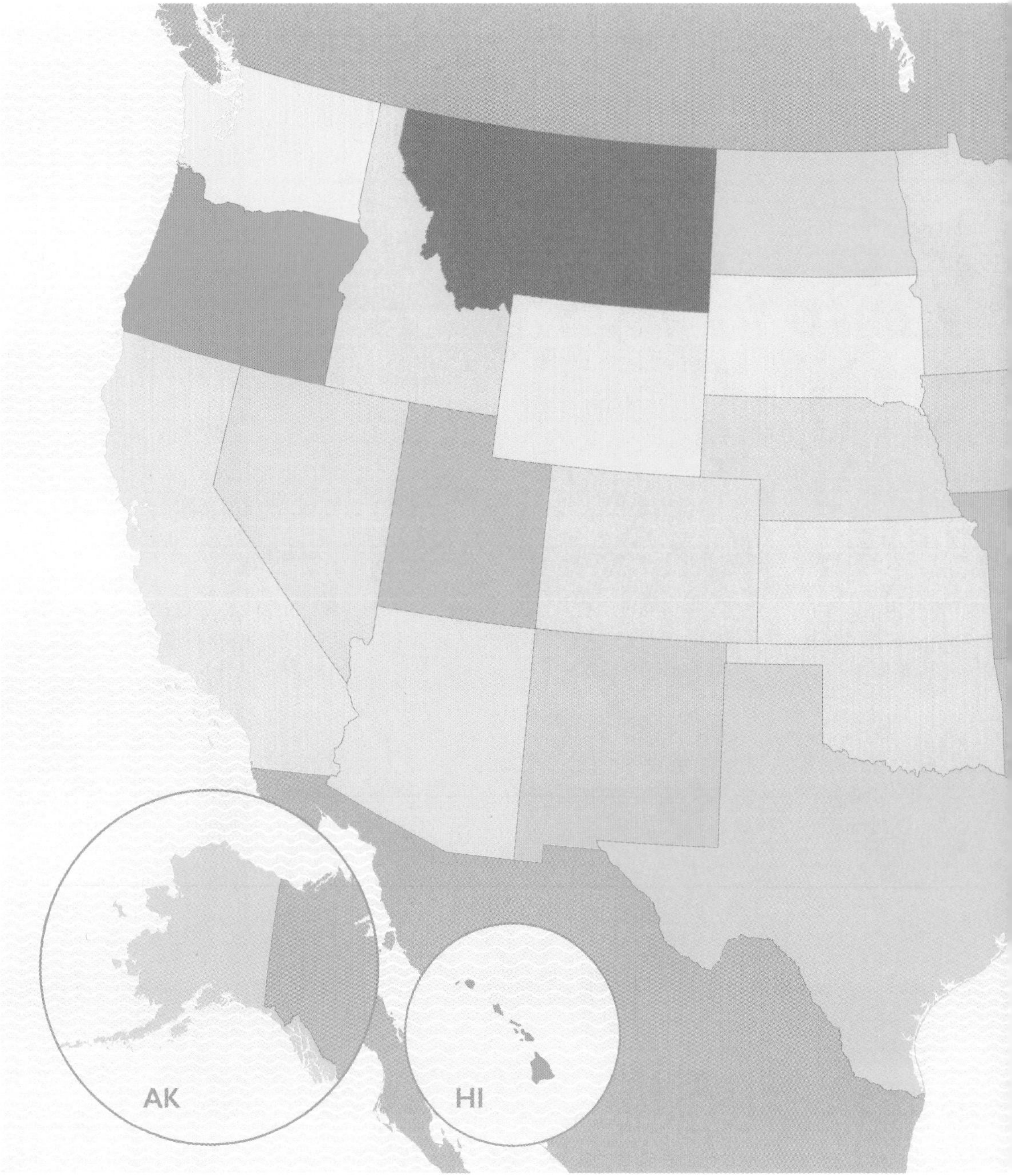

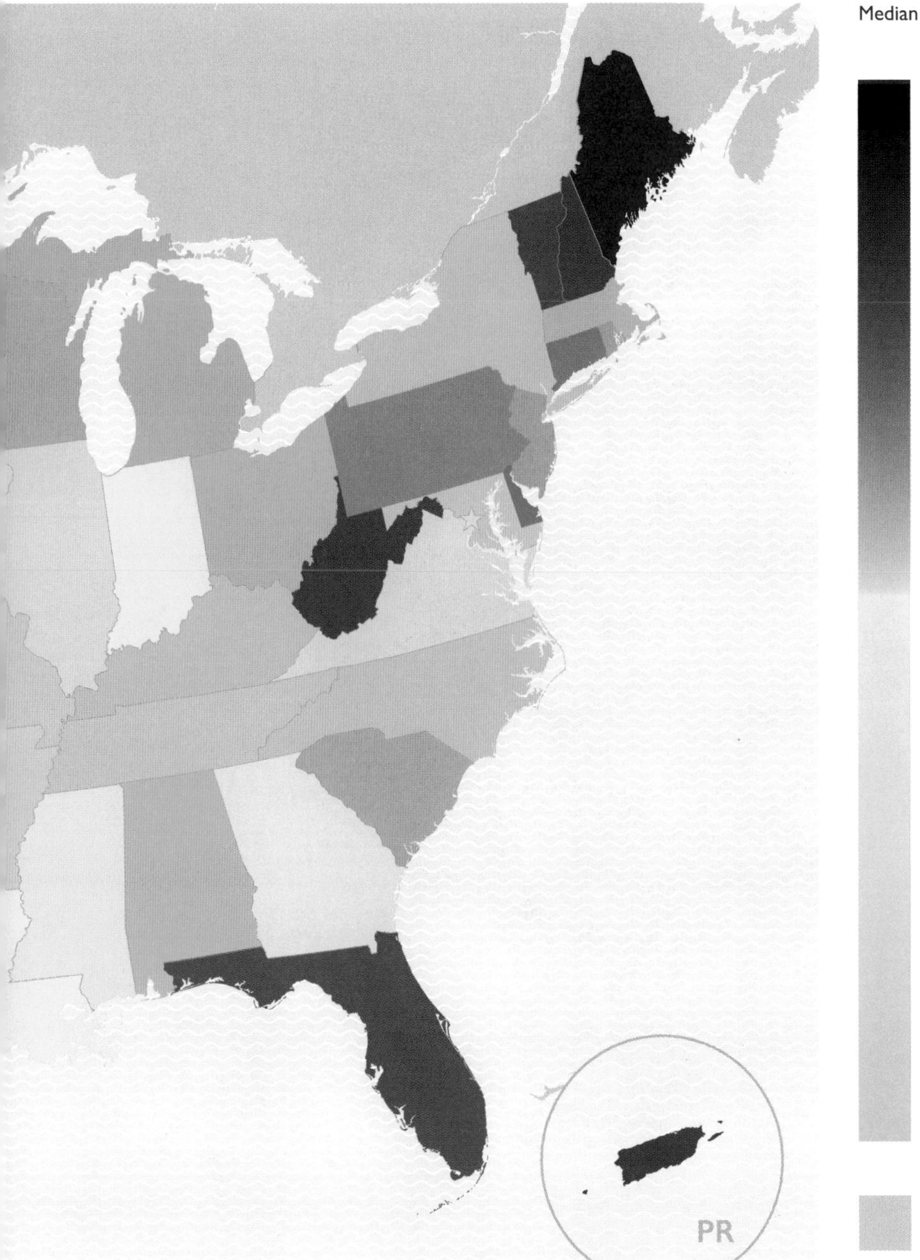
PR
Median age, in years
45
40
35
30
No data

80 Looking up to her:
Female height* in North America

4 ft, 11 in	5 ft, 1 in	5 ft, 3 in	5 ft, 5 in	
150 cm	155 cm	160 cm	165 cm	No data

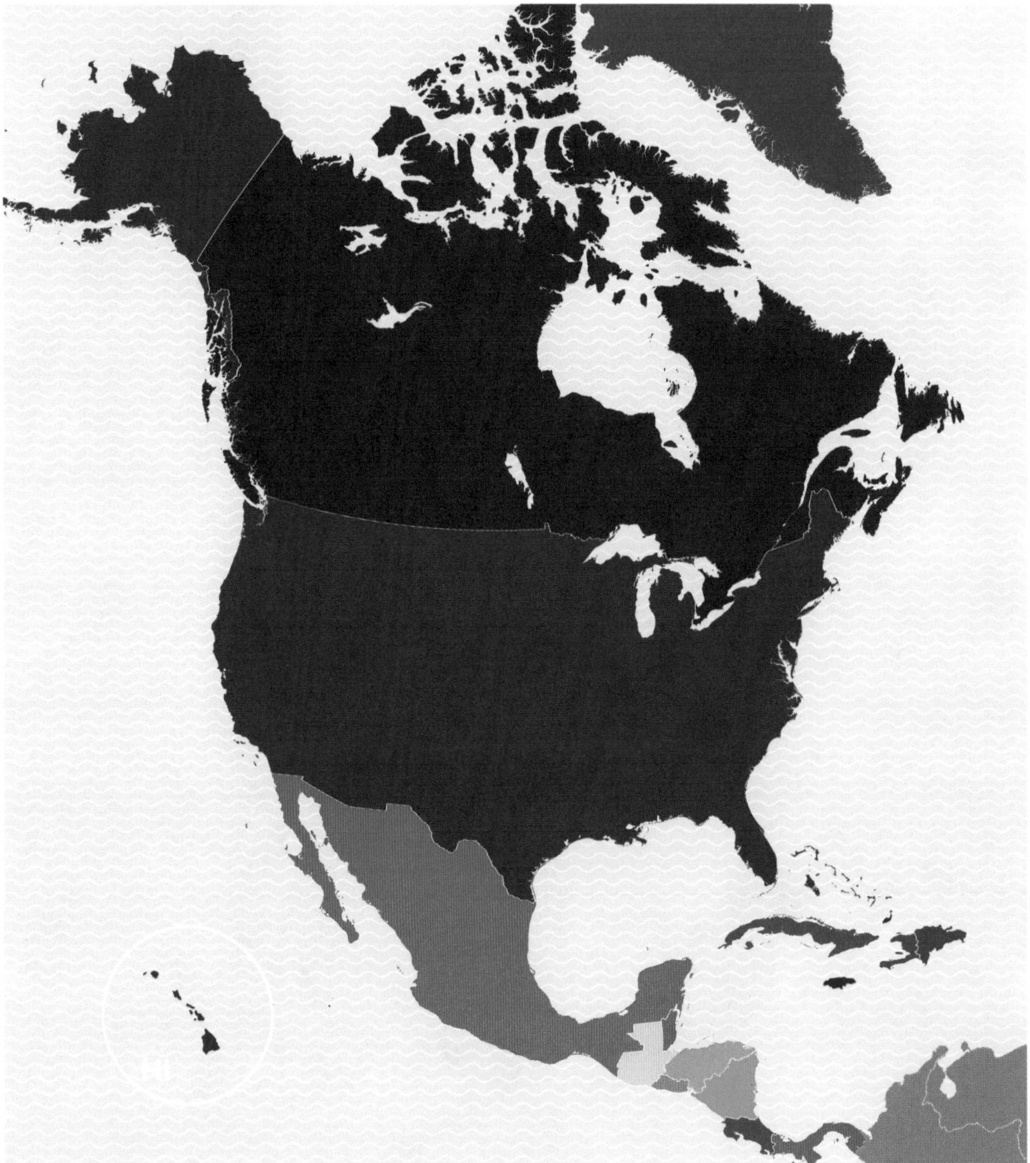

* Mean heights

81 He's the tops:
Male height* in North America

5 ft, 5 in | 5 ft, 7 in | 5 ft, 9 in | 5 ft, 11 in | No data

165 cm | 170 cm | 175 cm | 180 cm

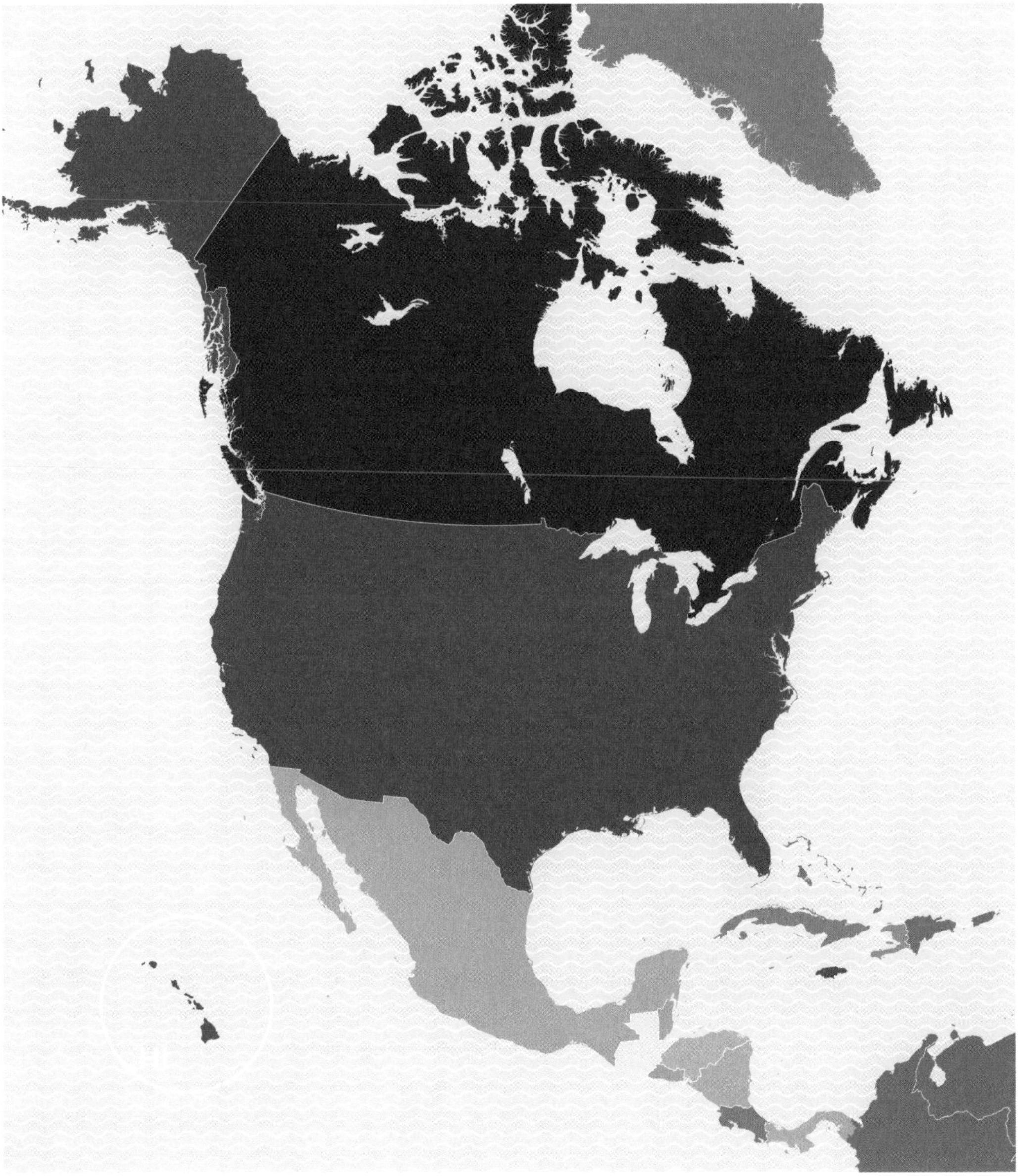

* Mean heights

82 Men are from Alaska, women are from Puerto Rico: Gender ratio by state

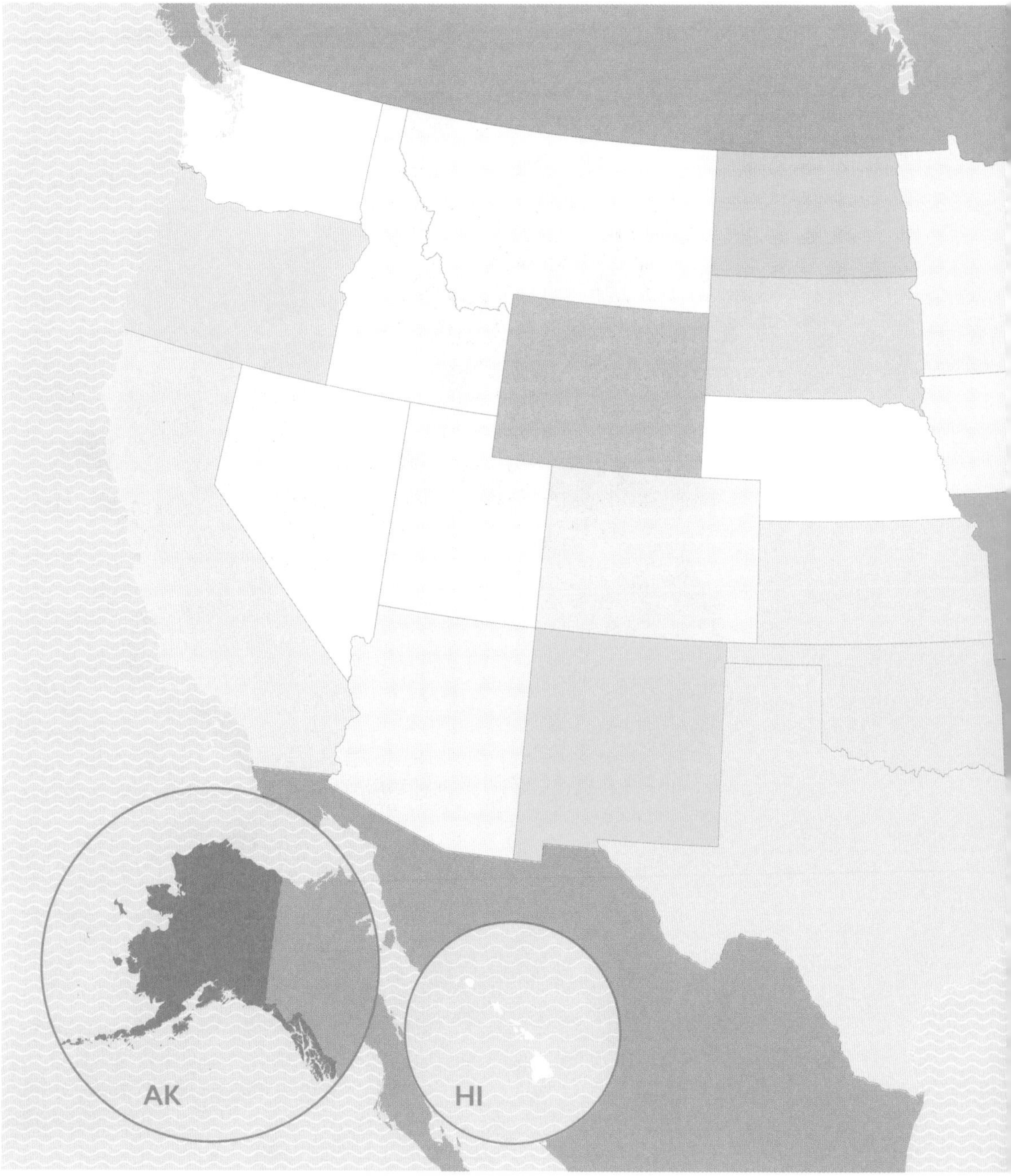

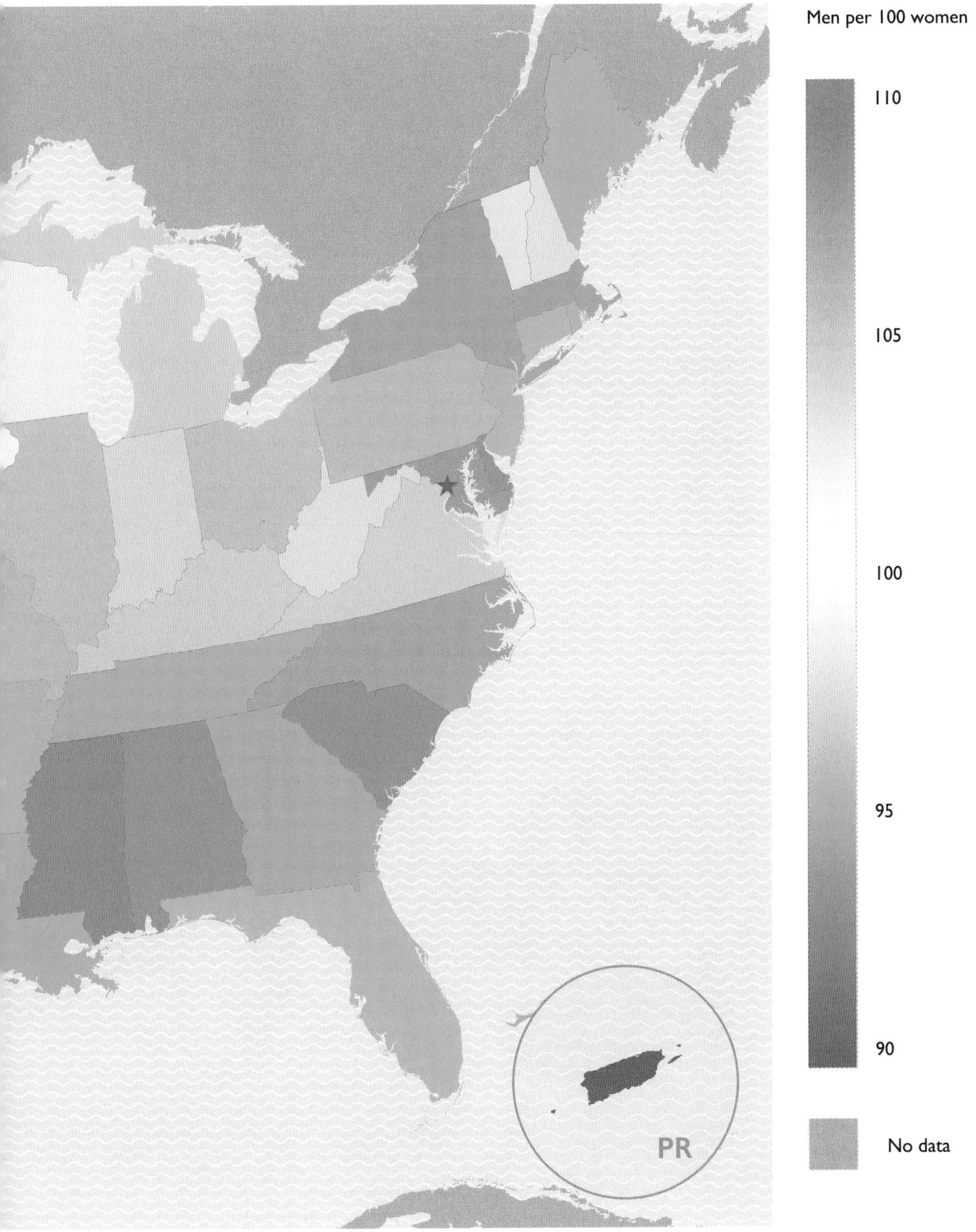
Men per 100 women
110
105
100
95
90
No data
PR

83 Running from the chapel:
Unmarried population by state

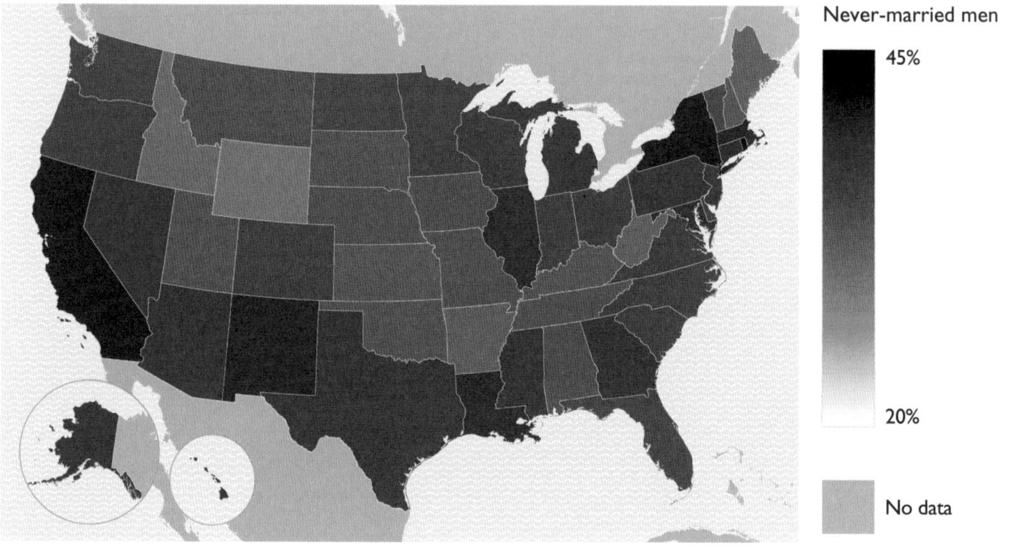

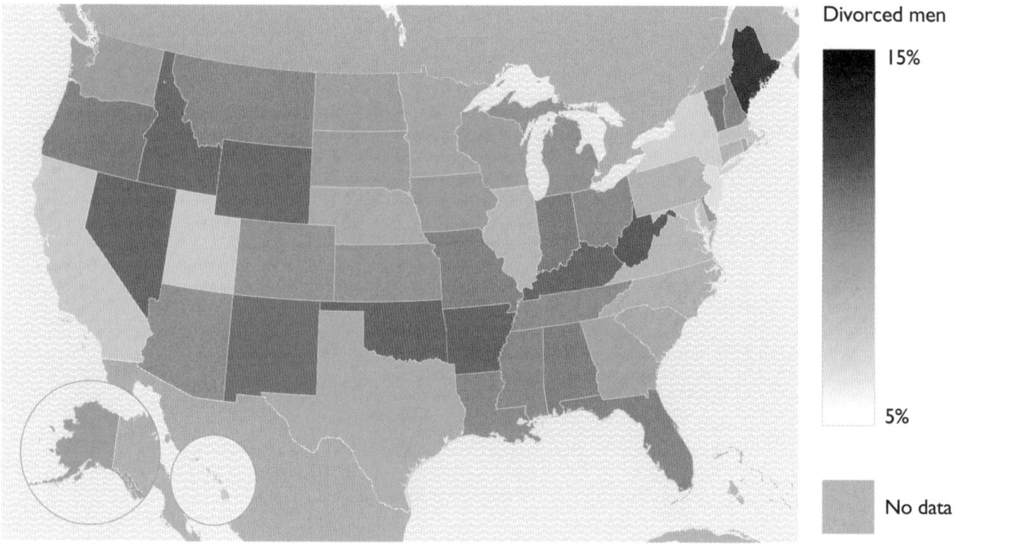

The percentages of men and women here are subsets of the entire population ages 15 and up. Everyone in that age group *not* depicted here is married, separated but not divorced, or widowed.

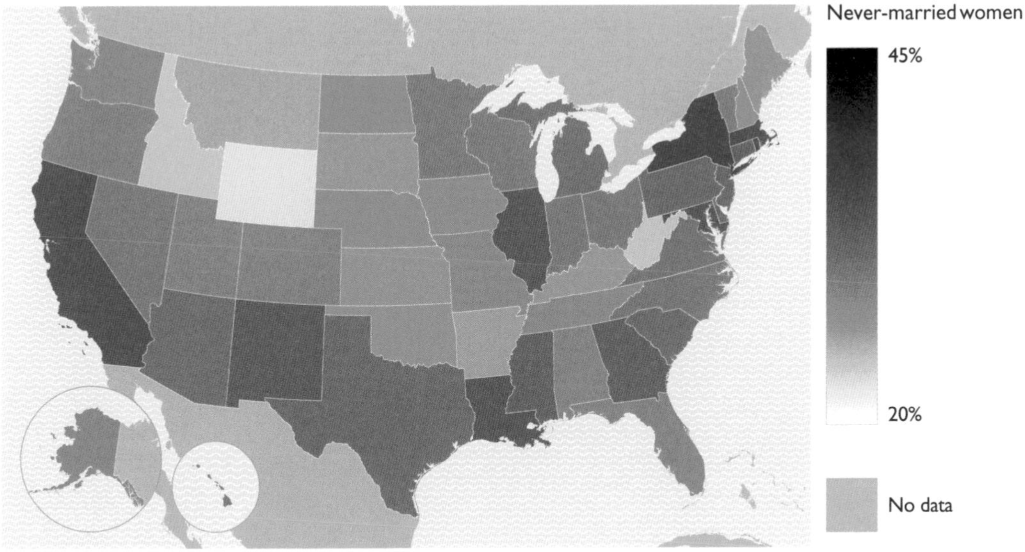

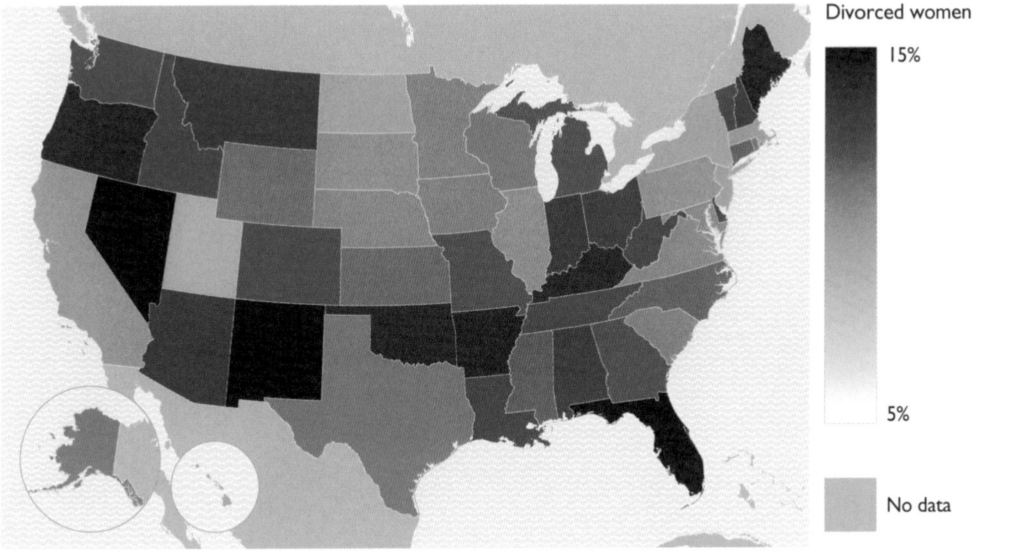

84 **Finding my happy place:** Quality of life, happiness, and well-being in North America

The happiness and well-being indexes we draw from here synthesize a wide range of factors—from social support to physical and mental health, from food access to financial security.

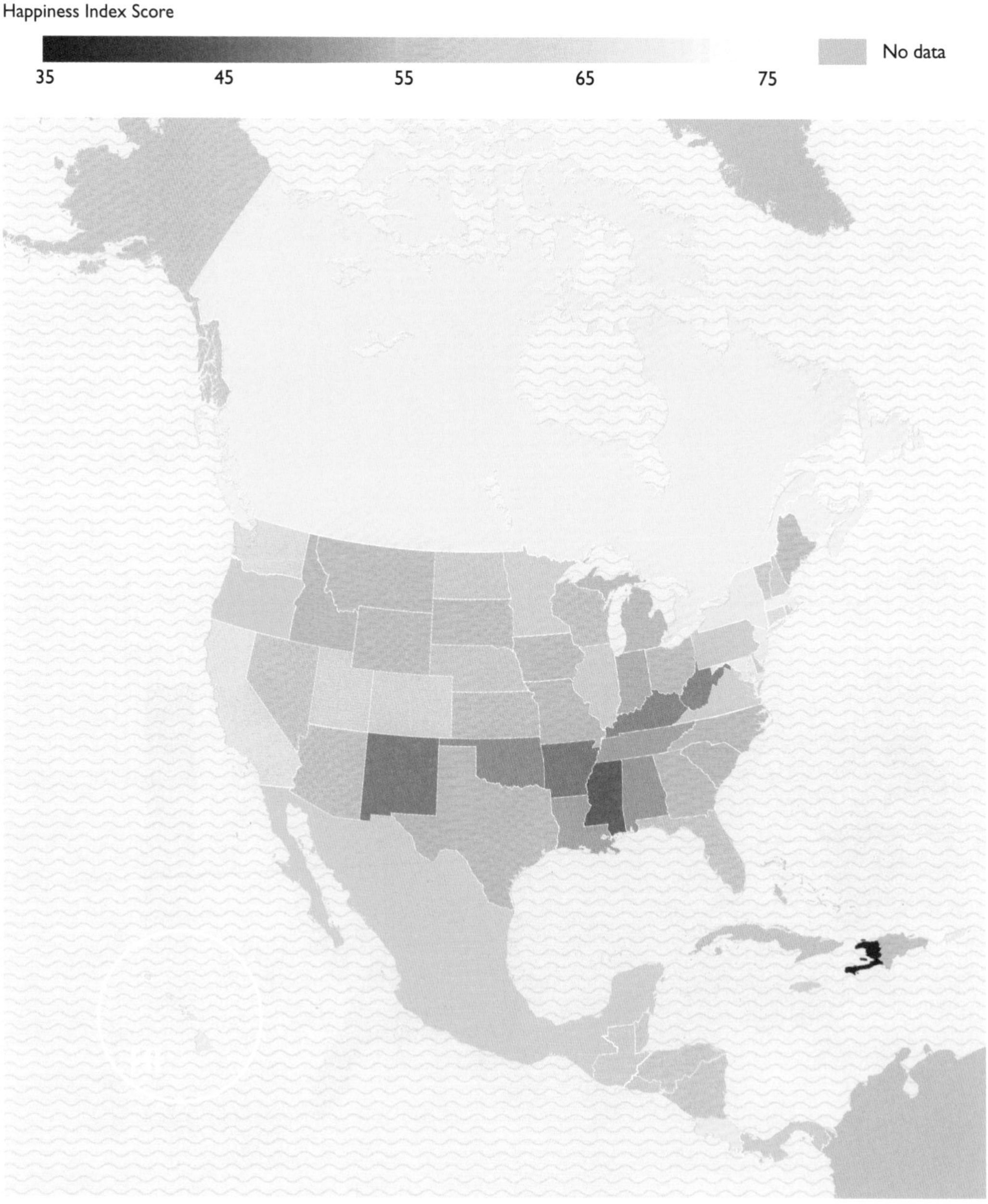

85 **I'm coming out:** LGBTQ acceptance in North America

The score for each state, province, or country's relative level of social acceptance of LGBTQ people reflects how inclusive its laws and services are. And while we know that placing this map side-by-side with the happiness map might imply causation, we want to remind readers that correlation does not, in fact, imply causation. But we can't stop readers from noticing the correlation.

Acceptance Index Score

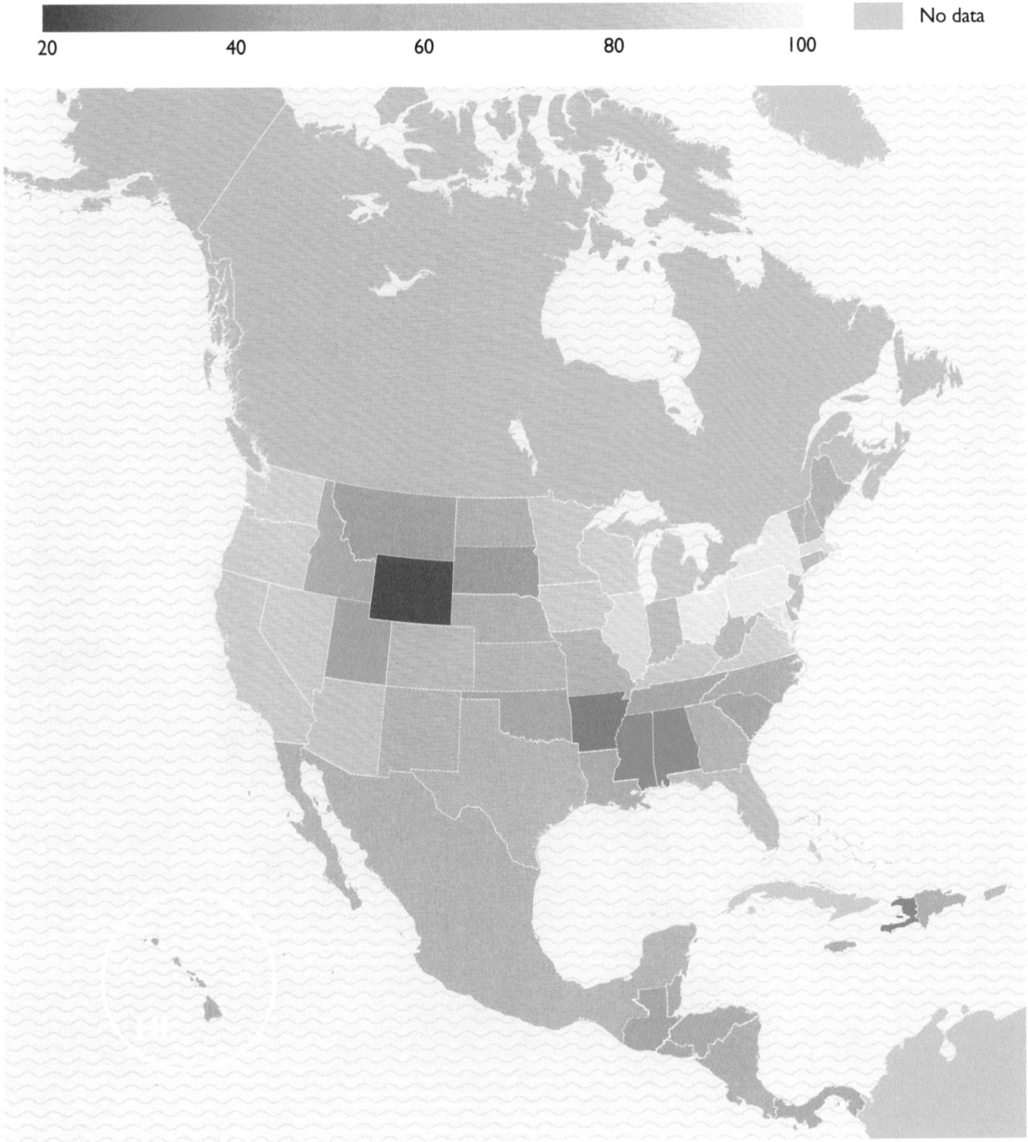

86 American tragedies:
Accident mortality by state

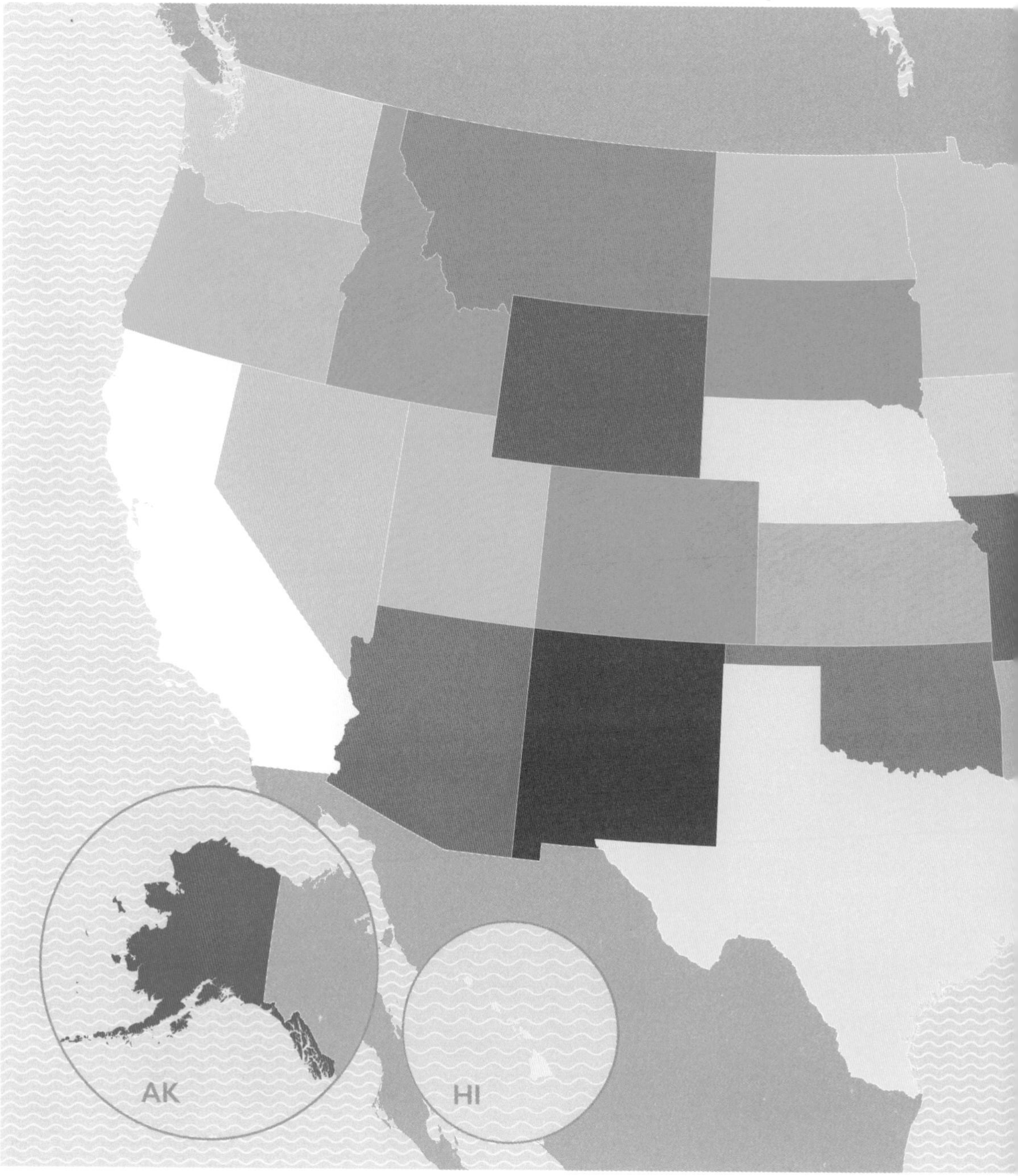

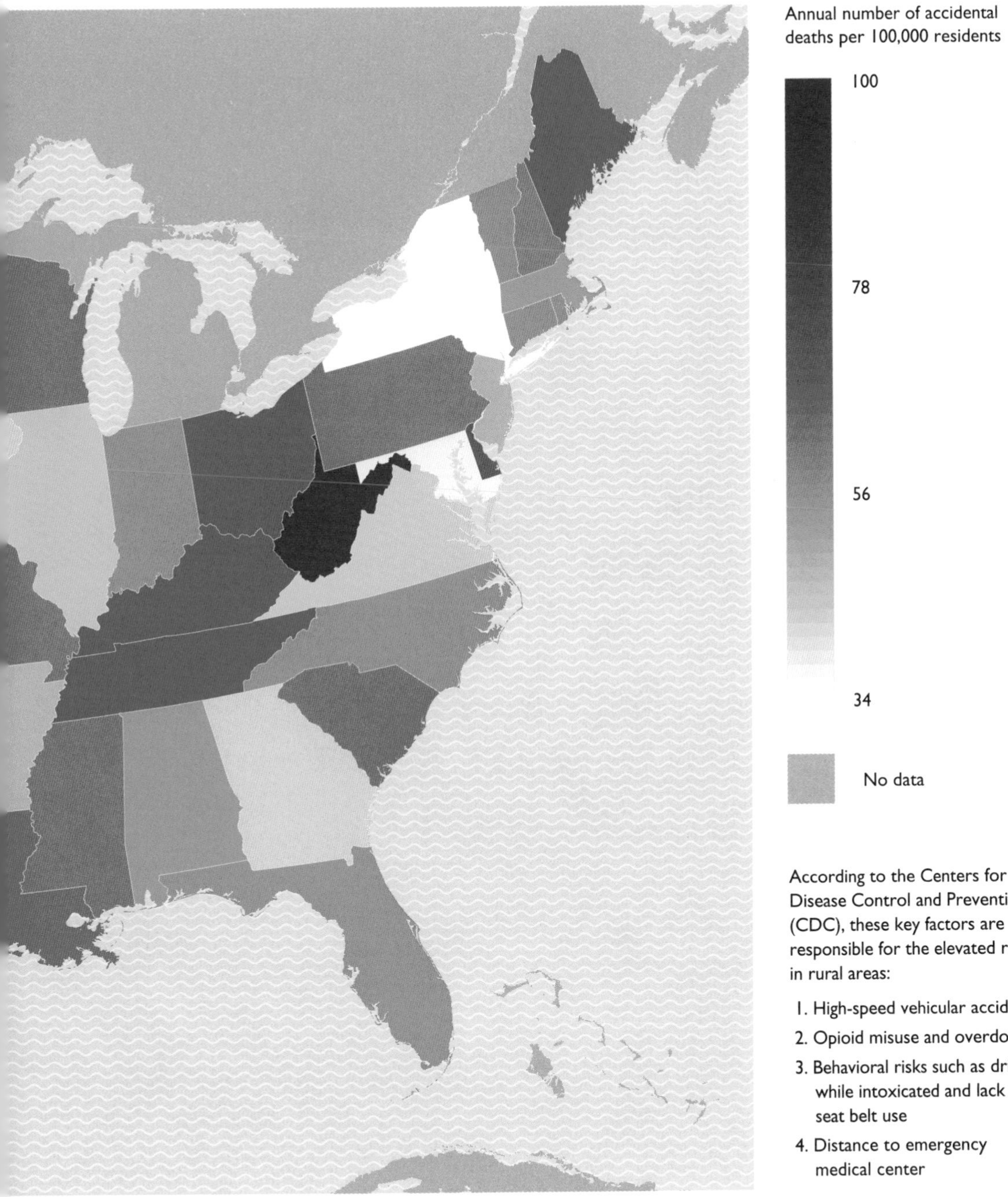

According to the Centers for Disease Control and Prevention (CDC), these key factors are responsible for the elevated risk in rural areas:

1. High-speed vehicular accidents
2. Opioid misuse and overdose
3. Behavioral risks such as driving while intoxicated and lack of seat belt use
4. Distance to emergency medical center

87 America, the next generation:
Birth rates by state

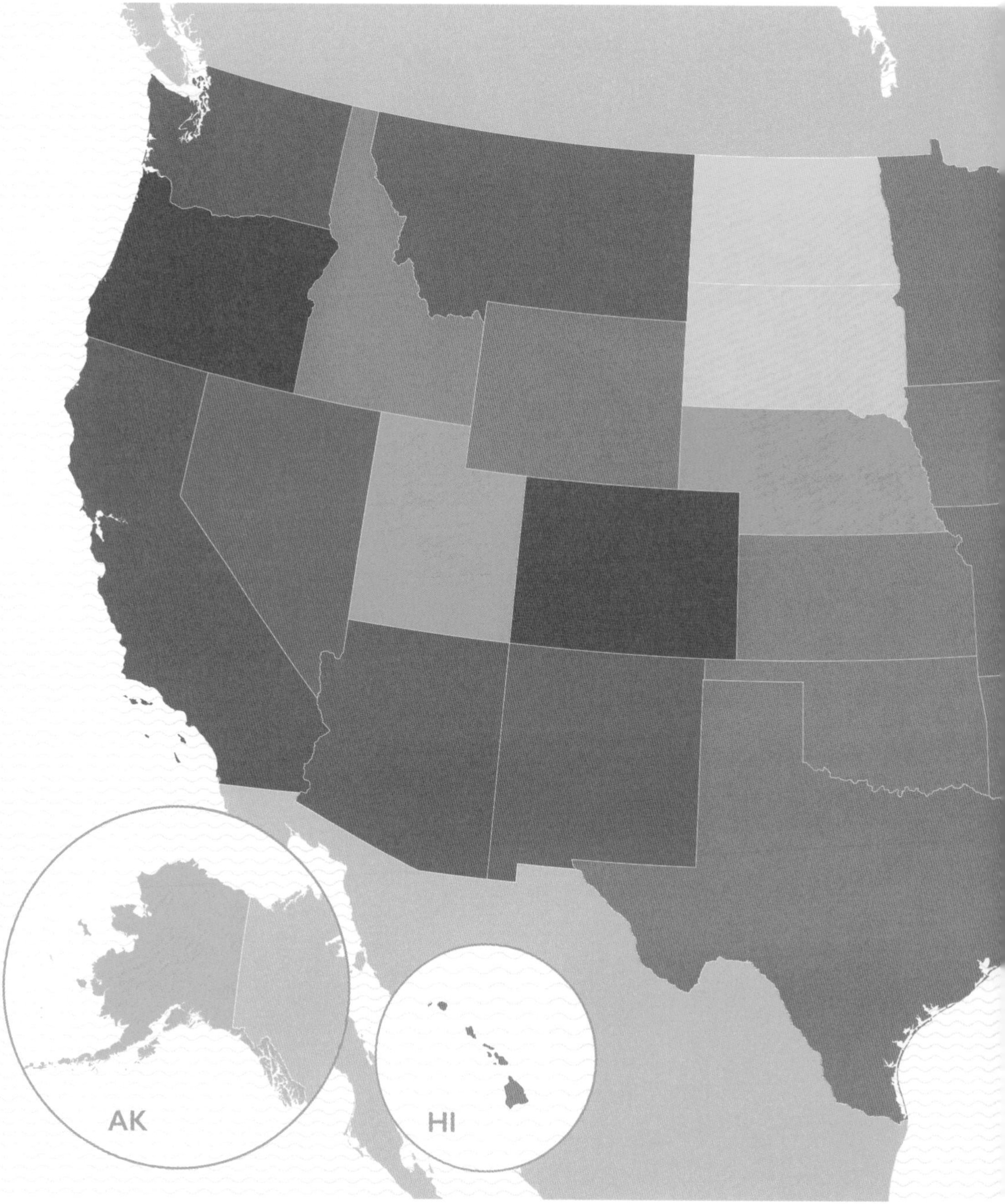

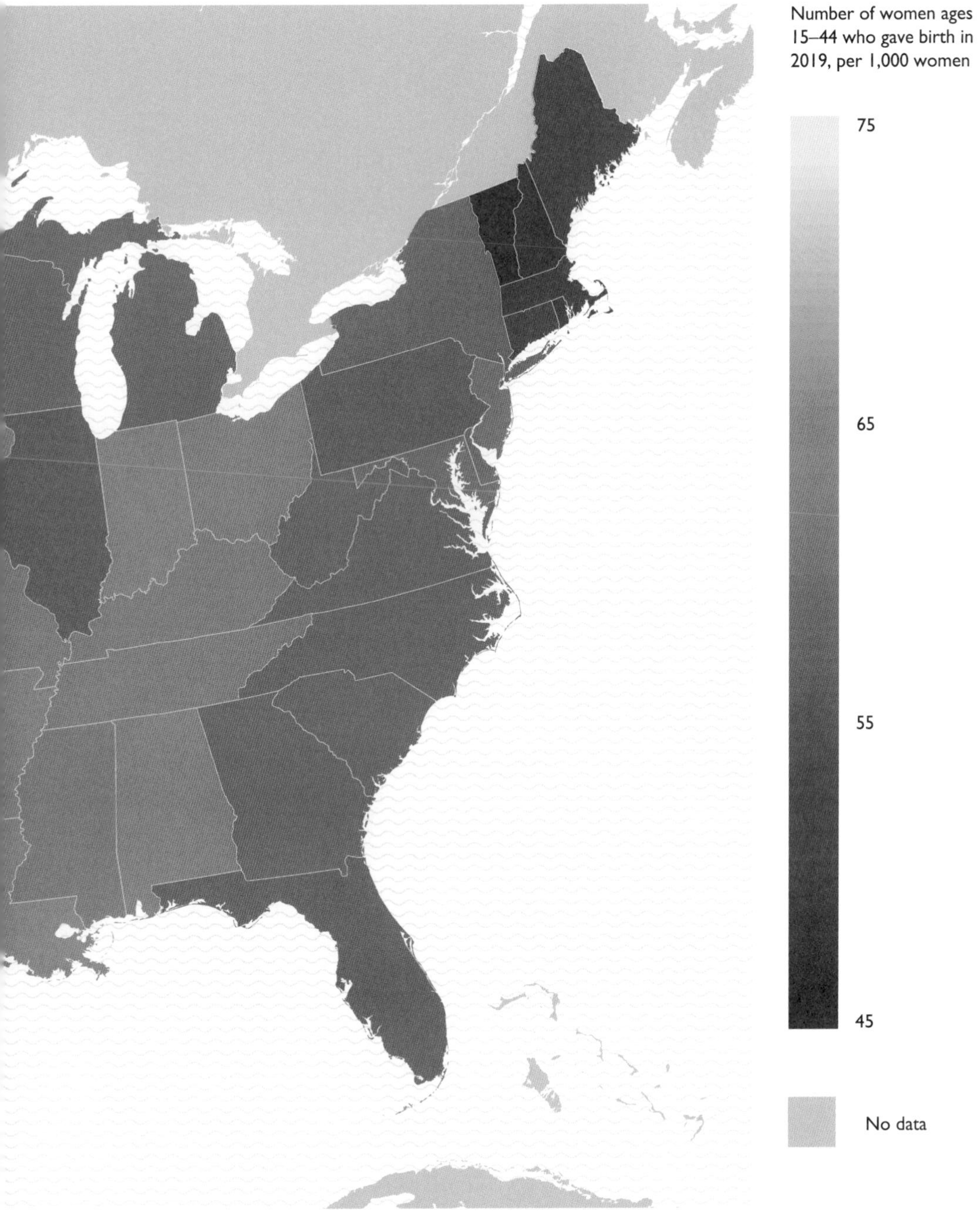
Number of women ages 15–44 who gave birth in 2019, per 1,000 women
75
65
55
45
No data

88 Down and out:
COVID-19 anxiety and depression

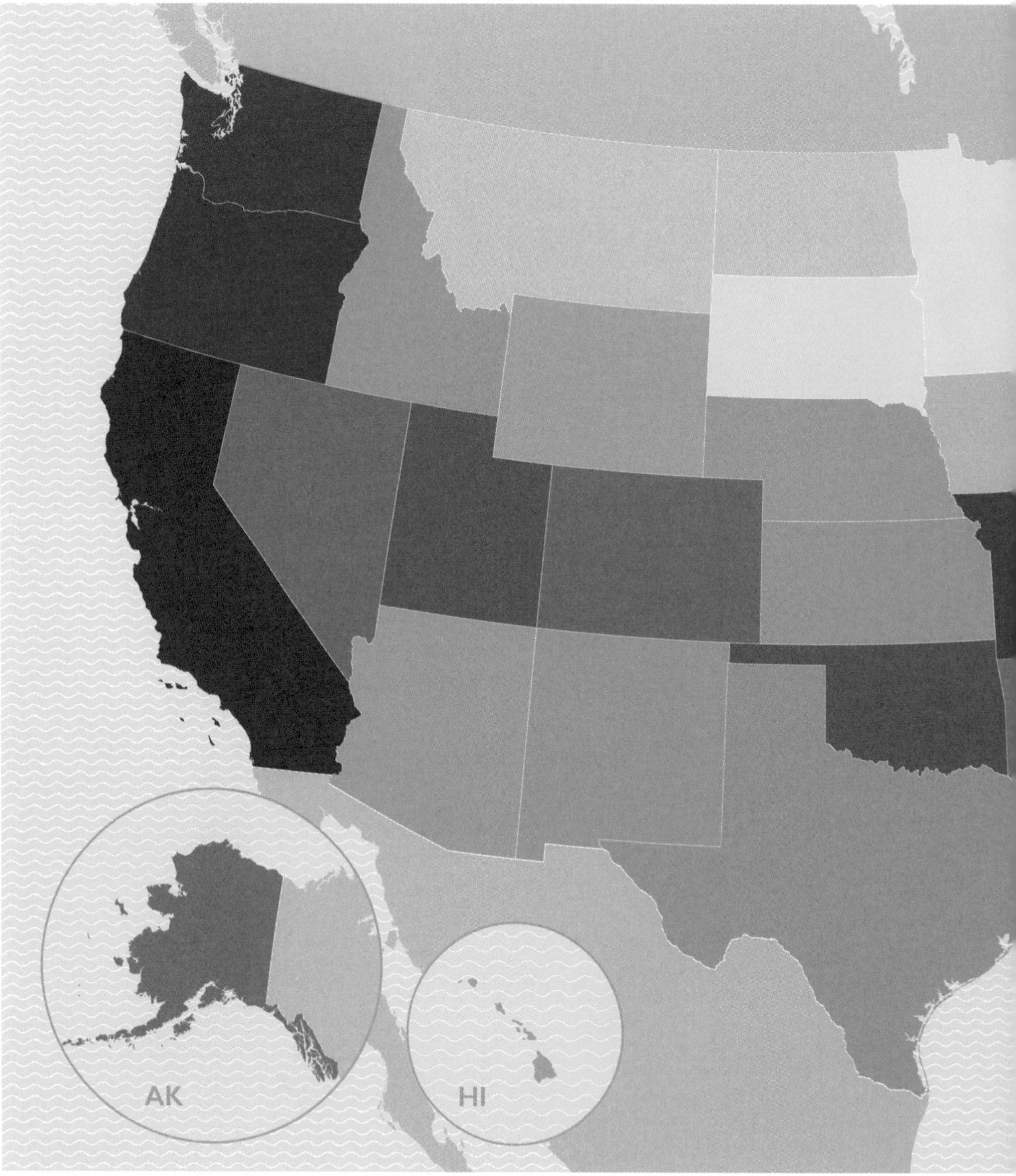

The National Center for Health Statistics (NCHS) and the Centers for Disease Control and Prevention (CDC) conducted a survey tracking the impact of the pandemic on anxiety and depression at regular intervals from April 2020 to July 2021, asking how people felt over the last seven days. We chose to map their data collected the weeks spanning January 6 through January 18, 2021, because during that time the US peaked in daily confirmed new cases of COVID-19 (on January 8).

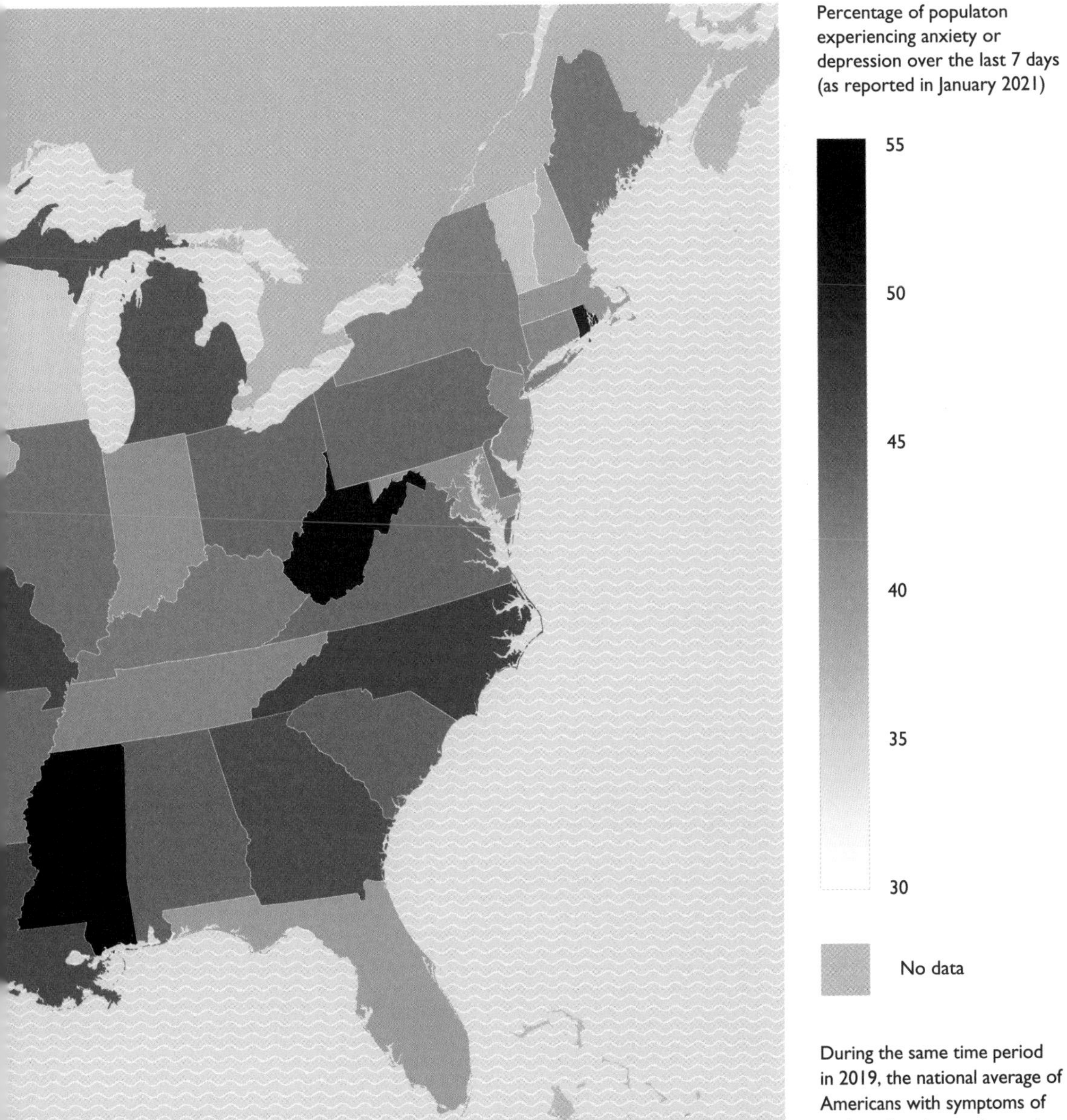

During the same time period in 2019, the national average of Americans with symptoms of anxiety disorder or depression stood at 11 percent. In January 2021, the national average was 41.1 percent.

7

INDUSTRY AND TRANSPORT

89 Expansive spans:
The 10 longest bridges in North America

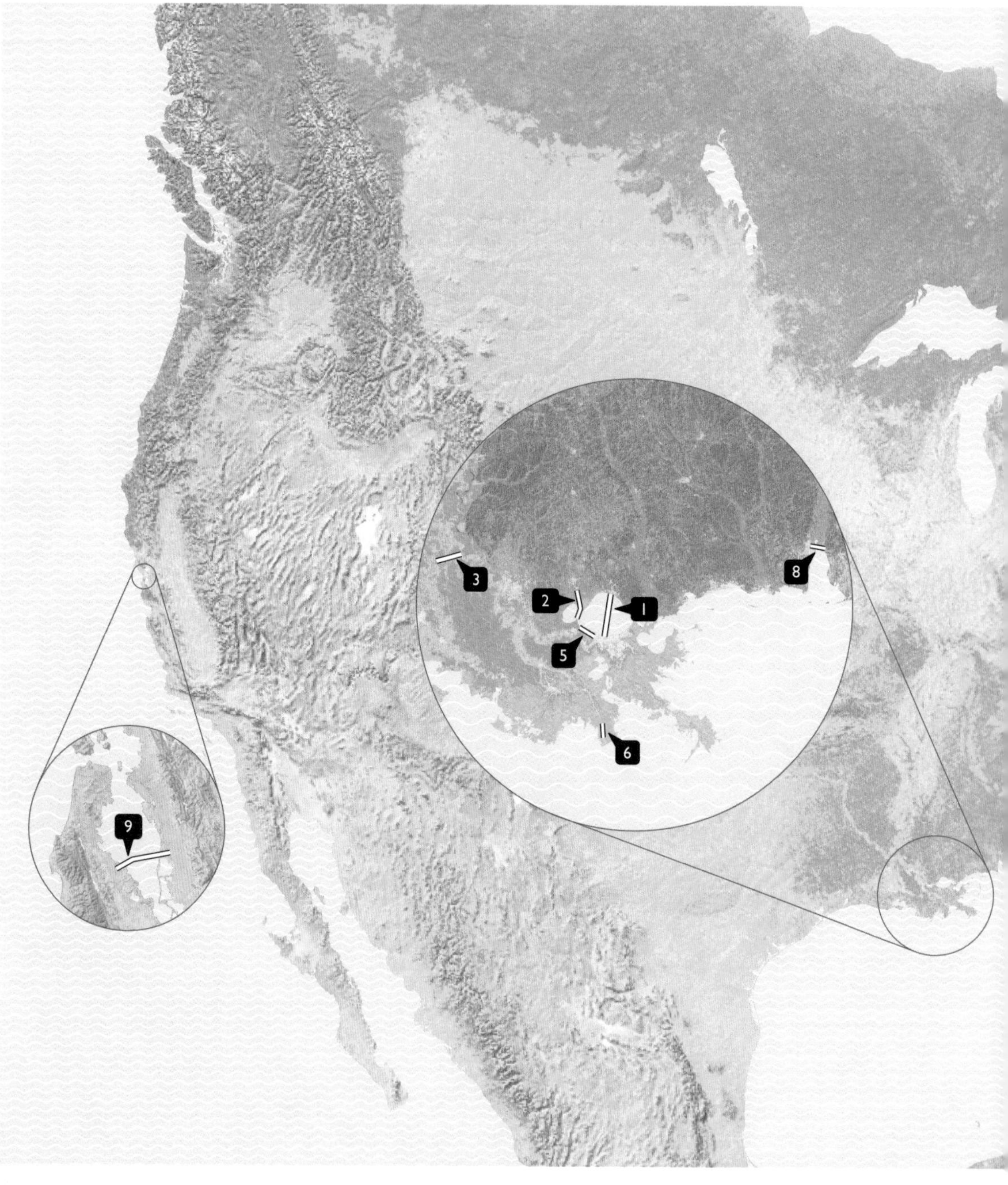

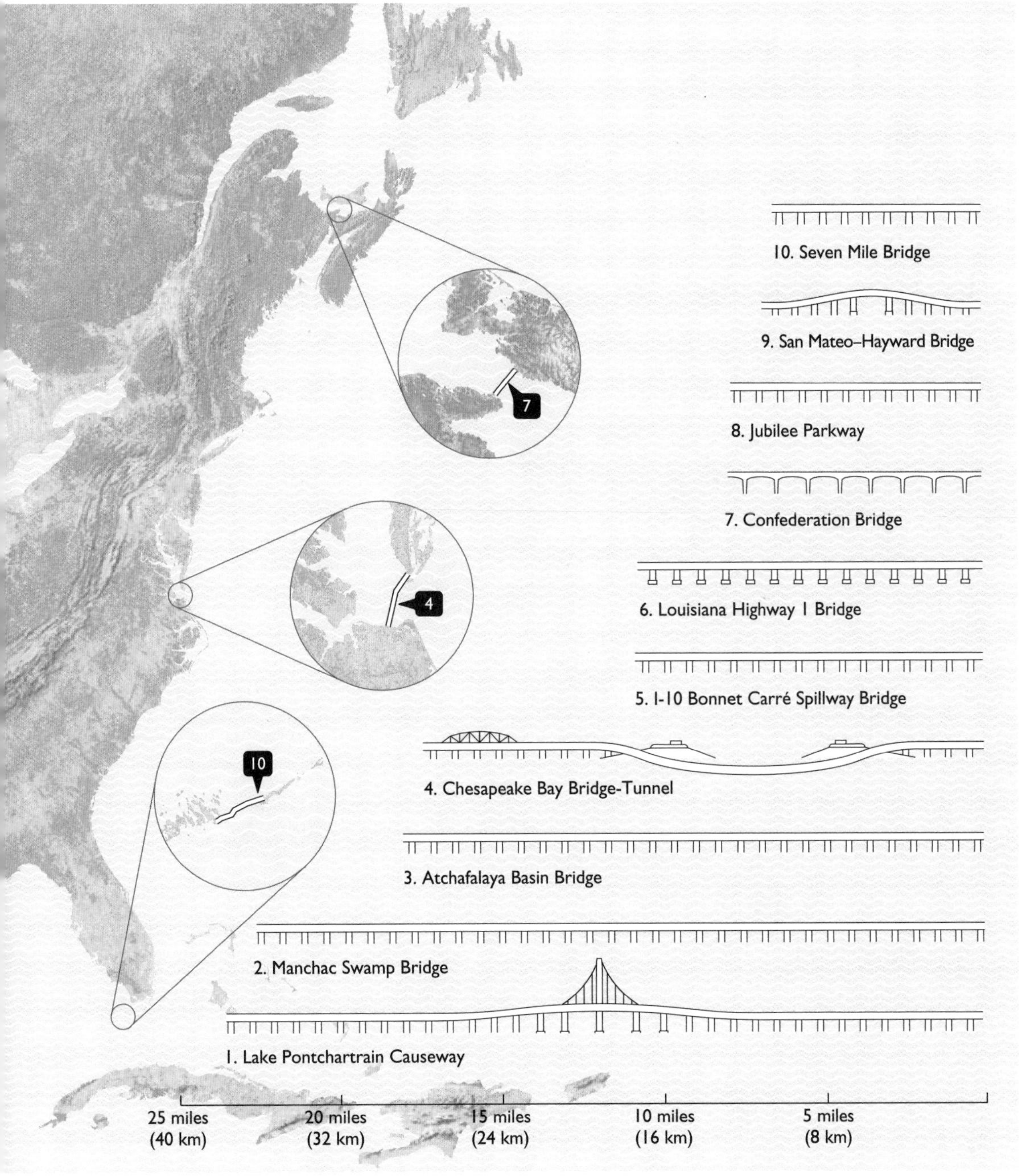
7
4
10
10. Seven Mile Bridge
9. San Mateo–Hayward Bridge
8. Jubilee Parkway
7. Confederation Bridge
6. Louisiana Highway 1 Bridge
5. I-10 Bonnet Carré Spillway Bridge
4. Chesapeake Bay Bridge-Tunnel
3. Atchafalaya Basin Bridge
2. Manchac Swamp Bridge
1. Lake Pontchartrain Causeway
25 miles (40 km)
20 miles (32 km)
15 miles (24 km)
10 miles (16 km)
5 miles (8 km)

90 **Lost at sea:**
Famous shipwrecks in North American waters

1. *Baychimo* | 1931
Beaufort Sea/Chukchi Sea/north Alaskan coast
Known as "the Ghost Ship of the Arctic"; caught in ice, abandoned by its crew; has been spotted drifting or frozen in ice; last alleged sighting was 1969

2. HMS *Terror* | 1848
King William Island's Terror Bay

3. HMS *Erebus* | 1848
Far eastern Queen Maud Gulf

Vessels from explorer Sir John Franklin's ill-fated attempt to complete the Northwest Passage; froze in ice at the top of Victoria Strait

4. *Exxon Valdez* | 1989
Prince William Sound, AK
Oil tanker that struck Bligh Reef, causing the worst oil spill in US history until Deepwater Horizon

5. *Eastland* | 1915
Chicago River, Chicago, IL
Deadliest Great Lakes–region maritime disaster; this passenger steamer capsized sitting in a wharf, killing 844 people

6. *Edmund Fitzgerald* | 1975
Off Whitefish Point, MI
Famous wreck popularized by Gordon Lightfoot's 1976 song

7. *Le Griffon* | 1679
Mississagi Lighthouse area
The "holy grail" of undiscovered Great Lakes shipwrecks, used by Robert de La Salle to search for a northwest passage to China

8. *Lady Elgin* | 1860
Near Winnetka, on Lake Michigan
More than 300 died when this steamer collided with the schooner *Augusta* in the deadliest open-water shipwreck on the Great Lakes

9. *General Slocum* | 1904
East River, NY
An unexplained fire quickly engulfed this charter boat, and the captain steered it to North Brother Island rather than directly to shore (possibly for insurance reasons); 1,021 people died.

10. HMS *Endeavour* | 1778
Outer Newport Harbor, RI
Captain Cook's ship, ostensibly to sail to Tahiti to observe the transit of Venus, but with a secret colonization mission; repurposed during the American Revolution and scuttled as part of a blockade before the Battle of Rhode Island

11. *Whydah Gally* | 1717
Off Cape Cod, MA
The first pirate shipwreck ever found, a former slave ship captured by the pirate Captain Samuel "Black Sam" Bellamy while returning from her maiden voyage on the triangular trade route

13. *Sultana* | 1865
Mississippi River, 4 mi (6 km) from Memphis
Worst maritime disaster in US history; carrying former Union prisoners of war back north, this overcrowded steamboat sank after its strained boilers exploded, taking more than 1,700 lives

14. *Clotilda* | 1860
Off Twelve Mile Island, AL
Last ship to carry captive Africans to the US, half a century after the US banned the transatlantic slave trade; charred, sunken remains were discovered in 2018

15. USS *Arizona* | 1941
Pearl Harbor, HI
American battleship bombed on December 7 by the Japanese, it exploded and sank, killing 1,177 crew members and officers; now a National Historic Landmark, the shipwreck can still be viewed

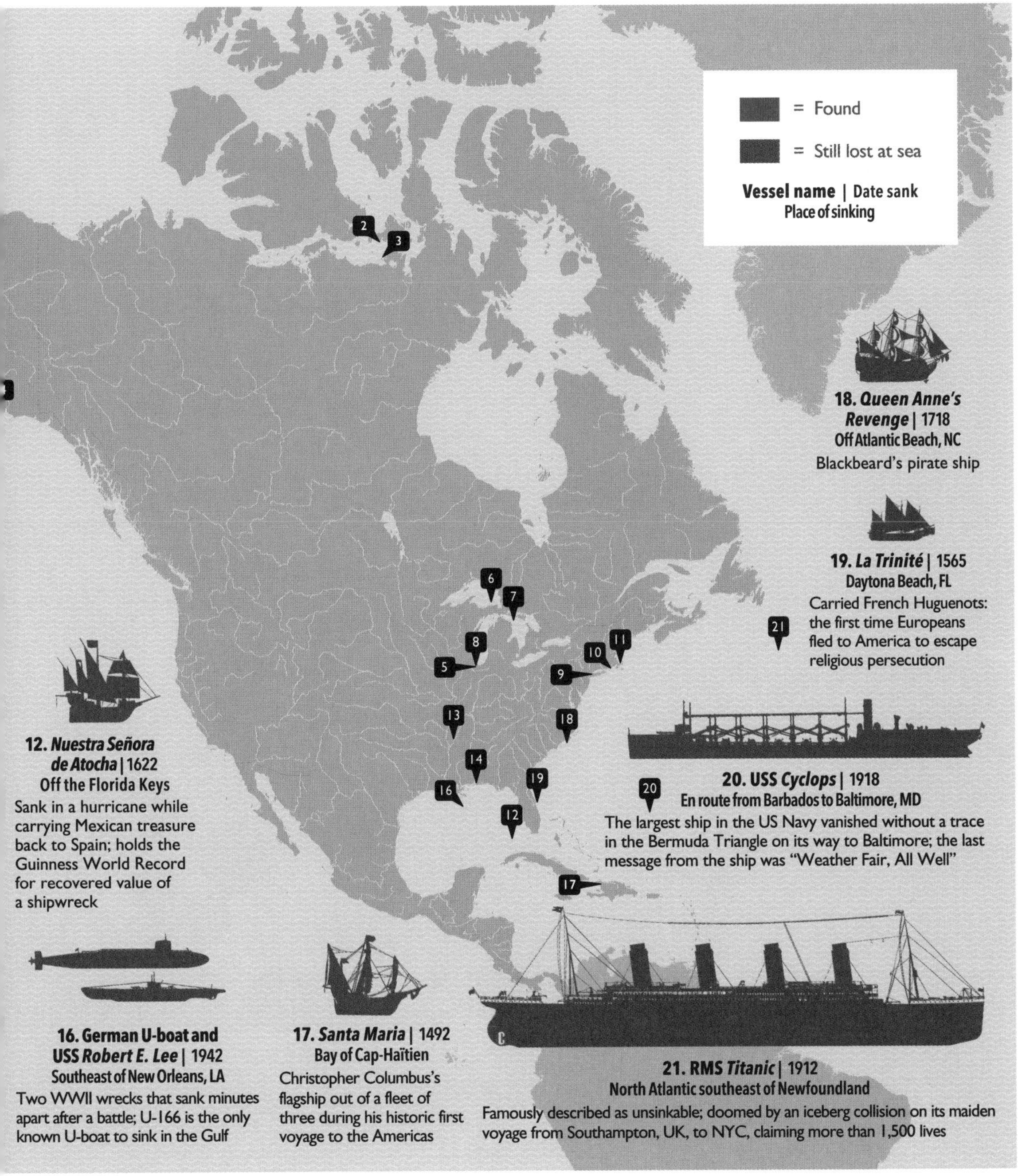
= Found
= Still lost at sea
Vessel name | Date sank
Place of sinking
18. *Queen Anne's Revenge* | 1718
Off Atlantic Beach, NC
Blackbeard's pirate ship
19. *La Trinité* | 1565
Daytona Beach, FL
Carried French Huguenots: the first time Europeans fled to America to escape religious persecution
12. *Nuestra Señora de Atocha* | 1622
Off the Florida Keys
Sank in a hurricane while carrying Mexican treasure back to Spain; holds the Guinness World Record for recovered value of a shipwreck
20. USS *Cyclops* | 1918
En route from Barbados to Baltimore, MD
The largest ship in the US Navy vanished without a trace in the Bermuda Triangle on its way to Baltimore; the last message from the ship was "Weather Fair, All Well"
16. German U-boat and USS *Robert E. Lee* | 1942
Southeast of New Orleans, LA
Two WWII wrecks that sank minutes apart after a battle; U-166 is the only known U-boat to sink in the Gulf
17. *Santa Maria* | 1492
Bay of Cap-Haïtien
Christopher Columbus's flagship out of a fleet of three during his historic first voyage to the Americas
21. RMS *Titanic* | 1912
North Atlantic southeast of Newfoundland
Famously described as unsinkable; doomed by an iceberg collision on its maiden voyage from Southampton, UK, to NYC, claiming more than 1,500 lives

91 Who's got the goods?: The 50 largest ports in the US

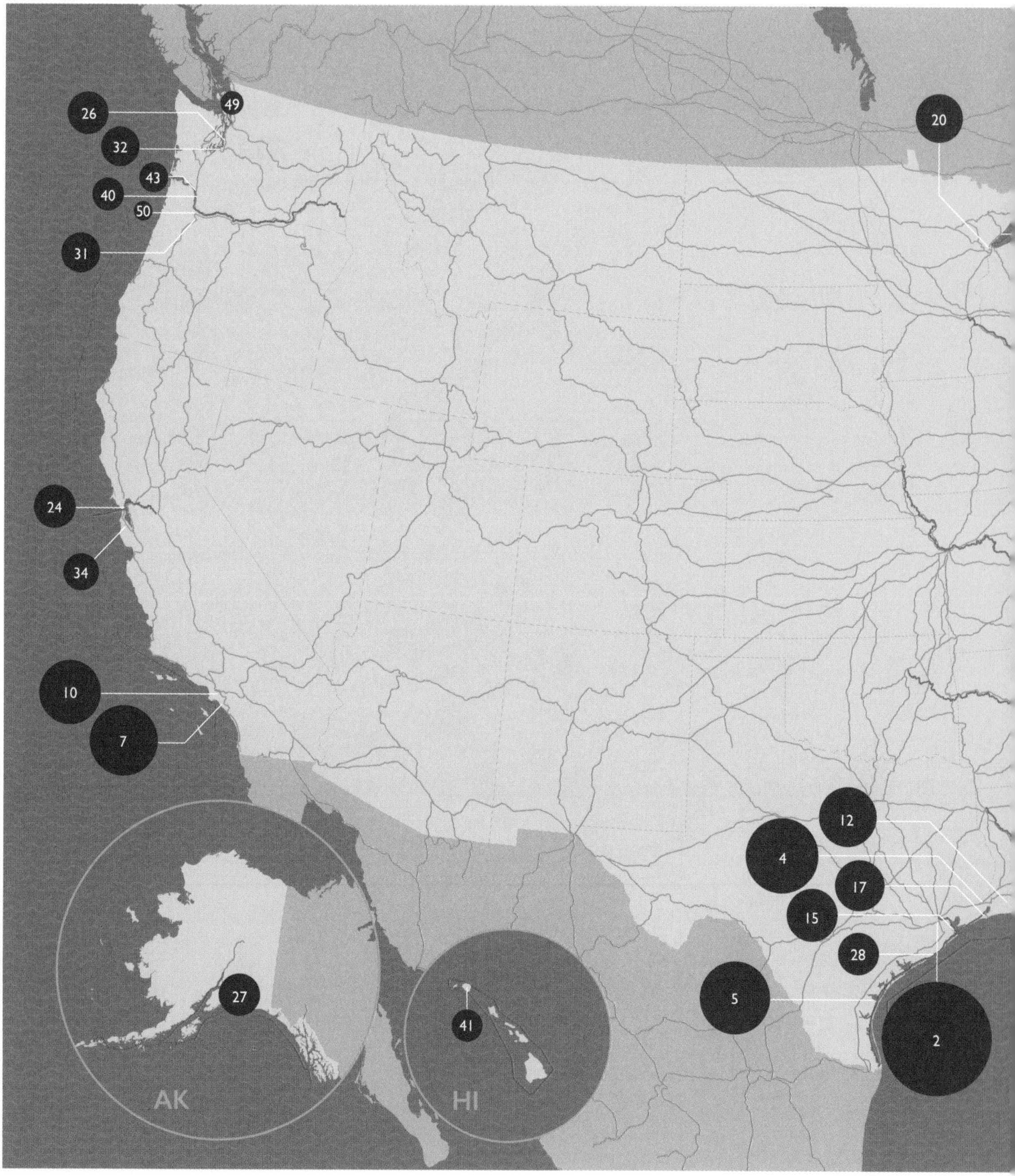

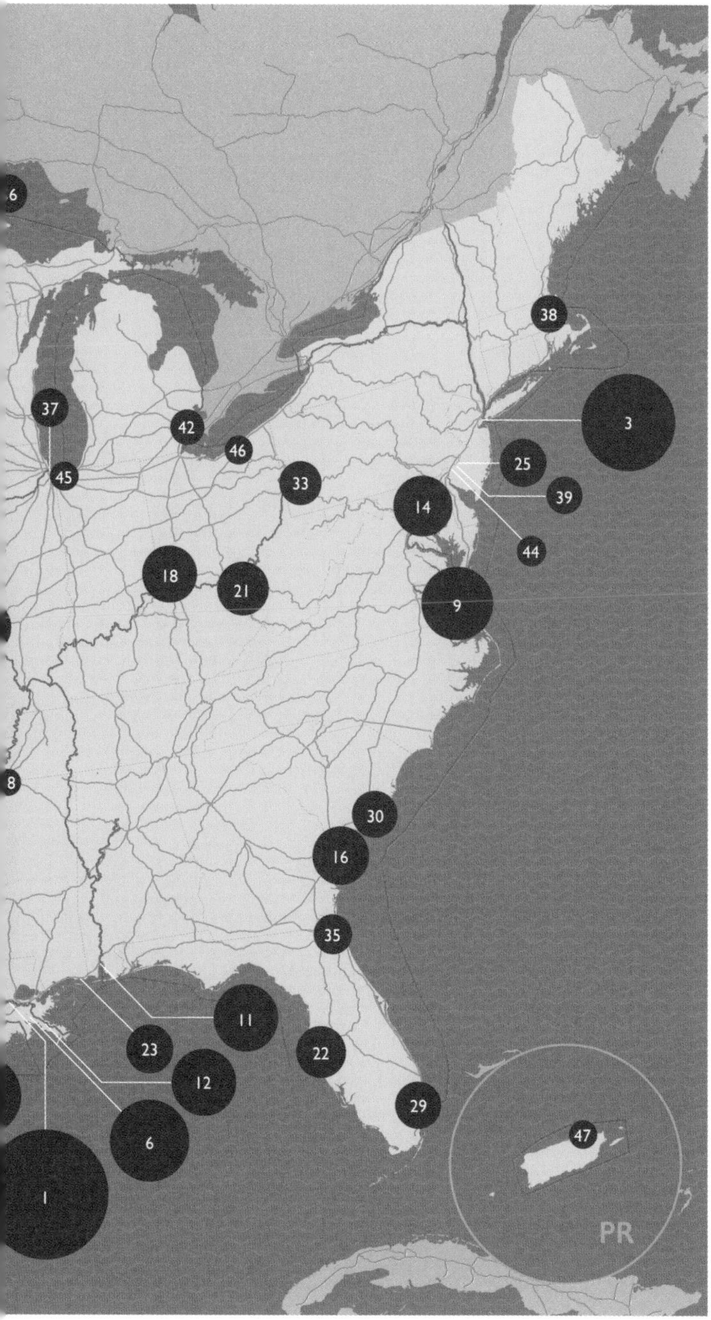

Port of . . .

1. South Louisiana
2. Houston
3. New York and New Jersey
4. Beaumont
5. Corpus Christi
6. New Orleans
7. Long Beach
8. Greater Baton Rouge
9. Virginia
10. Los Angeles
11. Mobile
12. Plaquemines ⎤—TIED
12. Lake Charles ⎦
14. Baltimore
15. Texas City
16. Savannah
17. Port Arthur
18. Cincinnati/Northern KY
19. Metropolitan St. Louis
20. Duluth-Superior
21. Huntington-Tristate
22. Tampa
23. Pascagoula
24. Richmond (CA)
25. Philadelphia
26. Seattle
27. Valdez
28. Freeport
29. Port Everglades
30. Charleston
31. Portland
32. Tacoma
33. Pittsburgh
34. Oakland
35. Jacksonville
36. Two Harbors
37. Chicago
38. Boston
39. Paulsboro
40. Kalama
41. Honolulu
42. Detroit
43. Longview
44. Marcus Hook
45. Indiana Harbor
46. Cleveland
47. San Juan
48. Memphis
49. Anacortes
50. Vancouver (WA)

Millions of metric tons annually

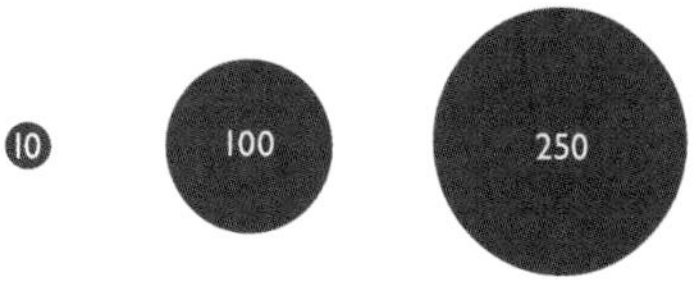

Railway

Marine highway

America's marine highways are a network of navigable waterways, an extension of the surface transportation system that includes rivers, bays, channels, coasts, the Great Lakes, open-ocean routes, and the Saint Lawrence Seaway System.

92 **Get out of town!:** How far one day of travel took you from NYC, since 1800

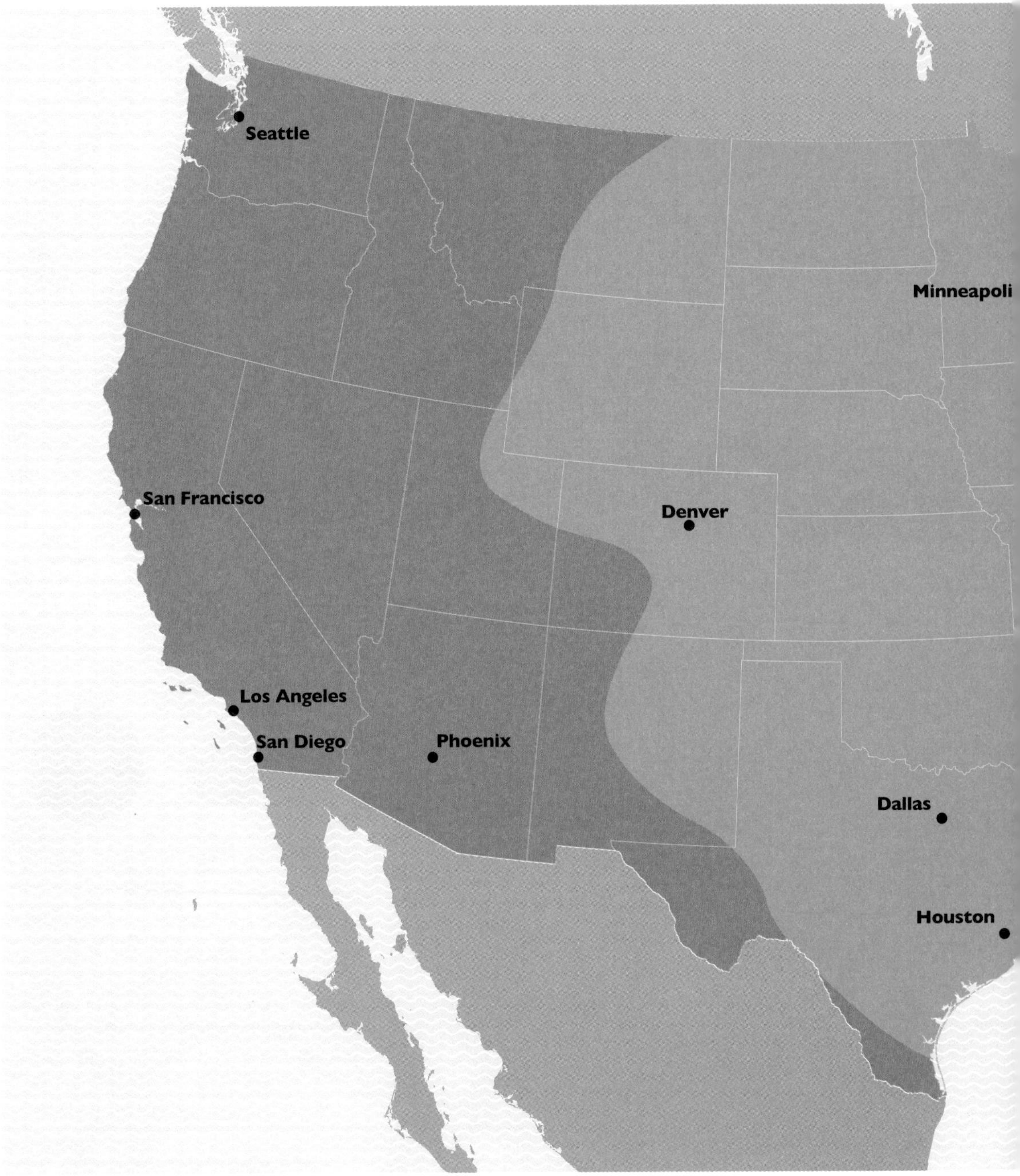

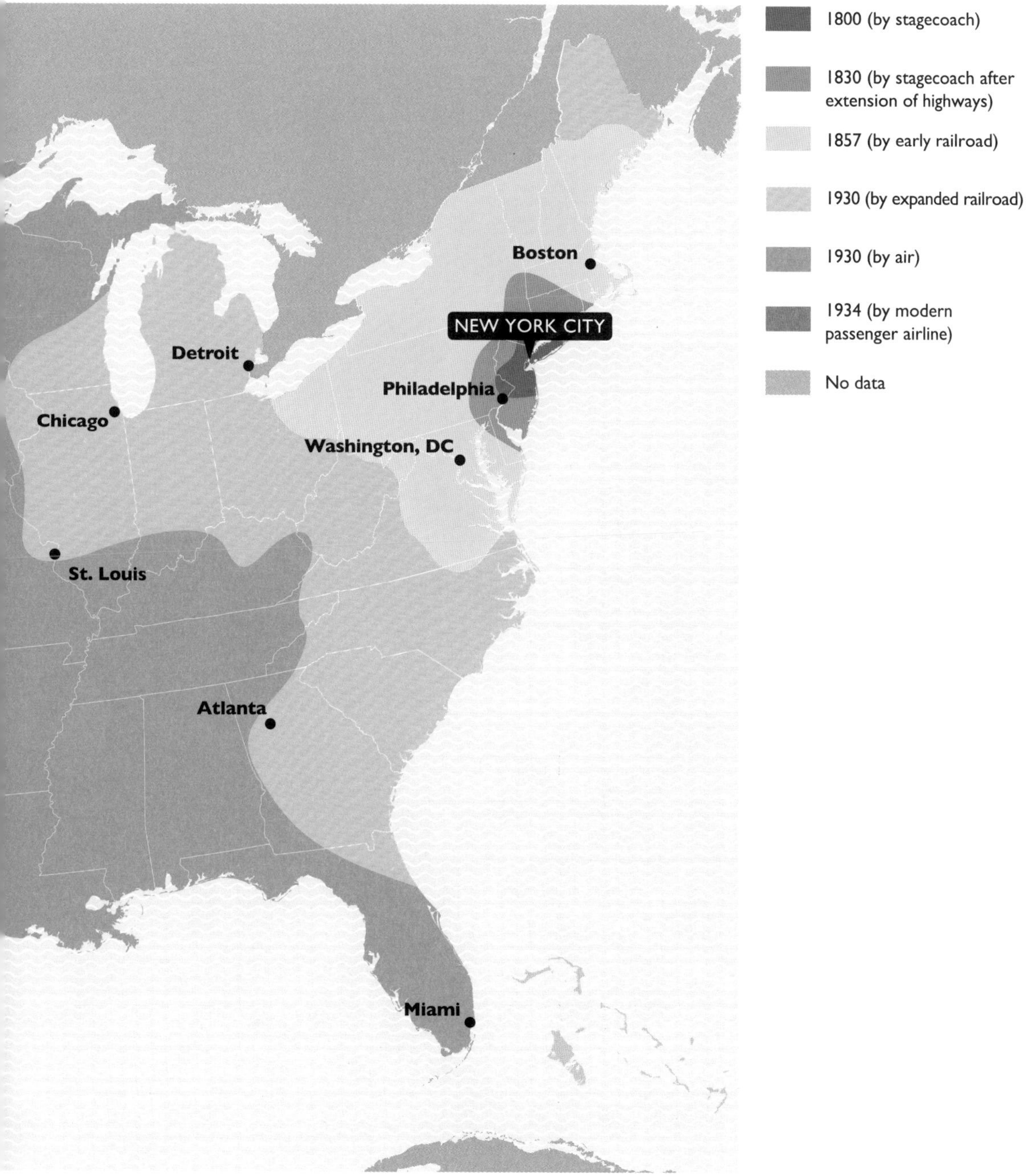
1800 (by stagecoach)
1830 (by stagecoach after extension of highways)
1857 (by early railroad)
1930 (by expanded railroad)
1930 (by air)
1934 (by modern passenger airline)
No data
Boston
NEW YORK CITY
Detroit
Philadelphia
Chicago
Washington, DC
St. Louis
Atlanta
Miami

93 **Moving on up:**
The 50 tallest buildings in North America

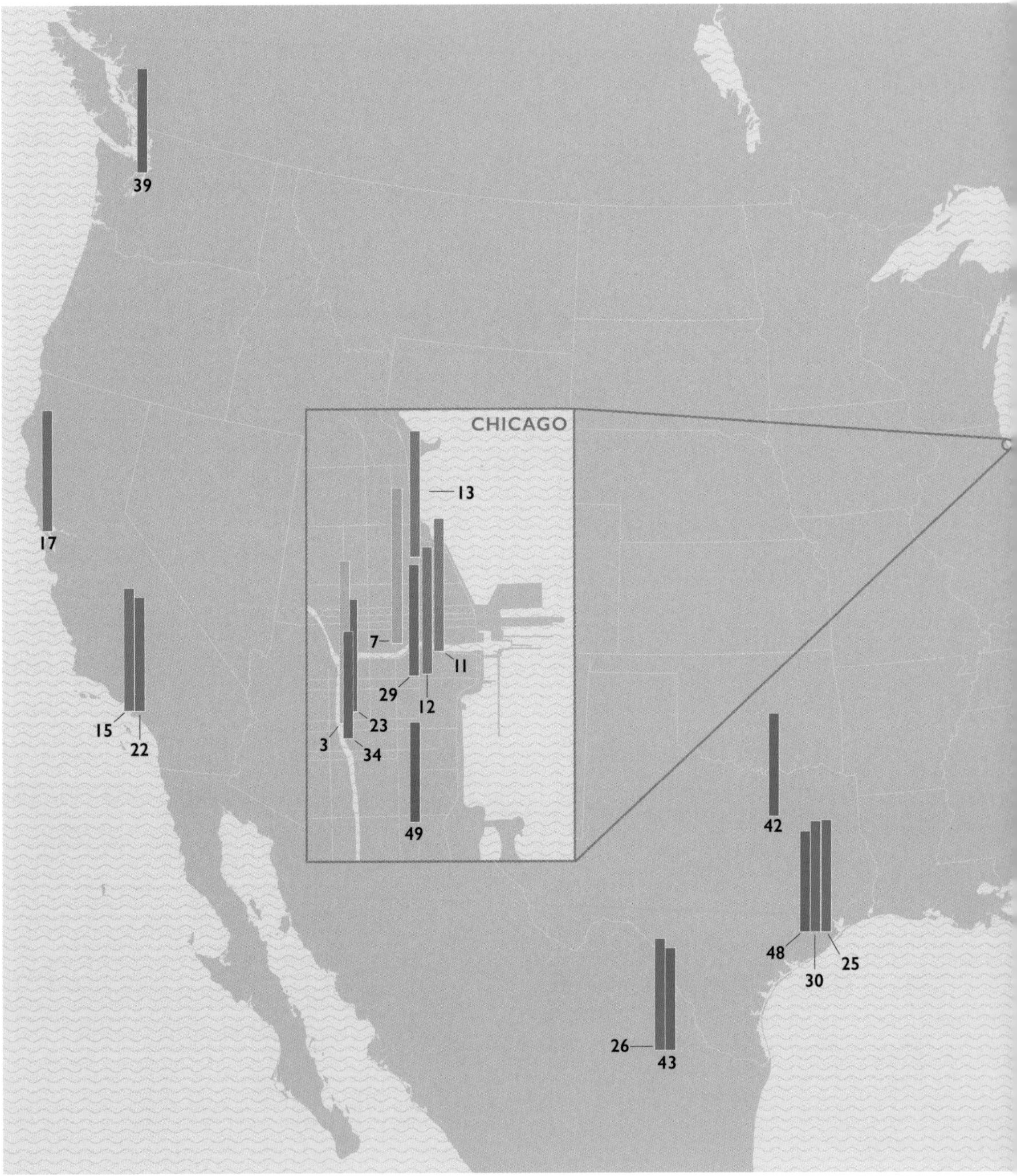

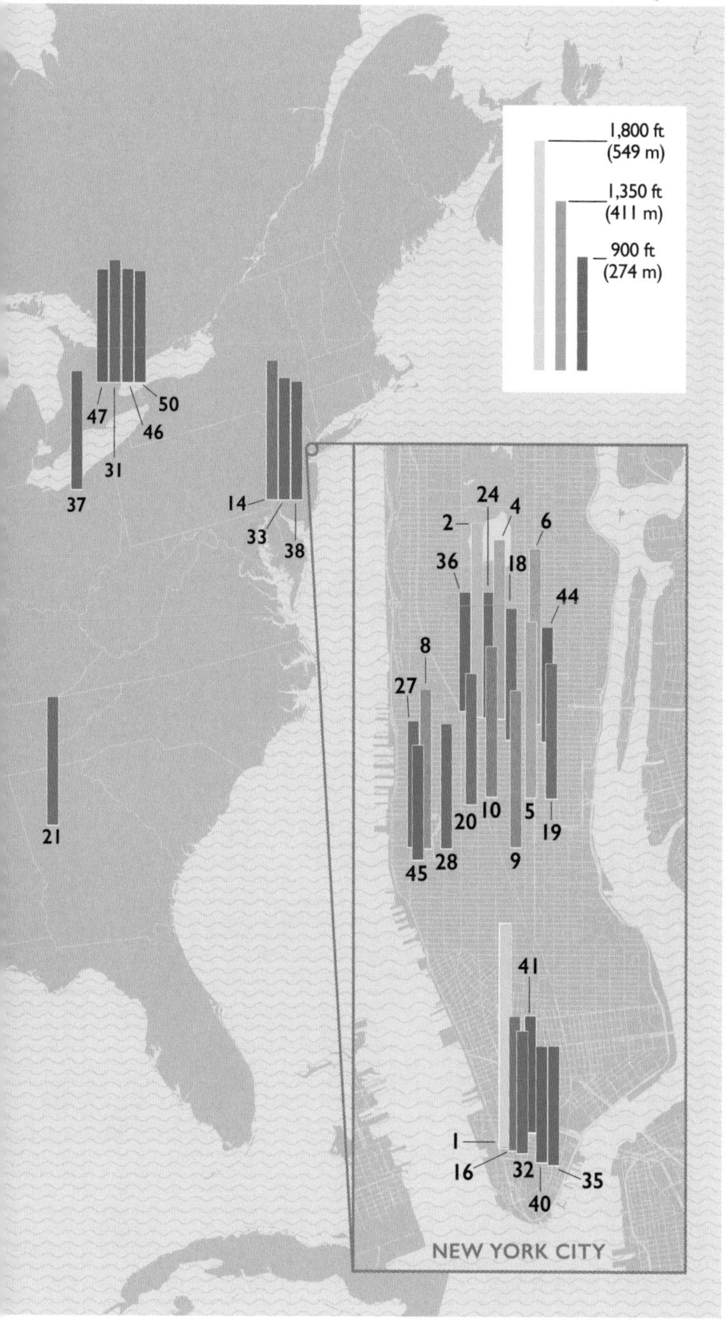

Building name	Height feet	meters
1. One World Trade Center	1,776	541.3
2. Central Park Tower	1,550	472.4
3. Willis Tower	1,451	442.1
4. 111 West 57th Street	1,428	435.3
5. One Vanderbilt	1,401	427.0
6. 432 Park Avenue	1,397	425.7
7. Trump International Tower	1,389	423.2
8. 30 Hudson Yards	1,270	387.1
9. Empire State Building	1,250	381.0
10. Bank of America Tower	1,200	365.8
11. St. Regis Chicago	1,191	362.9
12. Aon Center	1,136	346.3
13. 875 North Michigan Avenue	1,128	343.7
14. Comcast Technology Center	1,112	339.1
15. Wilshire Grand Center	1,100	335.3
16. 3 World Trade Center	1,079	328.9
17. Salesforce Tower	1,070	326.1
18. 53 West 53	1,050	320.1
19. Chrysler Building	1,046	318.9
20. New York Times Tower	1,046	318.8
21. Bank of America Plaza	1,023	311.8
22. U.S. Bank Tower	1,018	310.3
23. The Franklin - North Tower	1,007	306.9
24. One57	1,004	306.1
25. 600 Travis Street	1,002	305.4
26. T.Op Torre I	1,002	305.3
27. 35 Hudson Yards	1,000	304.8
28. One Manhattan West	996	303.6
29. Two Prudential Plaza	995	303.3
30. Wells Fargo Plaza	992	302.4
31. First Canadian Place	978	298.1
32. 4 World Trade Center	977	297.7
33. Comcast Center	974	296.7
34. 311 South Wacker Drive	961	292.9
35. 70 Pine	952	290.2
36. 220 Central Park South	950	289.6
37. Key Tower	947	288.7
38. One Liberty Place	945	288.0
39. Columbia Center	933	284.4
40. The Trump Building	927	282.6
41. 30 Park Place	926	282.2
42. Bank of America Plaza	921	280.7
43. Torre KOI	916	279.1
44. 601 Lexington	915	278.9
45. 15 Hudson Yards	914	278.6
46. The St. Regis Toronto	908	276.9
47. Scotia Tower	902	275.0
48. Williams Tower	901	274.6
49. NEMA Chicago	896	273.1
50. Aura at College Park	892	271.9

94 **Bumper to bumper, coast to coast:** How long it takes to get to work by county

Alaska has 16 of the 23 counties (technically, county equivalents) with average commutes under 10 minutes, with the high percentage of people walking to work as a major factor (see map on the next page). The others are in Hawaii, Colorado, Montana, Texas, and Kansas. None are in urban areas. Of the 22 counties with average commutes over 40 minutes, 7 are in the Washington, DC, metro area, and 5 are in and around New York City. Mora County, NM—one of the US's poorest and with few job prospects—has the longest commute of all (52.6 minutes).

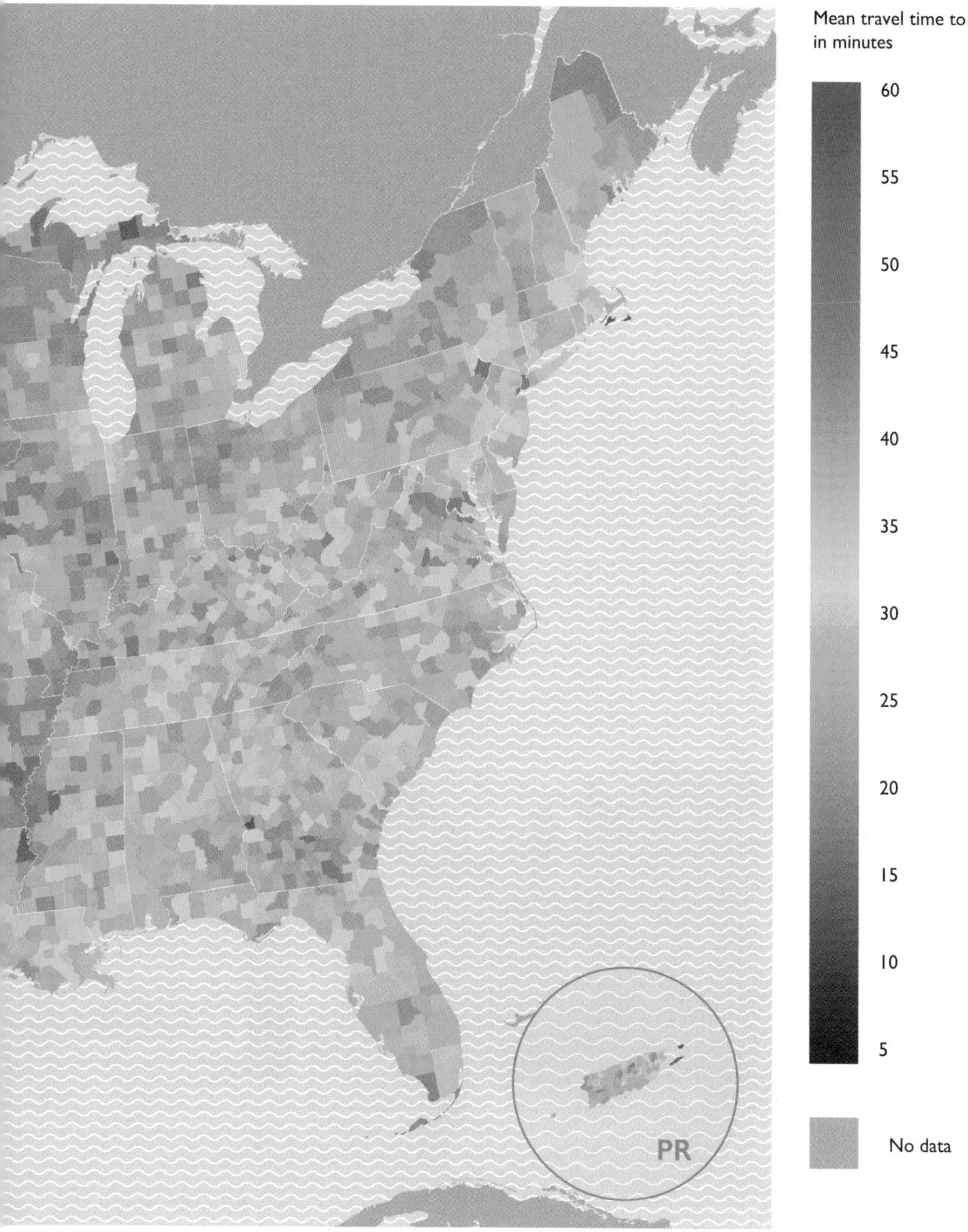

95 In the driver's seat:
How America gets to work

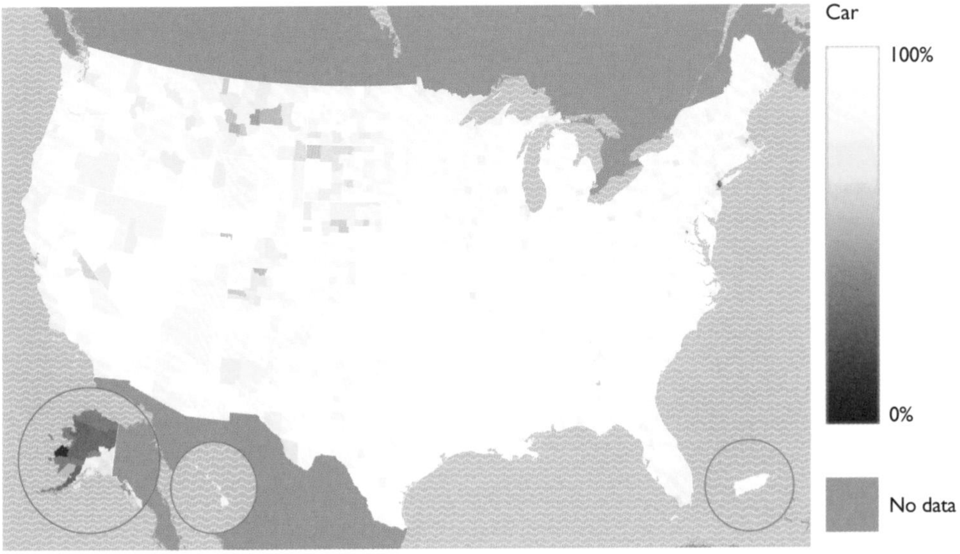

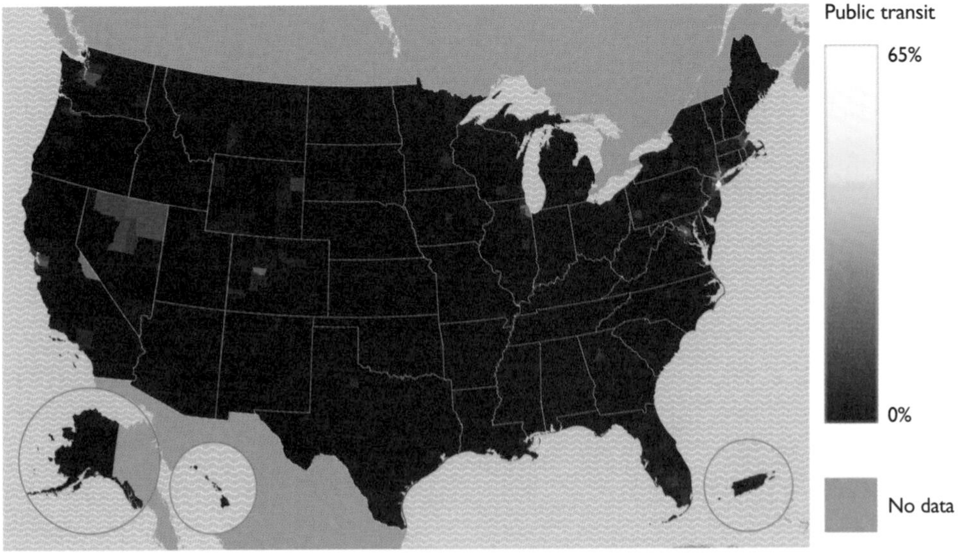

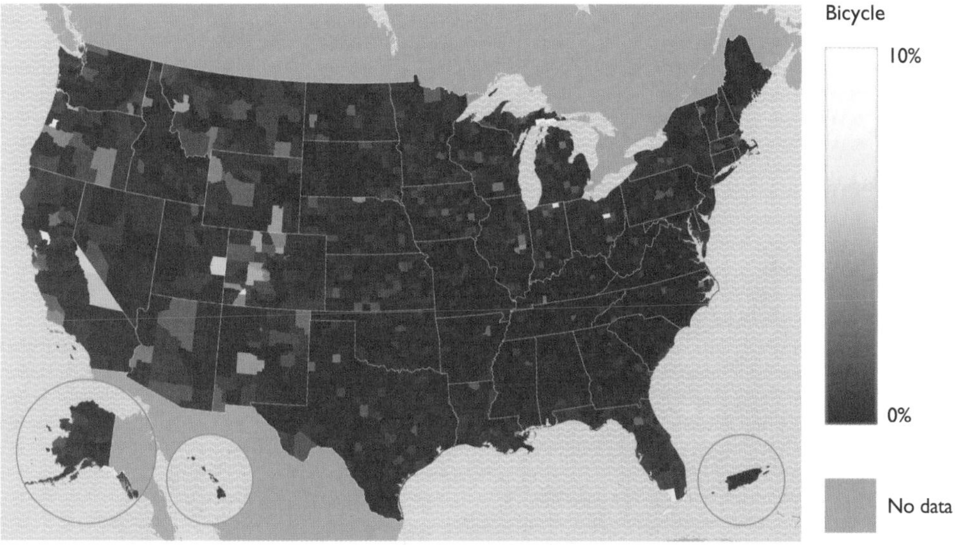
Bicycle
10%
0%
No data

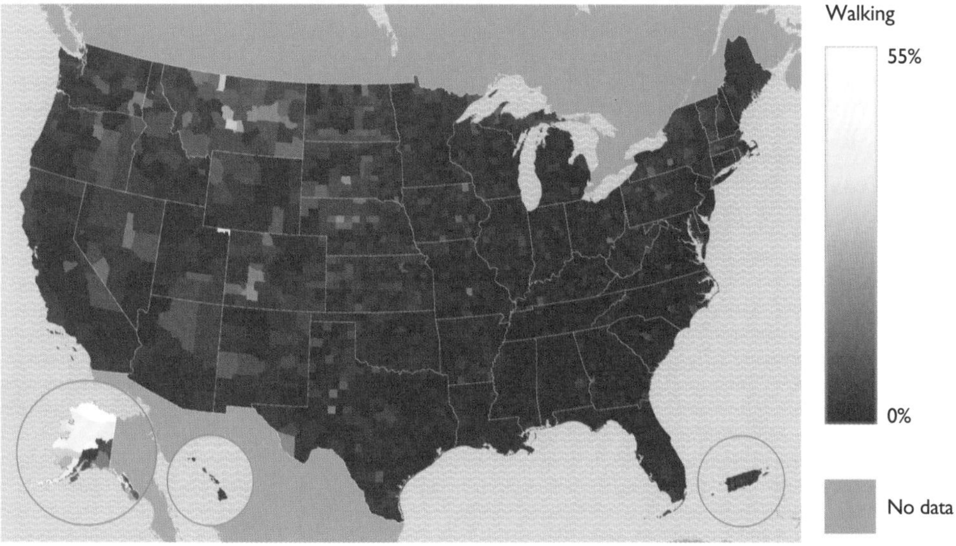
Walking
55%
0%
No data

96 **Don't bank on it:** The unbanked people of North America

Percentage of people age 15+ with an account at a bank or other financial institution

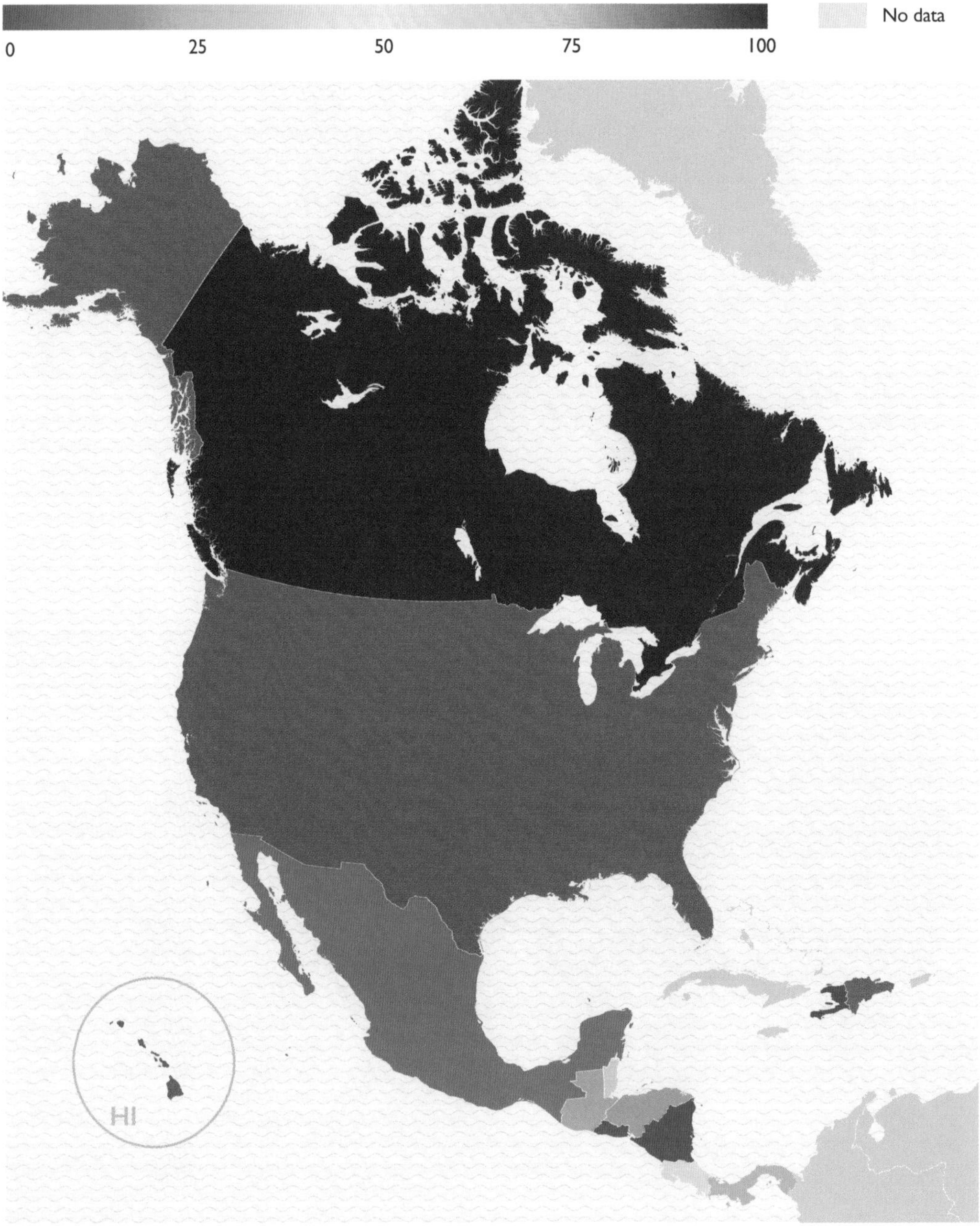

97 Road hogs:
The longest roads in the US, Canada, and Mexico

For drivers who love long road trips but hate navigating, you can't beat these roads. They take you the farthest you can go on one road without ever turning onto another.

Trans-Canada Highway (4,860 miles/7,820 km)

U.S. Highway 20 (3,237 miles/5,209 km)

Carretera Federal 15 (1,469 miles/2,364 km)

98 **Cattle call:** In which states are you likelier to run into a cow than a person?

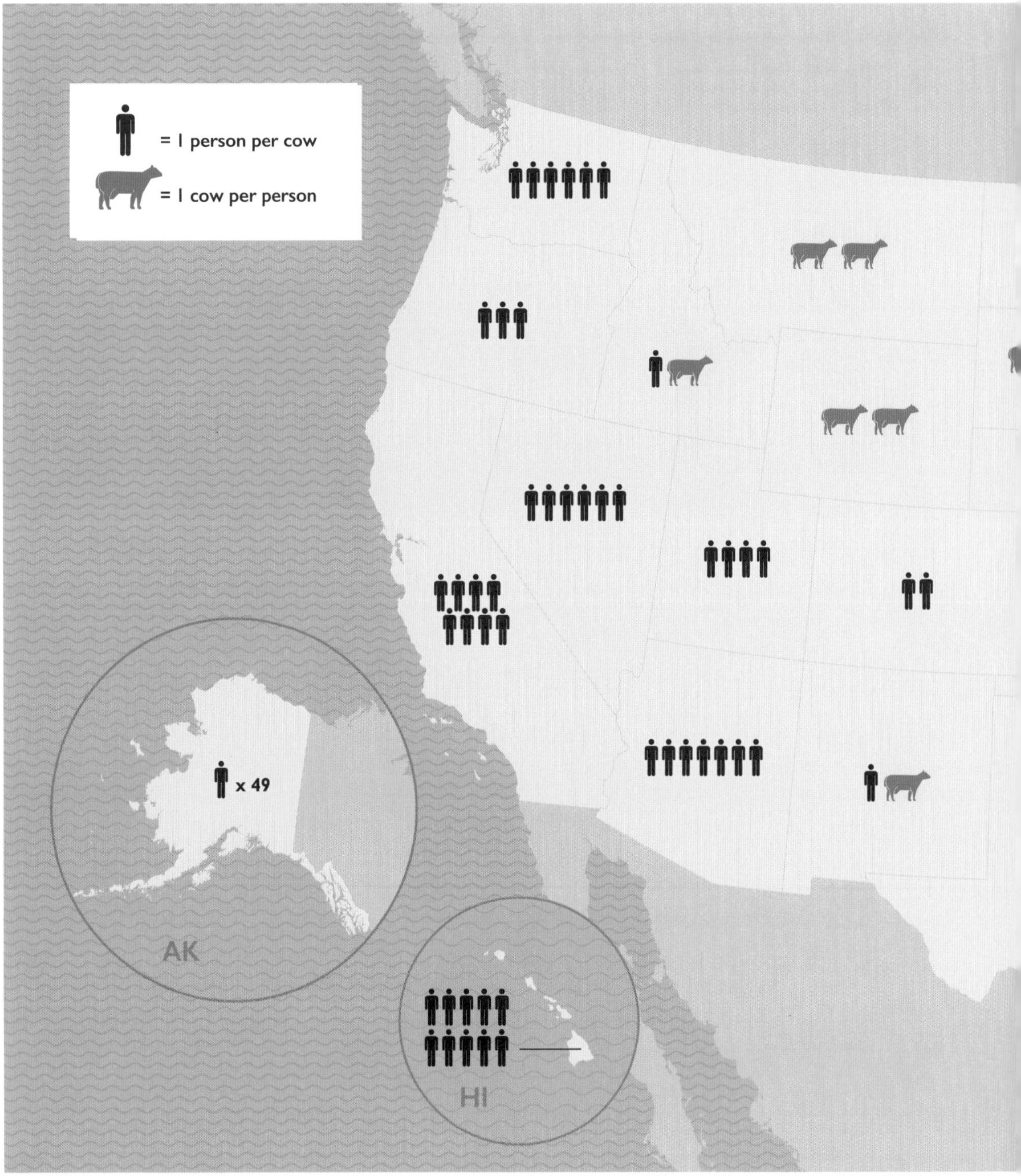

In six US states, the cattle (and calves) outnumber humans, led by South Dakota, with almost 4 million cattle grazing among its roughly 850,000 human inhabitants. And then there's New Jersey, whose cattle population of fewer than 30,000 is dwarfed by the nearly 9 million people who live there.

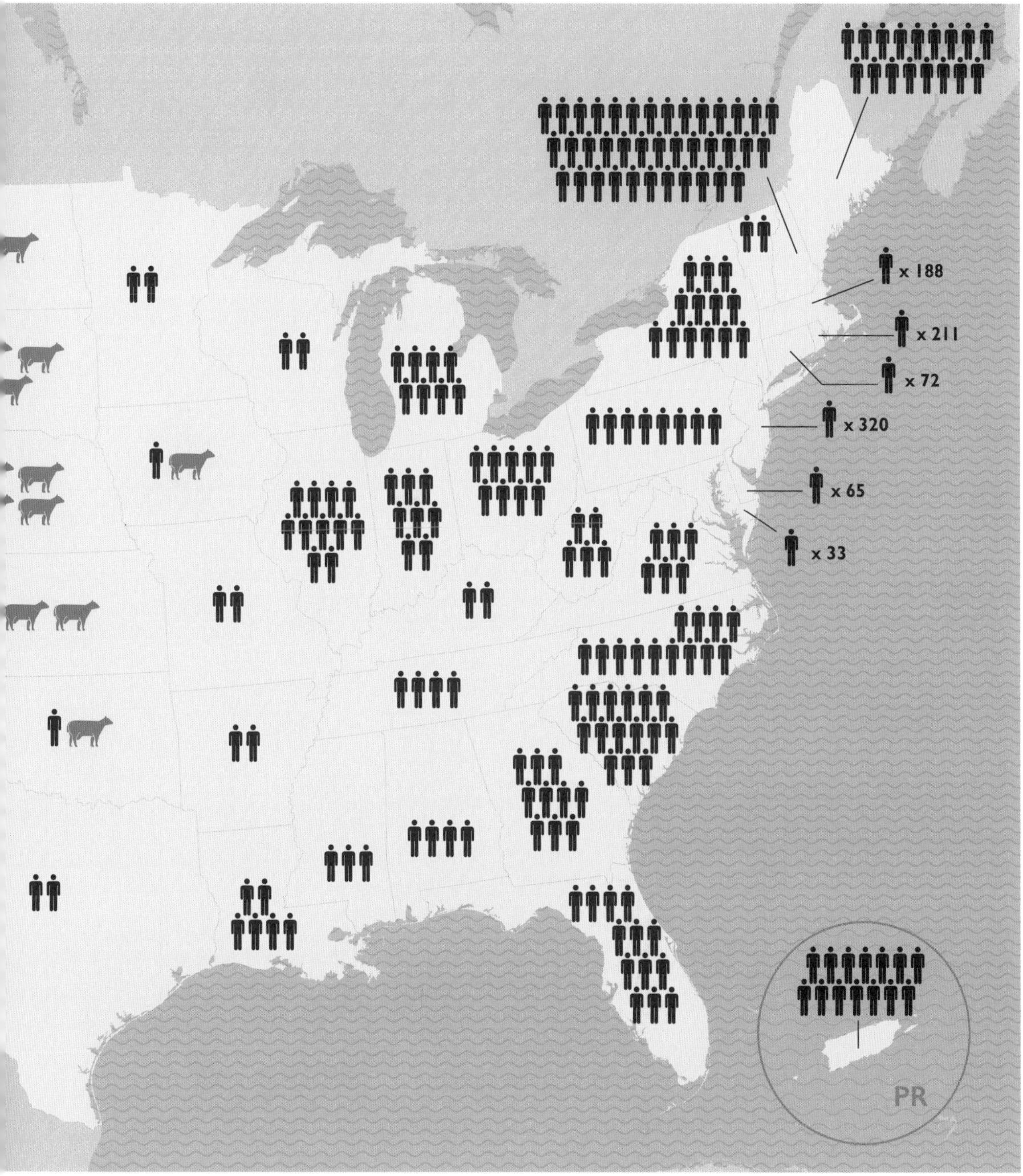

99 T-minus 15 spaceports:
Launchpads across the US

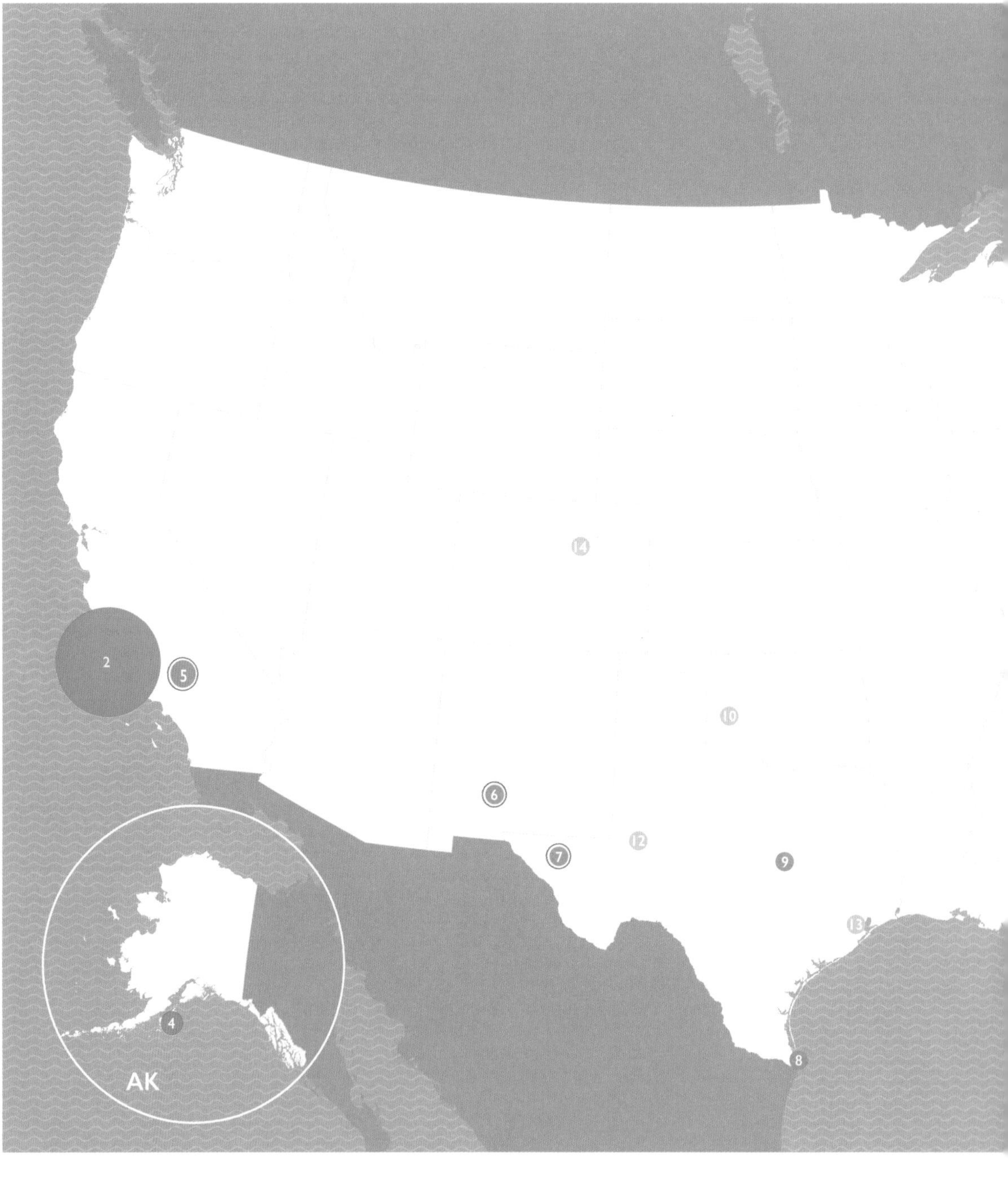

The Kennedy Space Center in Florida may be the most famous spaceport, having been the launch site for virtually all the US's most celebrated missions—but it's not the *only* spaceport, nor does it have a monopoly on crewed missions anymore. But it remains the granddaddy of orbital launch sites, which send spacecraft into orbit. In suborbital spaceflight, the craft goes up and then comes right back down (Blue Origin, for example, is suborbital, and the entire flight lasts about ten minutes). Very few spaceports can launch spacecraft at the altitude and horizontal velocity necessary to reach orbit.

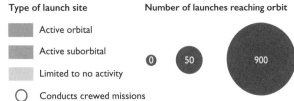

1. Cape Canaveral Space Force Station/Kennedy Space Center
First orbital launch 1/31/1958 (Explorer 1, the first satellite launched by the US); launch site for every crewed mission to reach orbit (162 total): Mercury (4), Gemini (10), Apollo (11), Shuttle (134), and SpaceX Crew Dragon (3); ideally situated for spacecraft requiring a west–east orbit

2. Vandenberg Air Force Base
First orbital launch 2/28/1959 (Discoverer 1, the world's first polar orbiting satellite); the nation's west coast space and missile facility; location of choice for spacecraft requiring a north–south polar Earth orbit and for safely testing intercontinental and intermediate range ballistic missiles

3. Wallops Flight Facility/Mid-Atlantic Regional Spaceport
First orbital launch 2/16/1961; NASA's principal facility for suborbital research programs using aircraft, scientific balloons, and sounding rockets; instrumental in testing the basic design and escape system of the Mercury space capsule; besides Cape Canaveral, the only other site providing cargo to the ISS

4. Pacific Spaceport Complex – Alaska
First orbital launch 9/30/2001; privately run since 2015; spaceport for small- and light-lift vertical rockets and stratospheric balloons; one of the best locations in the world for polar launch operations

5. Edwards Air Force Base/Mojave Air and Space Port
First orbital launch 4/5/1990; conducts horizontal launches (a carrier aircraft launches the spacecraft/rocket while airborne); used to send Pegasus (first privately developed space launch vehicle) into orbit and SpaceShipOne (first privately built crewed spacecraft) into suborbital space

6. Spaceport America
First crewed suborbital spaceflight May 2021 (SpaceShipTwo)

7. Blue Origin Launch Site, West Texas
Launch site for crewed Blue Origin, set for July 2021 liftoff

8. SpaceX Launch Site, Boca Chica
Test-launches regularly conducted for SpaceX

9. SpaceX Launch Site, McGregor
A SpaceX rocket development facility that tests structure and propulsion

10. Oklahoma Spaceport

11. Cecil Spaceport

12. Midland International Air and Space Port

13. Houston Spaceport

14. Colorado Air and Spaceport

15. Space Coast Regional Airport

Type of launch site

Active orbital

Active suborbital

Limited to no activity

Conducts crewed missions

Number of launches reaching orbit

0 50 900

100 **America offline:** Who doesn't have internet

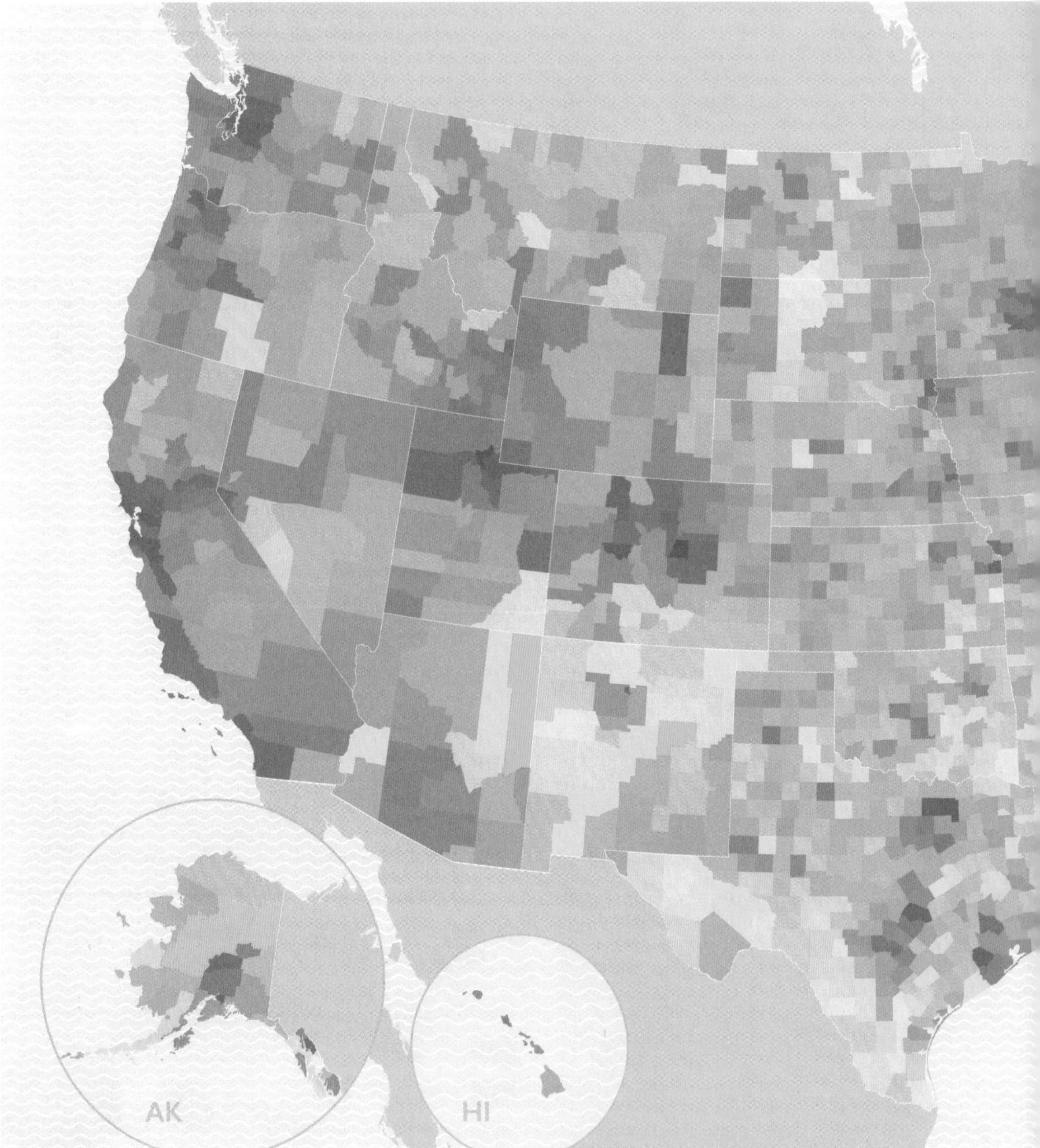

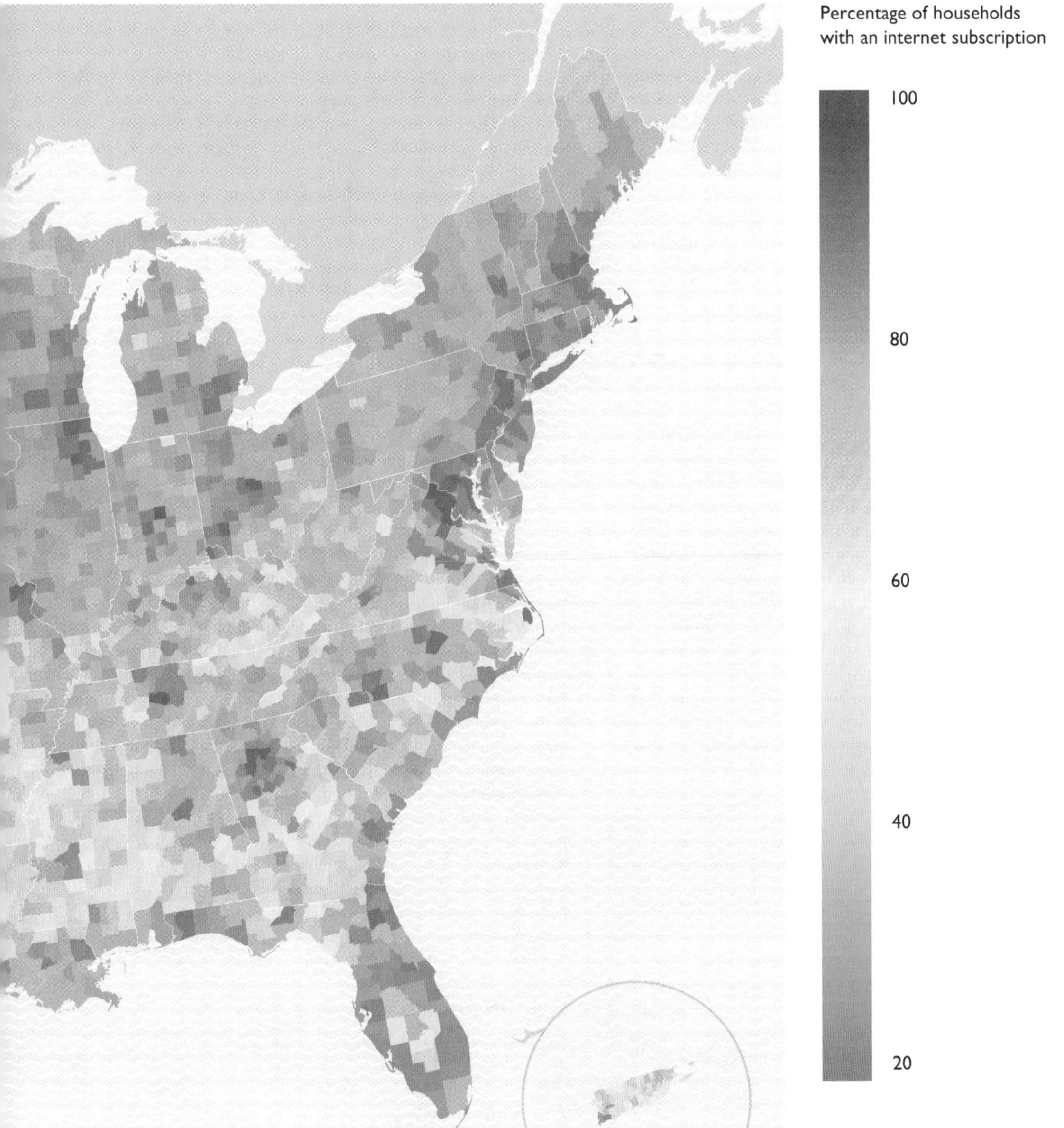
PR
Percentage of households
with an internet subscription
100
80
60
40
20
No data

SOURCES

Many of the maps in this book bring together data that otherwise can't be found in one place, and so our research required we consult a long list of sources—from newspapers to academic journals and more—to compile info for as many countries, states, provinces, and territories across the continent as possible. In these instances, we've marked "various" as a primary source to encompass our wide-ranging research. Many of the concepts behind our maps are original ideas, while some were inspired by others, in which cases we've provided credit here. For design, we heavily leaned on and express our utmost thanks to Natural Earth for data on map borders and to Tom Patterson for his land cover and shaded relief raster data.

Geography

1. Identity crisis: What even is North America?
Image: Copyright © Wikimedia 2015, "The North American Plate," under Creative Commons Attribution-ShareAlike 4.0 International.
Data: *The World Factbook 2021*. Washington, DC: Central Intelligence Agency, 2021.

2. Hawaii is a really long state: Comparing it to the West Coast puts it in perspective
Original concept.

3. Global scale: If your state or province were a country, what size by area would it be?
Data: "Información por entidad," Instituto Nacional de Estadística y Geografía.
"Land and freshwater area, by province and territory," Statistics Canada.
US Census Bureau. 2010 Census of Population and Housing, *Population and Housing Unit Counts*, CPH-2-1, United States Summary. Washington, DC: US Government Printing Office, 2012.
The World Factbook 2021. Washington, DC: Central Intelligence Agency, 2021.

4. Alas, Alaska: Big, weird, and wonderful
Image: Gaba, Eric. Copyright © Wikimedia 2008, "Alaska area compared to conterminous US," under Creative Commons Attribution-ShareAlike 3.0 Unported.
Data: "Alaska: About the Region," US National Park Service.
"Earthquake Risk in Alaska," Alaska Division of Geological and Geophysical Surveys.
"National Climate Extremes Committee (NCEC)," National Oceanic and Atmospheric Administration.

5. Counties that count: Giving a few across the US their due
Data: "County Population Totals: 2010–2019," US Census Bureau.
"Maps & GIS Data," State of Alaska Department of Labor and Workforce Development.
"Most unusual shaped counties," SkyscraperCity.
Stephen Hess, David Riordan, and John Oram. "Not all counties are equal," Mapzen.
Various.

6. Drawing outside the lines: Baffling North American borders
Data: Dempsey, Caitlin. "Interesting Geography Facts About the US–Canada Border," Geography Realm.
"List of enclaves and exclaves," Owlapps.
Nickum, Randy. "Interesting Exclaves of the United States," Google Sightseeing.
Selkirk, Diane. "The little-known US–Canada border war," BBC Travel.
"United Divide: A Linear Portrait of the USA/Canada Border," The Center for Land Use Interpretation.

7. Doing a double take: City pairs to leave you second-guessing
Data: Fuller, Gary. *The Trivia Lover's Guide to the World*. Plymouth: Rowman & Littlefield Publishers, 2012.

8. Just keep swimming: The first country you'll reach, coast to coast
Concept: Eric Odenheimer.
Weiyi Cai, Ana Swanson, and Laris Karklis. "What's across the ocean from you when you're at the beach, in 7 fascinating maps," *The Washington Post.*

9. The inner banks: Canada's great island lakes and lake islands
Data: "Nettilling Lake," Canadian Geographical Names Database.
"Sub-Sub-Sub Island on Victoria Island," Atlas Obscura.

10. On dry land: States and provinces not touching an ocean, gulf, or bay
Concept: Clive Newstead.

11. The 8 greatest Lakes in North America: Plus 7 pretty good ones
Data: Janssen, Sarah. *The World Almanac and Book of Facts 2018*. New York: World Almanac, 2018.

12. The United States of Washington: Every place named after the first president
Data: US Geological Survey.

13. Rhode nation: If every state were the size of Rhode Island, there would be 2,457 states
Original concept.

14. Name dropping: City names, then and now
Data: US National Park Service.
Various.

15. A river runs beside it: US natural borders
Data: Popelka, Sarah J. and Laurence C. Smith. "Rivers as political borders: a new subnational geospatial dataset," *Water Policy* 22, no. 3 (2020): 293–312.

Politics and Power

16. Madam Governor: Number of women governors by state or province
Original concept.

17. Belles of the ballot: When women gained the right to vote
Data: "A Brief History of Federal Voting Rights in Canada," Elections Canada.
"Map: States grant women the right to vote," National Constitution Center.
Walenta, Craig. "Ratification of Constitutional Amendments," USConstitution.net.
"Women and the vote: World suffrage timeline," New Zealand Government.

18. Engendering equality: The ongoing struggle to pass the Equal Rights Amendment
Data: "Commemorating the Nineteenth Amendment: Women's Suffrage at Home and Abroad," Council on Foreign Relations.
"Equal Rights Amendment," US National Archives and Records Administration.
EqualRightsAmendment.org, Alice Paul Institute.

19. Red California and blue Texas?: Presidential election results since 1856
Data: 270 to Win.
Dave Leip's Atlas of U.S. Presidential Elections.
J. Clark Archer, Stephen J. Lavin, Kenneth C. Martis, and Fred M. Shelley. *Historical Atlas of U.S. Presidential Elections 1788–2004*. Washington, DC: CQ Press, 2006.
Janssen, Sarah. *The World Almanac and Book of Facts 2018*. New York: World Almanac, 2018.

20. Worth the visit?: States the candidates made time for during the 2020 general election campaign
Data: "Map of General-Election Campaign Events and TV Ad Spending by 2020 Presidential Candidates," National Popular Vote.
"Out of 1,164 General-Election Campaign Events in the Past 4 Presidential Elections, 22 States Received 0 Visits and 9 More States Received Just 1," National Popular Vote.

21. "Did Not Vote" ends winning streak: A historic turnout finally beats the lack of turnout
Data: Fabina, Jacob. "Despite Pandemic Challenges, 2020 Election Had Largest Increase in Voting Between Presidential Elections on Record," US Census Bureau.
McDonald, Michael P. United States Elections Project.
"Official 2020 Presidential General Election Results," Federal Election Commission.

22. States that picked the greats: Predicting the pantheon of best presidents
Data: 270 to Win.
Dave Leip's Atlas of U.S. Presidential Elections.
J. Clark Archer, Stephen J. Lavin, Kenneth C. Martis, and Fred M. Shelley. *Historical Atlas of U.S. Presidential Elections 1788–2004*. Washington, DC: CQ Press, 2006.
Janssen, Sarah. *The World Almanac and Book of Facts 2018*. New York: World Almanac, 2018.

23. The price of leadership: Salaries of North American heads of state
Data: Copyright © Wikipedia 2020, "List of salaries of heads of state and government," under Creative Commons Attribution-ShareAlike 3.0 Unported License.

24. Land of equality?: If every state's population had Wyoming's representation in the Senate
Data: "Distribution of Electoral Votes," US National Archives and Records Administration.
"State Population Totals and Components of Change: 2010–2019," US Census Bureau.

25. Coaching all the way to the bank: The highest-paid public employee by state
Data: "NCAA Salaries," *USA Today*.
Open the Books © American Transparency.
"Search Federal Employee Salaries," FedsDataCenter.com.
Various.

26. Monroe's lament: The dependencies of North America
Data: "Monroe Doctrine (1823)," Our Documents.

27. Bang for the buck: Who pays most for their military?
Data: "SIPRI Military Expenditure Database," Stockholm International Peace Research Institute.

28. Living on the edge: 2 out of 3 Americans live within 100 miles of the border
Image: "The Constitution in the 100-Mile Border Zone," American Civil Liberties Union.

29. Where service comes first: Proportion of veterans by state
Data: "Percent of the Civilian Population 18 Years and Over Who Are Veterans," 2018 American Community Survey 1-Year Estimates, US Census Bureau.

Nature

30. Missing the forest for the trees: The can't-miss trees of North America
Data: US National Park Service.
Various.

31. Dog days: When does the hottest day of the year fall?
Image: "When to Expect the 'Warmest Day of the Year,'" National Oceanic and Atmospheric Administration.

32. Spring hasn't sprung: Punxsutawney Phil predicts late blooms
Image: "NPN Visualization Tool," USA National Phenology Network.
Data: "Groundhog Day Predictions," The Punxsutawney Groundhog Club.

33. Land before time: The greatest dinosaur finds in North America
Data: "Paper Dinosaurs 1824–1969," Linda Hall Library.
Various.

34. Our biggest hits: North America's largest impact craters
Data: "Earth Impact Database," Planetary and Space Science Centre University of New Brunswick.
"Shaping the Planets: Impact Cratering," Universities Space Research Association Lunar and Planetary Institute.

35. Above the fruited plain: The highest elevation in each state
Data: "Elevations and Distances in the United States," US Geological Survey.
"National Geodetic Survey," National Oceanic and Atmospheric Administration.

36. And now, the weather: North American weather extremes
Original concept.
Data: Various, including Clovered, Current Results, The DataFace, EarthSky, Farmers' Almanac, KGW-TV, Mount Washington Avalanche Center, Mountain Homies, NASA, National Climatic Data Center, Statista, The Travel, University Corporation for Atmospheric Research, *USA Today*, USA.com, and Vancouver is Awesome.

37. Caving the day: Cavernous wonders across North America
Data: US National Park Service.
Various.

38. Supervolcano: When Yellowstone blanketed the US and Canada in ash
Data: US Geological Survey.
US National Park Service.

39. Perma-lost: The Last Glacial Maximum compared to today
Data: "How to See a Glacier," US National Park Service.
"LGM Glaciation Extends," Project Z2, University of Cologne Department of Geography.
"Glaciated Areas," Natural Earth.
"Status of Glaciers in Glacier National Park," US Geological Survey.

40. Where darkness reigns: The dark skies still left for pristine stargazing
Data: Fabio Falchi, Pierantonio Cinzano, Dan Duriscoe, Christopher C. M. Kyba, Christopher D. Elvidge, Kimberly Baugh, Boris A. Portnov, Nataliya A. Rybnikova, and Riccardo Furgoni. "The new world atlas of artificial night sky brightness," Science Advances 2, no. 6 (2016): e1600377.

41. Singled out: Endangered or threatened species that live entirely within one state
Data: "Environmental Conservation Online System," US Fish and Wildlife Service.

42. A second act: North American species that have staged comebacks
Data: "Environmental Conservation Online System," US Fish and Wildlife Service.
The IUCN Redlist of Threatened Species.
Various.

43. Let there be light: Local time of sunset on the summer solstice
44. Don't let the sun go down on me: Local time of sunset on the winter solstice
Concept: Brian Brettschneider.
Data: "Sunrise/Sunset Calculator," National Oceanic and Atmospheric Administration.

45. Too much time on my hands: Hours of daylight on the summer solstice
46. Not enough time in the day: Hours of daylight on the winter solstice
Concept: Brian Brettschneider.
Data: "Solar Calculator," National Oceanic and Atmospheric Administration.

Culture and Sports

47. Tallest, fastest, steepest: Record-setting amusement parks and rides of North America
Data: Janssen, Sarah. *The World Almanac and Book of Facts 2018*. New York: World Almanac, 2018.
"More Roller Coaster Records," Ultimate Rollercoaster.
rcdb.
Various.

48. Ones for the books: Iconic moments in North American sports
Data: Mashraky, Reshef. "The 21 Greatest Moments in Sports History," Sports Retriever.
Tiedemann, James. "Top 10 Greatest U.S. Sports Moments," The Impact News.
Various.

49. Take me out to the crowd: The 20 largest North American stadiums by seating capacity
Concept: *Terra Maxima: The Records of Humankind*, ed. Wolfgang Kunth. Ontario: Firefly Books, 2013.

50. Topping the charts: The place names that appear in *Billboard* No. 1 song titles
Data: Copyright © Wikipedia 2020, "List of *Billboard* number-one singles," under Creative Commons Attribution-ShareAlike 3.0 Unported License.

51. Black American firsts: The places where they made their names
Data: "101 African American Firsts," BlackPast.org.
"Black people in history: Little-known figures to know," *CNN*.
Carney Smith, Jessie. *Black Firsts: 4,000 Ground-Breaking and Pioneering Historical Events*. Detroit: Visible Ink Press, 2012.

52. Don't drop the ball: The less-famous things that drop (or rise) on New Year's Eve
Data: Ahuja, Masuma. "16 weird things we drop to ring in the new year," *CNN*.
Morris, Seren. "Unusual New Year's Eve Events Across the U.S., From Possum to Potato Drops," *Newsweek*.
"New Year's Eve 2021 in the US," AstroSage.
"New Year's Eve: Weird Alternatives to the Times Square Ball," Spectrum News 1.
"Social and Cultural History - Installment #1: New Year's Eve Celebrations Around the U.S.," US History Teachers Blog.
Various.

53. How the stacks stack up: The 20 largest physical public library collections in the US
Data: "IMLS 2018 Public Libraries Survey," Institute of Museum and Library Services.

54. Magnificent museums: The 20 most popular museums in North America by attendance
Data: "TEA/AECOM 2019 Theme Index and Museum Index: The Global Attractions Attendance Report," Themed Entertainment Association.

55. Hometown heroes: Hall-of-famer birthplaces in the 4 major sports leagues
Original concept.
Data: "Legends of Hockey," Hockey Hall of Fame.
"Naismith Memorial Basketball Hall of Fame Inductees," Basketball Reference.
"Pro Football Hall of Fame Inductees," Pro Football Reference.
Stathead.

56. The Midas brush: The most expensive paintings of North America
Concept: Amoros, Raul. "The Most Expensive Paintings in the World, in One Map," HowMuch.net.

57. Go see a Frank Lloyd Wright building: Every one of his publicly accessible works
Data: "Frank Lloyd Wright's Work," Frank Lloyd Wright Foundation.
"Public Wright Sites," Frank Lloyd Wright Trust.

58. Bees expertise: Winning Scripps National Spelling Bee words by winner's home state
Data: Copyright © Wiktionary 2019, "Appendix: Scripps winning words," under Creative Commons Attribution-ShareAlike 3.0 Unported License.
Various newspapers covering the spelling bees (accessed through Newspapers.com).

59. All together now: The all-time largest gatherings these cities have ever seen
Data: Copyright © Wikipedia 2021, "List of largest peaceful gatherings" and "List of protests in the United States by size," under Creative Commons Attribution-ShareAlike 3.0 Unported License.
Various.

60. Where the stars are born: Where every Best Actor and Actress Oscar winner hails from
Data: "Academy Awards Best Actress and Best Supporting Actress Winners," Filmsite.
IMDb.
Janssen, Sarah. *The World Almanac and Book of Facts 2018*. New York: World Almanac, 2018.

61. What are you looking at?: The highest-rated programs on US TV
Data: Janssen, Sarah. *The World Almanac and Book of Facts 2018*. New York: World Almanac, 2018.

62. Ladies and gentlemen, the Beatles!: Every show they played in North America
Data: "Artists - The Beatles," *The Ed Sullivan Show*.
The Beatles Bible.
thebeatles.bizhat.com/tour_dates.htm.
Various.

People and Populations

63. His name is my name, too: Most common surnames by state or province
Data: Cohn, Angel. "What's the Most Common Surname in Your State?" Ancestry.com.
Forebears.
Gedeon, Joseph. "The 50 Most Common Last Names in America," 24/7 Wall Street.
Mexican Civil Registration Genealogy.
Various, including Whitepages.

64. Native lands in native hands: Indigenous population by state or province
Original concept.
Data: "American Indian Reservations, Trust Lands, and Native Hawaiian Home Lands," 2020 Gazetteer Files, US Census Bureau.
"Aboriginal Peoples Highlight Tables, 2016 Census," Statistics Canada.
"Indigenous Latin America in the Twenty-First Century: The First Decade," World Bank Group.
"Indigenous Peoples in Latin America: Statistical Information," Congressional Research Service.
"Panorama sociodemográfico de México 2015," Instituto Nacional de Estadística y Geografía.
US Census Bureau.

65. Where we're coming from: Most common countries of origin outside the US by state
Data: "Place of Birth for the Foreign-Born Population in Puerto Rico" and "Place of Birth for the Foreign-Born Population in the United States," American Community Survey 2019 5-Year Estimates, US Census Bureau.

66. American pie charts: Origin of residents by state
Data: "State of Residence by Place of Birth: 2018," US Census Bureau.

67. Birds of a feather: Most common cross-state migrations
Data: "State-to-State Migration Flows: 2019," US Census Bureau.

68. Brightest lights, biggest cities: The top 10 most populous US cities over time
Data: "Annual Estimates of the Resident Population for Incorporated Places of 50,000 or More, Ranked by July 1, 2019 Population: April 1, 2010 to July 1, 2019" and "Population of the 100 Largest Cities and Other Urban Places in the United States: 1790 to 1990," US Census Bureau.

69. Balancing act: America's population centers of gravity
Data: US Census Bureau.

70. Homes away from homes: Where people have second homes
Data: Cororaton, Gay. "Top Vacation Home Counties," National Association of Realtors.
"Selected Housing Characteristics" and "Vacancy Status," American Community Survey 2019 5-Year Estimates, US Census Bureau.

71. Homes all around, but not a place to stay: Number of second homes per unhoused person
Data: "The 2019 Annual Homeless Assessment Report (AHAR) to Congress," US Department of Housing and Urban Development.
"Selected Housing Characteristics" and "Vacancy Status," American Community Survey 2019 1-Year Estimates, US Census Bureau.

72. The extremely Big Apple: 38 states have smaller populations than NYC
Concept: Nicholas Gelos.

73. The language barrier: Percentage of households speaking only English by county
Data: "Household Language by Household Limited English Speaking Status," American Community Survey 2015 5-Year Estimates, US Census Bureau.

74. Turtle Island: Indigenous homelands in 1491
Image: Miller, Robert J. Arizona State University College of Law.
Sturtevant, William C. "Early Indian Tribes, Culture Areas, and Linguistic Stocks," Smithsonian Institution, 1967.

75. Strangers in their own land: Indigenous homelands today
Data: "American Indian Reservations, Trust Lands, and Native Hawaiian Home Lands," 2020 Gazetteer Files, US Census Bureau.
"Intergovernmental Affairs: Tribal Affairs – American Indian and Alaska Native (AIAN)," US Census Bureau.
Miller, Robert J. Arizona State University College of Law.
"TIGER/Line Shapefile, 2018, nation, U.S., Current American Indian/Alaska Native/Native Hawaiian Areas National (AIANNH) National," US Census Bureau.

Lifestyle and Health

76. Wafflography: Number of Waffle Houses by latitude
Data: "Find a Store," Waffle House.

77. Fast food diet: Which states are a drive-through lover's paradise?
Data: "Ranking Cities With the Most and Least Fast Food Restaurants," Datafiniti.

78. Getting the runaround: How fast recreational runners finish a marathon by state
Data: McLoughlin, Danny. "Who's faster? The Ultimate State Comparison for Marathons," RunRepeat.

79. One for the ages: Median age by state
Data: "Age and Sex," American Community Survey 2019 1-Year Estimates, US Census Bureau.

80. Looking up to her: Female height in North America

81. He's the tops: Male height in North America

Data: NCID Risk Factor Collaboration. "Height and body-mass index trajectories of school-aged children and adolescents from 1985 to 2019 in 200 countries and territories: a pooled analysis of 2181 population-based studies with 65 million participants." *The Lancet* 396, no. 10261 (2020): 1463–1534.

82. Men are from Alaska, women are from Puerto Rico: Gender ratio by state

Data: "Age and Sex," American Community Survey 2019 1-Year Estimates, US Census Bureau.

83. Running from the chapel: Unmarried population by state

Data: "Selected Social Characteristics in the United States," American Community Survey 2019 1-Year Estimates, US Census Bureau.

84. Finding my happy place: Quality of life, happiness, and well-being in North America

Data: "Community Well-Being Index," Sharecare.
"World Happiness Report 2021," World Happiness Report.

85. I'm coming out: LGBTQ acceptance in North America

Data: "City and Town Population Totals: 2010–2019,"
Flores, Andrew R. "Social Acceptance of LGBT People in 174 Countries: 1981 to 2017," UCLA School of Law Williams Institute.
"Population and Housing Unit Counts: 2021," and "QuickFacts: United States," US Census Bureau.
"Municipal Equality Index 2020," Human Rights Campaign.

86. American tragedies: Accident mortality by state

Data: "Accident Mortality by State," Centers for Disease Control and Prevention, 2019.

87. America, the next generation: Birth rates by state

Data: "Fertility Rates by State," Centers for Disease Control and Prevention, 2019.

88. Down and out: COVID-19 anxiety and depression

Data: "Household Pulse Survey," Centers for Disease Control and Prevention, National Center for Health Statistics.

Industry and Transport

89. Expansive spans: The 10 longest bridges in North America

Data: *Terra Maxima: The Records of Humankind*, ed. Wolfgang Kunth. Ontario: Firefly Books, 2013.
Various.

90. Lost at sea: Famous shipwrecks in North American waters

Data: "National Register of Historic Places," US National Park Service.
NOAA Office of National Marine Sanctuaries.
"Science and shipwrecks: Preserving America's maritime history," National Oceanic and Atmospheric Administration.
"Shipwrecks," Monitor National Marine Sanctuary.
Various.

91. Who's got the goods?: The 50 largest ports in the US

Data: "Interim Multimodal Freight Network" and "Tonnage of Top 50 U.S. Water Ports, Ranked by Total Tons," US Department of Transportation.

92. Get out of town!: How far one day of travel took you from NYC, since 1800

Data: Paullin, Charles O. *Atlas of the Historical Geography of the United States*, ed. John K. Wright. Washington, DC: Carnegie Institution, 1932. Digital edition edited by Robert K. Nelson et al., 2013.

93. Moving on up: The 50 tallest buildings in North America

Data: The Skyscraper Center by The Council on Tall Buildings and Urban Habitat.

94. Bumper to bumper, coast to coast: How long it takes to get to work by county

95. In the driver's seat: How America gets to work

Data: "Commuting Characteristics by Sex," American Community Survey 2019 5-Year Estimates, US Census Bureau.

96. Don't bank on it: The unbanked people of North America

Data: World Bank 2017 Global Findex Database.

97. Road hogs: The longest roads in the US, Canada, and Mexico

Data: "Ask the Rambler: What Is the Longest Road in the United States?" US Department of Transportation Federal Highway Administration.
"Binational Freight Corridor Study," Arizona Department of Transportation.
"Recopilación de Información de Carreteras, Puentes y Estaciones Meteorológicas, Para el Desarrollo del Proyecto de Vulnerabilidad de Estructuras de Puentes en Zonas de Gran Influencia de Ciclones Tropicales," Instituto de Ingeniería.
"Trans-Canada Highway," The Canadian Encyclopedia.
"Trans-Canada Highway," Encyclopedia Britannica.
Various.

98. Cattle call: In which states are you likelier to run into a cow than a person?

Data: "2017 Census of Agriculture," USDA National Agricultural Statistics Service.

99. T-minus 15 spaceports: Launchpads across the US

Data: "Licenses" and "Spaceports by State," US Federal Aviation Administration
Roberts, Thomas G. "Spaceports of the World," Center for Strategic and International Studies.
Space-Track.org.

100. America offline: Who doesn't have internet

Data: "Types of Computers and Internet Subscriptions," American Community Survey 2019 5-Year Estimates, US Census Bureau.

ACKNOWLEDGMENTS

We'd like to thank Ian Wright and Granta Books for doing an inspirational job with the first Brilliant Maps book, and The Experiment for bringing it to North America—all paving the way for our own book. Matthew thanks his research team of KB, ZB, AB, and CD. He also thanks Gerald and Charlie Bucklan for helping to keep him grounded. Victor would like to thank his parents, grandparents, siblings Steph, James, Jeff, and Natalie, and friend Mike Hammond for all their feedback. Thanks to Robert J. Miller at the Arizona State University College of Law for his expertise on Indigenous Americans, Fabio Falchi of the Universidade de Santiago de Compostela for his expertise on dark skies, Danny McLoughlin of RunRepeat, and many others who provided insights on a variety of topics. At The Experiment, thanks to everyone guiding this book to publication, including the publicity team of Jennifer Hergenroeder, Will Rhino, and Cliff Robbins; Margie Guerra for contracts help; Zach Pace, Hannah Matuszak, Karen Giangreco, and Pamela Schechter for managing and organizing the project and getting it to the printer (Hannah, thanks also for your editorial feedback throughout); Nick Cizek, our editor and, for all intents and purposes, coauthor; Matthew Lore and Peter Burri for running the show and taking a chance on us; and our spectacular illustrator, Jack Dunnington, who not only designed everything you've seen in this book but doubled as another editor, too, as he helped shape our datasets to bring them to life in heroic fashion. The simple fact is, without Jack, this book would not have been possible.

ABOUT THE AUTHORS

As a boy in the 1990s, **MATTHEW BUCKLAN** eagerly awaited each new issue of *National Geographic*. There were just so many places, and they were all uniquely captivating. His favorite part was the centerfold map, which he would explore for hours. As he grew older, the nascent internet opened up a whole new world of geography, letting him see people and news from all of the places that had previously been mere points on a map. This love of human geography eventually propelled him to the state geography bee; today, he still can't look away from a great map. He lives in the Milwaukee area.

VICTOR CIZEK still owns a picture atlas he got as a kid on a trip to Washington, DC, and sees it as a source of inspiration for him to make this book, along with his visual arts background and geographically minded coauthor Matt. He grew up in and lives in northeast Ohio, represented by a steel mill on that old atlas.

JACK DUNNINGTON is a Brooklyn-based artist and designer. More work can be found at jackdunnington.com.

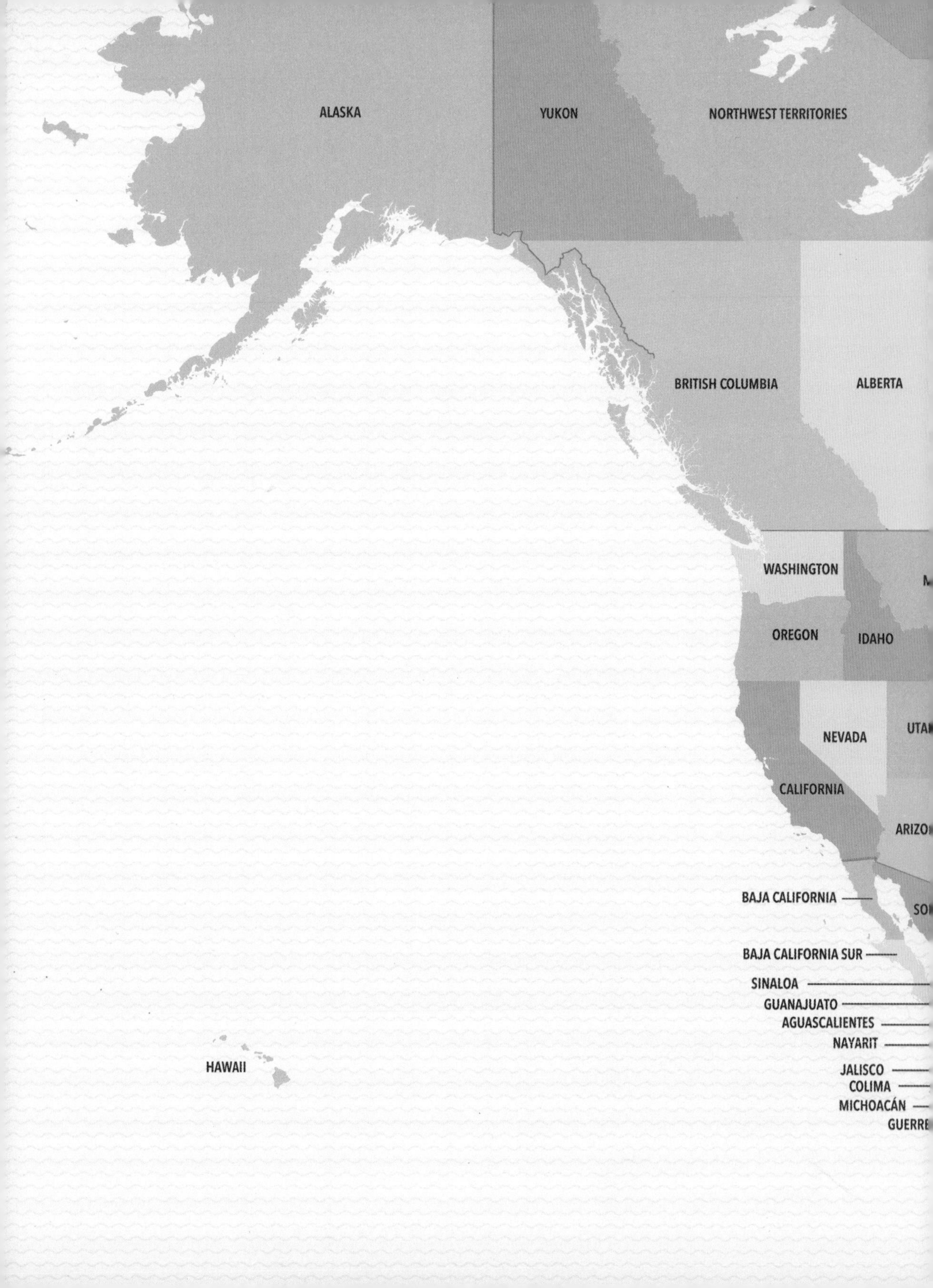
ALASKA
YUKON
NORTHWEST TERRITORIES
BRITISH COLUMBIA
ALBERTA
WASHINGTON
OREGON
IDAHO
NEVADA
CALIFORNIA
BAJA CALIFORNIA
BAJA CALIFORNIA SUR
SINALOA
GUANAJUATO
AGUASCALIENTES
NAYARIT
JALISCO
COLIMA
MICHOACÁN
HAWAII